Mathematics for Digital Science 1

SCIENCES

Computer Science,
Field Director – Jean-Charles Pomerol

Operational Research and Decision,
Subject Head – Patrick Siarry

Mathematics for Digital Science 1

Fundamentals

Gérard-Michel Cochard
Mhand Hifi

ISTE

WILEY

First published 2025 in Great Britain and the United States by ISTE Ltd and John Wiley & Sons, Inc.

ISTE Ltd
27-37 St George's Road
London SW19 4EU
UK

www.iste.co.uk

John Wiley & Sons, Inc.
111 River Street
Hoboken, NJ 07030
USA

www.wiley.com

Library of Congress Control Number: 2024946485

British Library Cataloguing-in-Publication Data
A CIP record for this book is available from the British Library
ISBN 978-1-78945-194-8

ERC code:
PE1 Mathematics
PE1_17 Mathematical aspects of computer science
PE1_20 Control theory, optimisation and operational research

Contents

Preface

The term *digital* relates to information and communication sciences and technologies, covering areas such as computing, digital electronics and telecommunications. Advances in these fields, in terms of discoveries, developments and applications, have grown exponentially over the last 50 years. Such progress has profoundly transformed human activities, making the transition to "all digital" a significant issue not only economically, but also politically. Concurrently, professions related to digital technology are constantly evolving. It is noteworthy that these technological advances are based on a substantial mathematical foundation. Thus, any engineer or researcher aiming to drive innovation must possess a strong knowledge of mathematics.

Many students opt for computer science-related courses early on without first acquiring the essential mathematical foundations they might need in the future. While they can certainly excel as technicians, they may not be as well prepared to become effective engineers or researchers.

It should be recognized that computer science cannot be conceived without a solid mathematical basis. Advances in digital technologies have been closely linked to the support of mathematics. The pioneers of computing were above all mathematicians: Alan Turing, Claude Shannon, John Von Neumann and even Charles Babbage, who is nicknamed the grandfather of computing. They could never have designed computers without their mathematical expertise. Other historical figures such as Euclid, who formulated the first algorithm in history, George Boole, father of binary algebra, Ada Lovelace, pioneer in the creation of the first computer programs, Grace Hopper, to whom the first language compiler is attributed, and Margaret Hamilton, a key figure in software engineering, were also mathematicians. Today, areas such as artificial intelligence, managing large datasets, and information security are at the heart of computer science research. Once again, it

Mathematics for Digital Science 1,
by Gérard-Michel COCHARD and Mhand HIFI.

is evident that without a solid mathematical foundation, it is difficult to innovate in these areas.

The digital mathematics courses offered aim to meet the need for fundamental mathematical knowledge essential for mastering and advancing digital technologies. While the book primarily targets university and engineering school students, it also serves as a resource for working IT professionals seeking to enhance their mathematical skills as part of continuing education in their respective field.

The three-volume work compiles the lessons taught to numerous generations of students in the first two university cycles, specifically in Bachelor's/Master's degrees in Computer Science or in Computer Methods Applied to Business Management. The first – and present – volume covers the essential mathematical foundations for approaching digital technologies and is described below, while the second volume focuses on digital information, and the third volume is devoted to data analysis and optimization.

Chapter 1 of the present (first) volume presents the fundamental concepts of logic and establishes the foundations of deductive reasoning.

Chapter 2 covers the fundamentals of sets and binary relations, while Chapter 3 focuses on combinatorial analysis.

Chapter 4 covers the definition and properties of the structure of Boolean algebra, as well as the manipulation of Boolean functions. Subsequently, Chapter 5 addresses the theoretical aspects of electronic circuits and their practical implementation for the manufacture of electronic modules.

Chapter 6 revisits arithmetic concepts that were once foundational in mathematics education in primary and secondary schools, now highlighting their profound relevance in the field of information security. Chapters 7 and 8 explore data protection and confidentiality, addressing error detection and correction codes, as well as encryption systems.

The last three chapters, namely 9, 10 and 11, focus on data processing and the predictions that can be derived from it. They re-examine the fundamental notions of probability and statistics, crucial for data analysis. More specifically, the final chapter illustrates how chance can be exploited to conduct experiments through simulations.

To achieve the goal of mastering the application of mathematical results, demonstrations are presented whenever they are accessible to the reader, who is assumed to be familiar with basic knowledge. In some cases, the validity of the results is trusted, while pointing out that detailed demonstrations can be found in more advanced works.

For each chapter, numerous examples are provided to illustrate the presented concepts. Generally, these examples are elaborated in detail to ensure better understanding on the part of the reader.

Finally, for each addressed theme, the choice is made to present them exhaustively, starting from the initial definitions and hypotheses. These basic notions will certainly be familiar to many readers. The intention behind such an approach is to spare those with knowledge gaps the need to look up these concepts in other literature.

October 2024

1

The Concept of Logic

CONCEPTS COVERED IN THIS CHAPTER.–

Logic is essential, not only for computer operations, but also for artificial intelligence. The concepts of logic outlined in this chapter are therefore of fundamental importance.

Beginning with syllogistic reasoning, the exploration examines propositional calculus and its fundamental operations: negation, conjunction, disjunction, conditional, biconditional, as well as their combinations.

The last section focuses on the tools of deductive reasoning.

References: [EXE 59, MEN 64, QUI 84, WOO 80, LIP 83].

1.1. Syllogisms

Syllogisms are a type of logical reasoning that dates back to the time of Aristotle in Ancient Greece. They are constructed from logical propositions, known as premises, which are linked by a conclusion. The conclusions are derived from the logical of these propositions, allowing the deduction of a conclusion from the given premises. In what follows, the example of a classic syllogism will be considered:

(i) all men are mortal,

(ii) all Greeks are men,

(iii) therefore, all Greeks are mortal.

The representation of such a logic is termed "Barbara's syllogism", named after a figure in medieval logic who popularized it. In the above example, there are two

Mathematics for Digital Science 1,
by Gérard-Michel COCHARD and Mhand HIFI.

premises and a conclusion. The first premise is represented by (i), the second by (ii) and the conclusion by (iii).

The described syllogism is considered valid because it respects the rules of formal logic. The conclusion follows necessarily from the two premises and cannot be false if the premises are true. Indeed, each premise of a syllogism must be evaluated according to its truth value to determine whether the conclusion that follows from it is also true or false. In the example, TRUE or FALSE (a truth value) can be assigned to each premise; then it is said that a proposition is a statement which takes one and only one truth value.

EXAMPLE 1.1.– Some illustrations:

"The sky is blue,"
"Money cannot buy happiness,"
"The house is burning and the firefighters are not there,"
"Computer scientists are all crazy,"

are "propositions".

"Eat your soup!,"
"Good heavens!,"
"Where are they all running?"

do not correspond to propositions.

A syllogism, such as the one considered above, can be represented by the following notation:

all Xs are Ys (major premise),
all Zs are Xs (minor premise),
so, all Zs are Ys (conclusion).

And especially,

$$Premises\ \{Every\ X\ is\ Y, Every\ Z\ is\ X)$$

$$Conclusion: every\ Z\ is\ Y$$

The structure is also known as the "canonical form" of a syllogism. The terms X, Y and Z represent categories or classes of things (identified objects), and the syllogism explores the logical relationship between these categories. The canonical form of a syllogism facilitates highlighting the logical structure of reasoning and verifying its validity using the rules of formal logic. Note that a proposition can be placed in one of the following four categories, as shown in Table 1.1.

A	Universal and affirmative	e.g. "all men are mortals"
E	Universal and negative	e.g. "no man is mortal"
I	Particular and affirmative	e.g. "some men are mortal"
O	Particular and negative	e.g. "some men are not mortal"

Table 1.1. *Categories of propositions*

Note that a syllogism is composed of three propositions, each belonging to one of the four classes of propositions (A, E, I and O). The middle term connects the two premises of the syllogism, and there are four possible figures depending on the position of this term. The number of possible categories for a syllogism is $4^3 = 64$ possible combinations (four categories for each of the three propositions). Furthermore, depending on the position of X, Y, Z in the premises, there are four classes of syllogisms (see Figure 1.1). The number of categories of syllogisms then becomes $4^3x4 = 256$. For each possible combination of premises, there are two possible formulations (A implies B or B implies A), effectively doubling the number of categories of syllogisms: 512.

category	I	II	III	IV
premise	XY	YX	XY	YX
premise	ZX	ZX	XZ	XZ
conclusion	ZY	ZY	ZY	ZY

Figure 1.1. *Syllogism classes*

In summary, there are 512 possible categories of syllogisms, but the number of admissible syllogisms is smaller due to the presence of "rules of reasoning".

EXAMPLE 1.2.– Some illustrations:

– A syllogism with two negative premises is not valid. In a valid syllogism, if one of the two premises is negative, the conclusion must be negative.

– In a valid syllogism, if both premises are affirmative, then the conclusion must be affirmative.

– A syllogism is not valid if it contains two particular premises.

These rules, some of which are presented above, actually reduce the number of valid forms of syllogisms to 19.

As previously discussed, syllogisms follow a general form:

(premise 1, premise 2) implies (conclusion).

We can generalize and acknowledge that syllogisms can involve more than two premises, as demonstrated in Lewis Caroll's "kangaroo argument".

EXAMPLE 1.3.– The premises:

1. The only animals in this house are cats.
2. Any animal can become a favorite animal that loves to watch the moon.
3. When I hate an animal, I avoid it.
4. No animal is carnivorous unless it prowls at night.
5. No cat fails to kill mice.
6. No animal pleases me, except those in this house.
7. Kangaroos cannot become pets.
8. Only carnivores kill mice.
9. I hate animals that I do not like.
10. Animals that roam at night always like to look at the moon.

Conclusion: I always avoid kangaroos.

The objective of the chapter is to provide an overview of the mechanisms for reasoning based on the manipulation of mathematical concepts.

According to N. Bourbaki (Elements of the History of Mathematics):

> We consider that the intervention of metamathematics in the presentation of logic and mathematics can and must be reduced to the very elementary part which deals with the handling of operator symbols and deductive criteria.

1.2. Elementary operations of propositional calculus

1.2.1. *Negation*

Consider the following propositions P and Q:

P: Amiens is in France.

Q: it is false that Amiens is in France.

The proposition Q is derived from the proposition P by prefixing the phrase "it is false that" to it. Q is regarded as the negation of P, denoted as ¬P. The symbol "¬" represents negation and can thus be interpreted as "it is false that". The truth table illustrating negation is simple (see Figure 1.2).

P	¬P
T	F
F	T

Figure 1.2. *Negation truth table*

Expressing negation in a literary manner often poses difficulties. While the previous example, where ¬P translates to "Amiens is not in France" is straightforward, more complex cases exist. To illustrate such complexity, consideration will be given to a few examples.

Beginning with the first:

EXAMPLE 1.4.–

P: all integers are real numbers.

Q_1: some integers are not real numbers.

Q_2: no integer is a real number.

Which of Q_1 and Q_2 represents the negation of P?

After reflection, it becomes evident that P and Q_2 can be false simultaneously. Thus, Q_2 is not the negation of P. Conversely, Q_1 represents ¬P.

Here is a second example:

EXAMPLE 1.5.–

P: all angles can be trisected (i.e. divided into three equal parts) using only the ruler and the compass (notoriously wrong!).

Q_1: no angle can be trisected using only the ruler and compass.

Q_2: there is at least one angle that cannot be trisected using only the ruler and compass.

What is the negation of P, Q_1 or Q_2? P is false because it is impossible to trisect an angle of 37° with only a ruler and compass.

Q_1 is false because we can trisect an angle of 180° with a ruler and compass (see Figure 1.3).

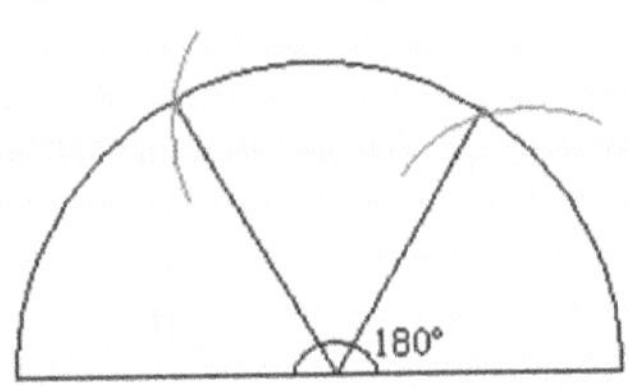

Figure 1.3. *Trisection of an angle of 180°*

It is therefore Q_2 which is the negation of P.

It should be noted that difficulties often arise when propositions contain words such as "All," "Some" or similar expressions. However, as will be discussed later in this chapter, these difficulties can be resolved by using quantifiers and the rules associated with them.

1.2.2. *Conjunction*

A conjunction is a logical connector that combines two propositions to form a third one, using the word "and" to link clauses. The conjunction of two propositions P and Q is denoted by $P \wedge Q$.

EXAMPLE 1.6.–

P: the number 12 is rational.

Q: the number 12 is positive.

P∧Q: the number 12 is rational and positive.

Conjunction is a logical operation that can be represented by a truth table, as shown in Figure 1.4.

P	Q	P∧Q
T	T	T
T	F	F
F	T	F
F	F	F

Figure 1.4. *Conjunction truth table*

It is important to emphasize that the conjunction P ∧ Q is true only if both propositions P and Q are true. In other words, the conjunction is true only when both initial propositions are simultaneously true. If either proposition is false, then the conjunction will be false.

1.2.3. *Disjunction*

Disjunction is a logical connector that combines two propositions to form a third, using the word "or" to link clauses. The disjunction of two propositions P and Q is denoted by P ∨ Q.

EXAMPLE 1.7.–

P: a price reduction is granted to women.

Q: a price reduction is granted to soldiers.

P∨Q: a price reduction is granted to women or soldiers.

By definition, the disjunction truth table is given in Figure 1.5.

P	Q	P∨Q
T	T	T
T	F	T
F	T	T
F	F	F

Figure 1.5. *Disjunction truth table*

It is important to emphasize that the disjunction P ∨ Q is false only if both propositions P and Q are simultaneously false. The disjunction, which is defined as the *inclusive or* and is represented by the symbol ∨, comes from the Latin "vel" and means "either or both". In other words, if either of the initial propositions is true, or if both propositions are true, then the disjunction will be true.

There is another type of "or", called *exclusive or*, which comes from the Latin "aut" and which means "either one or the other". In the case of the logical operation of "exclusive or", the first row of the truth table would be T, T, F. However, there is no need to create a special symbol for "exclusive or", because it can be expressed using the already defined symbols of conjunction (∧), disjunction (∨) and negation (¬), as shown in Figure 1.6.

P	Q	P∨Q	P∧Q	¬(P∧Q)	(P∨Q)∧(¬(P∧Q))
T	T	T	T	F	F
T	F	T	F	T	T
F	T	T	F	T	T
F	F	F	F	T	F

Figure 1.6. *"Exclusive or" truth table*

Therefore, the expression (P ∨ Q) ∧ (¬(P ∧ Q)) represents the "exclusive or", where P and Q are two logical propositions. There are other ways of expressing "exclusive or" using different combinations of conjunction, disjunction and negation, which will be discussed later.

1.2.4. *Conditional*

Common language often uses the conditional form: "if P then Q".

This proposition is an implication constructed from the propositions P and Q, denoted as P → Q or "if P then Q". It can be associated with a truth table, which is, by definition, that of Figure 1.7

P	Q	P→Q
T	T	T
T	F	F
F	T	T
F	F	T

Figure 1.7. *Conditional truth table*

Note that the implication P → Q is always true, except when P is true and Q is false. In all other cases, the implication is considered true.

EXAMPLE 1.8.– Consider x and y as angles, along with the following propositions:

P: x and y have their parallel sides.

Q: x = y.

The four rows of the truth table are shown in Figure 1.8.

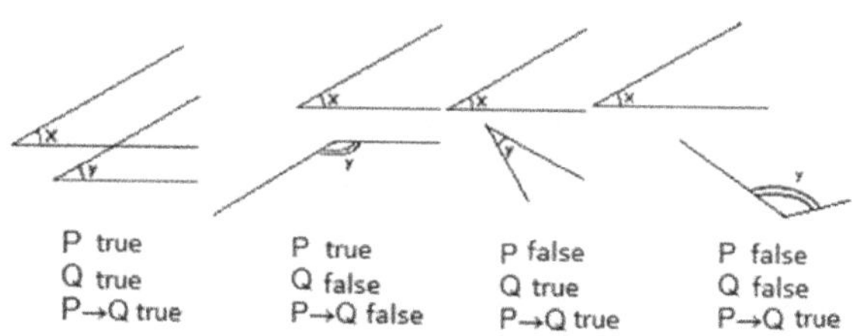

Figure 1.8. *Truth table*

EXAMPLE 1.9.– Consider the following propositions:

P: it is raining.

Q: I will drive you home.

P→Q: if it rains, then I will drive you back.

Note that if P is true (it is raining), this implies that if the proposition Q (I will drive you home) is false, then the conditional P→Q is false. Indeed, if P is true and

Q is false, this means that although it is raining, I am not taking you back by car, which contradicts the implication of the conditional. This situation is "abnormal" and justifies that P→Q is false in this specific case.

1.2.5. *Biconditional*

P	Q	P→Q	Q→P	(P→Q)∧(Q→P)
T	T	T	T	T
T	F	F	T	F
F	T	T	F	F
F	F	T	T	T

Figure 1.9. *Truth table*

Let us examine the proposition (P→Q) ∧ (Q→P) whose truth table (see Figure 1.9) is constructed from the previous truth tables (see Figure 1.7). This new operation is denoted as P↔Q and is called *biconditional*. It can be expressed by "P if and only if Q".

The biconditional is true if P and Q have the same truth value, and false otherwise. Therefore, the truth table in Figure 1.10 illustrates this relationship.

P	Q	P↔Q
T	T	T
T	F	F
F	T	F
F	F	T

Figure 1.10. *Biconditional truth table*

By comparing the truth table of the "biconditional" to that of the "exclusive or", we can observe that the "exclusive or" can be represented by ¬(P↔Q).

1.2.6. *Tautologies*

Given two propositions P and Q, let us examine all the binary operations that can be considered and draw up the corresponding truth tables (see Figure 1.11):

– The right half of Figure 1.11 is obtained by taking the negation of the left part.

– Table 14 in Figure 1.11 corresponds to a result which is always true; this is a *tautology*.

– Table 16 corresponds to a result which is always false; this is *a contradiction*. The negation of a tautology is a contradiction.

1

P	Q	P
T	T	T
T	F	T
F	T	F
F	F	F

2

P	Q	Q
T	T	T
T	F	F
F	T	T
F	F	F

3

P	Q	$\neg P$
T	T	F
T	F	F
F	T	T
F	F	T

4

P	Q	$\neg Q$
T	T	F
T	F	T
F	T	F
F	F	T

5

P	Q	$P \wedge Q$
T	T	T
T	F	F
F	T	F
F	F	F

6

P	Q	$P \vee Q$
T	T	T
T	F	T
F	T	T
F	F	F

7

P	Q	$\neg(P \wedge Q)$
T	T	F
T	F	T
F	T	T
F	F	T

8

P	Q	$\neg(P \vee Q)$
T	T	F
T	F	F
F	T	F
F	F	T

9

P	Q	$P \to Q$
T	T	T
T	F	F
F	T	T
F	F	T

10

P	Q	$Q \to P$
T	T	T
T	F	T
F	T	F
F	F	T

11

P	Q	$\neg(P \to Q)$
T	T	F
T	F	T
F	T	F
F	F	F

12

P	Q	$\neg(Q \to P)$
T	T	F
T	F	F
F	T	T
F	F	F

13

P	Q	$P \leftrightarrow Q$
T	T	T
T	F	F
F	T	F
F	F	T

14

P	Q	T
T	T	T
T	F	T
F	T	T
F	F	T

15

P	Q	$\neg(P \leftrightarrow Q)$
T	T	F
T	F	T
F	T	T
F	F	F

16

P	Q	$\neg T$
T	T	F
T	F	F
F	T	F
F	F	F

Figure 1.11. *Possible binary operations*

From the propositions P, Q, …, and the elementary operations $\wedge$, $\vee$, $\neg$, it is possible to construct propositional forms f(P, Q, …). Depending on the values of P, Q, …, the form can be true or false. If f(P, Q, …) is always true independently of the truth values of P, Q, …, the form f is a tautology. Conversely, if f(P, Q, …) is always false independently of the truth values of P, Q, …, the form f is a contradiction.

The notion of tautology is important in propositional logic because it simplifies complex logical expressions and verifies the validity of reasoning by establishing that a proposition is a tautology.

EXAMPLE 1.10.– $P\vee(\neg P)$, $\neg(P\wedge(\neg P))$, $(\neg(P\wedge Q))\leftrightarrow((\neg P)\vee(\neg Q))$ are tautologies. This can be simply by drawing up the truth tables (see the different truth tables grouped in Figure 1.12).

The following propositions are contradictions: $\neg(P\vee(\neg P))$, $P\wedge(\neg P)$, $(P\wedge Q)\leftrightarrow((\neg P)\vee(\neg Q))$.

The first two are contradictions because they are the negations of the first two previously presented tautologies. The third proposition can be verified by setting up a truth table.

P	¬P	P∨(¬P)
T	F	T
F	T	T

P	¬P	¬(P∧(¬P))
T	F	T
F	T	T

P	Q	P∧Q	¬(P∧Q)	¬P	¬Q	(¬P)∨(¬Q)	(¬(P∧Q))↔((¬P)∨(¬Q))
T	T	T	F	F	F	F	T
T	F	F	T	F	T	T	T
F	T	F	T	T	F	T	T
F	F	F	T	T	T	T	T

Figure 1.12. *Representation of the different truth tables*

A tautology of the form $P\leftrightarrow Q$ is a logical equivalence. According to the definition of $\leftrightarrow$, this means that P and Q have the same truth tables. Therefore, it is noted that

$$P \Leftrightarrow Q \text{ or more simply } P \equiv Q$$

Thus, in the previous examples, the logical proposition $\neg(P\wedge Q)$ is equivalent to the logical proposition $(\neg P)\vee(\neg Q)$, denoted as $\neg(P\wedge Q) \equiv (\neg P)\vee(\neg Q)$. It means that these two propositions have the same truth tables and are therefore logically equivalent.

The following table lists the main basic logical equivalences:

Idempotence	$P \vee P \equiv P$		$P \wedge P \equiv P$	
Associativity	$(P \vee Q) \vee R \equiv P \vee (Q \vee R)$		$(P \wedge Q) \wedge R \equiv P \wedge (Q \wedge R)$	
Commutativity	$P \vee Q \equiv Q \vee P$		$P \wedge Q \equiv Q \wedge P$	
Distributivity	$P \vee (Q \wedge R) \equiv (P \vee Q) \wedge (P \vee R)$		$P \wedge (Q \vee R) \equiv (P \wedge Q) \vee (P \wedge R)$	
Identify	$P \vee f \equiv P$	$P \wedge v \equiv P$	$P \vee v \equiv v$	$P \wedge f \equiv f$
Complementarity	$P \vee (\neg P)$	$P \wedge (\neg P) \equiv f$	$\neg v \equiv f$	$\neg f \equiv v$
Involution	$\neg(\neg P) \equiv P$			
Morgan's laws	$\neg(P \vee Q) \equiv (\neg P) \wedge (\neg Q)$		$\neg(P \wedge Q) \equiv (\neg P) \vee (\neg Q)$	

In this table, t denotes a tautology and f denotes a contradiction. The table only shows the symbols $\wedge$, $\vee$, $\neg$ which is sufficient since the other symbols can be expressed in terms of these. Logical equivalences can contain all symbols.

Note that these equivalences are fundamental in propositional logic and can be used to simplify complex logical expressions.

EXAMPLE 1.11.– The equivalence $P \rightarrow Q \equiv (\neg P) \vee Q$ is evident. This means that $P \rightarrow Q$ can be expressed as $(\neg P) \vee Q$ using negation and disjunction. This equivalence is commonly known as the law of implication or law of detachment (see the truth table shown in Figure 1.13).

P	Q	P→Q	¬P	(¬P)∨Q
T	T	T	F	T
T	F	F	F	F
F	T	T	T	T
F	F	T	T	T

Figure 1.13. *Truth table*

Consequences:

– If it rains, I open my umbrella ⇔ It does not rain or I open my umbrella.

– If the triangle is equilateral, it is isosceles ⇔ The triangle is not equilateral or it is isosceles.

– If the man is single, he is unhappy ⇔ The man is not single or he is unhappy.

Another interesting tautological expression is represented by the form P→Q. When P→Q holds as a tautology, it signifies that P logically implies Q, symbolized as P ⇒ Q, indicating that Q holds true whenever P does. This translates to "If P is true, then Q is also true." Consequently, if P is false, no information can be drawn regarding Q. Fundamentally, P can be viewed as a sufficient condition for Q.

P	Q	R	P→Q	Q→R	(P→Q)∧(Q→R)	P→R	(P→Q)∧(Q→R)→(P→R)
T	T	T	T	T	T	T	T
T	T	F	T	F	F	F	T
T	F	T	F	T	F	T	T
T	F	F	F	T	F	F	T
F	T	T	T	T	T	T	T
F	T	F	T	F	F	T	T
F	F	T	T	T	T	T	T
F	F	F	T	T	T	T	T

Figure 1.14. *Truth table*

A syllogism is a specific instance of logical implication. Consider, for example, the following syllogism (transitivity): P → Q, Q → R, therefore P → R which can be represented as follows: ((P→Q)∧(Q→R)) → (P→R).

Examining the truth table in Figure 1.14, it is evident that (P→Q)∧(Q→R) ⇒ (P→R).

1.3. Tools of deductive reasoning

In the domain of mathematics, the primary objective is to move from one true proposition to another through *deductive reasoning,* leading to a *demonstration.* This process is facilitated by employing various tools, which can be categorized into three main categories:

– *The replacement rule*: this rule permits the substitution of a proposition with another that is logically equivalent, maintaining the validity of the demonstration.

– *The substitution principle*: this principle enables the replacement of a variable with an equivalent expression, preserving the integrity of the demonstration.

– *Inference rules*: these rules facilitate the derivation from one or more propositions to a new proposition, employing logical forms such as implication or contrapositive.

These fundamental tools in mathematics are instrumental in constructing rigorous and precise demonstrations.

The *replacement rule* can be formulated as:

REPLACEMENT RULE. *In a tautology, the replacement rule permits the substitution of a proposition with another having the same truth value.*

This is evident because, in a tautology, the truth values of the component propositions do not determine the truth value of the form, which is consistently TRUE. Thus, one proposition can be substituted with another of the same truth value without altering the truth value of the tautology. Hence, the tautology $[(P\rightarrow(Q\rightarrow R)]\leftrightarrow[(P\wedge Q)\rightarrow R)]$ transforms, by replacing R with P, into a new tautology: $[(P\rightarrow(Q\rightarrow P)]\leftrightarrow[(P\wedge Q)\rightarrow P]$.

The principle of substitution operates differently:

PRINCIPLE OF SUBSTITUTION.– *For a propositional form* f (P, Q, …), *if* P *and* P' *are logically equivalent,* then f(P, Q, …) *and* f(P', Q, …) *are also logically equivalent, i.e.* $P \equiv P' \Rightarrow f(P, Q, \ldots) \equiv f(P', Q, \ldots)$.

EXAMPLE 1.12.– Illustrating the principle of substitution with the following example, consider the propositional form $\neg(P\wedge Q) \rightarrow P$ (which is false). Using Morgan's law, knowing that $\neg(P\wedge Q) \equiv (\neg P)\vee(\neg Q)$, hence deducing $(\neg P)\vee(\neg Q)\rightarrow P$.

EXAMPLE 1.13.– To demonstrate the distinction between the replacement rule and the substitution principle, Morgan's law $\neg(P\vee Q) \equiv (\neg P)\wedge(\neg Q)$ can be derived from Morgan's law $\neg(P\wedge Q) \equiv (\neg P)\vee(\neg Q)$. Begin with the hypothesis that $\neg(P\wedge Q)\leftrightarrow(\neg P)\vee(\neg Q)$ is a tautology. Replace P by $\neg P$: this simplifies to the tautology $\neg((\neg P)\wedge Q)\leftrightarrow(\neg(\neg P))\vee(\neg Q)$. Replacing Q by $\neg Q$: this provides the tautology $\neg((P)\wedge(\neg Q))\leftrightarrow(\neg(\neg P))\vee(\neg(\neg Q))$.

With $\neg(\neg P) \equiv P$, and $\neg(\neg Q) \equiv Q$, and according to the substitution principle and the above result, thus $\neg((\neg P)\wedge(\neg Q)) \equiv P\vee Q$.

Let us then consider the propositional form $\neg(P\vee Q)$. According to the principle of substitution and the previous line: $\neg(P\vee Q) \equiv (\neg((\neg P)\wedge(\neg Q)))$, or again by substitution $\neg(P\vee Q) \equiv (\neg P)\wedge(\neg Q)$.

The *rules of inference* are logical implications listed as follows:

– *Modus ponens* rule: if P is true and (P→Q) is true, then Q is true.

– *Conjunctive inference* rule: if P is true and Q is true, then P∧Q is true.

– *Conjunctive simplification* rule: if P∧Q is true, then P is true.

– *Hypothetical syllogism* rule: if (P→Q) is true and (Q→R) is true, then (P→R) is true.

– *Modus tollens* rule: if (P→Q) is true and ¬Q is true, then ¬P is true.

– Rule *of inference by case*: if (A→Q) is true and (B→Q) is true, then (A∨B)→Q is true.

All these rules are easily demonstrated using truth tables.

EXAMPLE 1.14.– Demonstrations employing truth tables can be found in Figures 1.15–1.17.

P	Q	P→Q	(P∧(P→Q))→Q
T	T	T	T
T	F	F	T
F	T	T	T
F	F	T	T

Figure 1.15. *Modus ponens*

P	Q	P→Q	¬Q	(P→Q)∧(¬Q)	¬P	((P→Q)∧(¬Q))→¬P
T	T	T	F	F	F	T
T	F	F	T	F	F	T
F	T	T	F	F	T	T
F	F	T	T	T	T	T

Figure 1.16. *Modus tollens*

A	B	Q	A→Q	B→Q	A∨B	(A∨B)→Q	(A→Q)∧(B→Q)	((A→Q)∧(B→Q))→((A∨B)→Q)
T	T	T	T	T	T	T	T	T
T	T	F	F	F	T	F	F	T
T	F	T	T	T	T	T	T	T
T	F	F	F	T	T	F	F	T
F	T	T	T	T	T	T	T	T
F	T	F	T	F	T	F	F	T
F	F	T	T	T	F	T	T	T
F	F	F	T	T	F	T	T	T

Figure 1.17. *Inference by case*

1.4. Quantification

Consider the following syllogism:

– All men are mortal.

– Socrates is a man.

– So, Socrates is mortal.

It is practical to use functional symbolism:

– H(Socrates) will mean "Socrates is a man."

– M(Socrates) will mean "Socrates is mortal."

– H(x) will mean "x is a man."

– M(x) will mean "x is mortal."

Consider that for all the values of the variable x, H(x) represents a proposition that can be either true or false:

– H(Napoleon) is true.

– H(Eiffel Tower) is false.

H(x) is termed a *predicate,* specifically *monadic* as it contains only one variable.

EXAMPLE 1.15.– P(x): $2x + 1 = 7$

If $x = 3$, P(x) is true.

If $x = 8$, P(x) is false.

EXAMPLE 1.16.– Q(x) : $(x+1)^2 = x^2 + 2x + 1$

Q(x) is always true, regardless of the value assigned to x.

In the last example, Q(x) is true for all x. This fact can be expressed using the *universal quantifier* $\forall x$.

$\forall x$ Q(x) signifies "for all x, Q(x) holds true"

Using the universal quantifier, the syllogism concerning Socrates can be expressed as follows:

$\forall x\ H(x) \rightarrow M(x)$,
H(Socrates),
So M(Socrates).

Another quantifier is the *existential quantifier*, denoted by ∃, which means "there is at least one". Thus, in the context of example 1.15, $\exists\ x\ P(x)$ could be used to express "there exists at least one x such that $2x + 1 = 7$".

Similarly, the assertion "Some men are intelligent" can be expressed using predicates:

H(x): x is a man.
I(x): x is intelligent.
under the form $\exists x\ [H(x) \wedge I(x)]$.

Here is a suggestion for improvement:

Note the presence of brackets (or parentheses or braces) in the application of logical quantifiers, indicating their domain of action or scope. By using the universal (∀) and existential (∃) quantifiers, we can succinctly express the four categorical forms A, E, I and O of traditional logic. Now, examine the following predicate: B(x): "x is white". Suppose the universe of x is finite, composed of three objects a, b and c for simplicity. In such a scenario, it becomes straightforward to verify the following equivalences:

$$\forall x\ B(x) \equiv B(a) \wedge B(b) \wedge B(c) \quad [1.1]$$

$$\exists x\ B(x) \equiv B(a) \vee B(b) \vee B(c) \quad [1.2]$$

A	Universal and affirmative	e.g. all men are intelligent	$\forall x\ [H(x) \rightarrow I(x)]$
E	Universal and negative	e.g. no man is intelligent	$\forall x\ [H(x) \rightarrow \neg I(x)]$
I	Particular and affirmative	e.g. some men are intelligent	$\exists x\ [H(x) \wedge I(x)]$
O	Particular and negative	e.g. some men are not intelligent	$\exists x\ [H(x) \wedge \neg I(x)]$

Table 1.2. *Categories of propositions*

Having established this equivalence, taking the negation of both sides of equation [1.1] yields:

$$\neg[\forall x\, B(x)] \equiv \neg[B(a) \wedge B(b) \wedge B(c)]$$

$$\text{Or } \neg[\forall x\, B(x)] \equiv (\neg B(a)) \vee (\neg B(b)) \vee (\neg B(\chi)) \qquad \text{(Morgan's law)}$$

or by replacing B by $\neg B$,

$$\neg[\forall x\, \neg B(x)] \equiv B(a) \vee B(b) \vee B(c) \equiv \exists x\, B(x) \text{ according to [1.2]}$$

Thus, $\exists x\, B(x) \equiv \neg[\forall x\, \neg B(x)]$ implies, on the one hand, $\exists x\, \neg B(x) \equiv \neg[\forall x\, B(x)]$ and, on the other hand, $\neg(\exists x\, B(x)) \equiv \forall x\, \neg B(x)$. This demonstrates the process of negating of a logical expression with a quantifier. Additionally, this result holds true for any universe

$$\neg[\forall x\, B(x)] \equiv \exists x\, \neg B(x)$$

$$\neg[\exists x\, B(x)] \equiv \forall x\, \neg B(x)$$

Observing the effect of negation indicates the exchange of the symbols $\forall$ and $\exists$. This observation enables us to obtain the negation of propositions containing "All", "Some", "None", etc.

To illustrate the application of these exchange rules, consider the following examples:

EXAMPLE 1.17.– Let P represent the proposition: "all that glitters is not gold", which can be expressed as $\forall x\, [B(x) \rightarrow \neg O(x)]$, where $B(x)$ denotes that the object x shines, and $O(x)$ denotes that object x is gold. The negation of P can be derived by applying the rules of negation and quantifier exchange:

$$\neg\{\forall x\, [B(x) \rightarrow \neg O(x)]\} \equiv \exists x\, \neg[B(x) \rightarrow \neg O(x)] \equiv \exists x\, [B(x) \wedge O(x)]$$

Thus, the negation of P becomes the proposition: "there exists at least one object which shines, and which is made of gold". In this case, $\neg P$ can then be expressed as $\exists x\, [B(x) \wedge O(x)]$, signifying that some shining objects are made of gold.

EXAMPLE 1.18.– Let Q represent the proposition: "the rule applies to everyone", expressed as $\forall x\, [B(x) \rightarrow \neg O(x)]$, where $R(x)$ denotes that the rule applies to x. The negation of Q can be derived by applying the rules of negation and quantifier exchange: $\neg\forall x\, R(x) \equiv \exists x\, \neg R(x)$. Thus, the negation of Q becomes the proposition: "there exists at least one person to whom the rule does not apply". Hence, $\neg Q$ can then be represented as $\exists x\ \neg R(x)$, indicating that the rule does not apply to certain individuals.

EXAMPLE 1.19.– Let S represent the proposition: "Every polygon has an exterior and an interior," expressed as $\forall x\ [P(x) \rightarrow I(x) \wedge E(x)]$, where P(x) denotes that object "x" is a polygon, I(x) denotes that object "x" has an interior and E(x) denotes that object "x" has an exterior. The negation of S is formulated as follows:

$$\neg[\forall x\ [P(x) \rightarrow I(x) \wedge E(x)]] \equiv \exists x\ \neg[P(x) \rightarrow I(x) \wedge E(x)] \equiv \exists x\ [\neg P(x) \vee (I(x) \wedge E(x))]$$

$$\equiv \exists x\ [P(x) \wedge (\neg(I(x) \wedge E(x)))] \equiv \exists x\ [P(x) \wedge ((\neg I(x)) \vee (\neg E(x)))]$$

$$\equiv \exists x\ [[P(x) \wedge (\neg I(x))] \vee [P(x) \wedge (\neg E(x))]]$$

Thus, $\neg S$ implies that there exist polygons with either no interior or no exterior.

Consider the expression $\exists x\ [F(x) \vee G(x)]$ where, for example, F(x) represents "*x* is a saint" and G(x) represents "x is a genius". It is evident that there are individuals who are either saints or geniuses, indicating the existence of individuals who are saints or individuals who are geniuses, such that:

$$\exists x\ [F(x) \vee G(x)] \equiv [\exists x\ F(x)] \vee [\exists x\ G(x])$$

the generalization to be accepted. Similarly,

$$\forall x\ [F(x) \wedge G(x)] \equiv [\forall x\ F(x)] \wedge [\forall x\ G(x)]$$

1.5. Prenex forms and pure forms

A compound proposition may contain forms such as $\exists x\ [P \wedge F(x)]$, where P is a proposition unrelated to the object x and F(x) is a predicate. Clearly, the scope of the quantifier can be restricted, allowing us to write, for example:

$$\exists x\ [P \wedge F(x)] \equiv P \wedge [\exists x\ F(x)]$$

Similarly,

$$\forall x\ [P \wedge F(x)] \equiv P \wedge [\forall x\ F(x)]$$
$$\exists x\ [P \vee F(x)] \equiv P \vee [\exists x\ F(x)]$$
$$\forall x\ [P \vee F(x)] \equiv P \vee [\forall x\ F(x)]$$

Then, the following deduction can be made:

$$\exists x\,[P \to F(x)] = \exists x\,[(\neg P) \vee F(x)] = (\neg P) \vee [\exists x\,F(x)] = P \to \exists x\,F(x)$$

$$\forall x\,[P \to F(x)] = \forall x\,[(\neg P) \vee F(x)] = \neg P \vee [\forall x\,F(x)] = P \to \forall x\,F(x)$$

$$\exists x\,[P \to F(x)] \equiv P \to \exists x\,F(x)$$
$$\forall x\,[P \to F(x)] \equiv P \to \forall x\,F(x)$$

Nevertheless, the following results are less evident:

$$\exists x\,[F(x) \to P] \equiv \forall x\,F(x) \to P$$
$$\forall x\,[F(x) \to P] \equiv \exists x\,F(x) \to P$$

Indeed, $\exists x\,[F(x) \to P] \equiv \exists x\,[\neg F(x) \vee P) \equiv [\exists x\,\neg F(x)] \vee P \equiv \neg[\forall x\,F(x)] \vee P \equiv \forall x\,F(x) \to P$, with a similar demonstration for the second result.

A compound proposition can also involve multiple predicates, necessitating the distinction of several variables. For instance, $\forall x\,[F(x) \to \exists y[G(y) \wedge H(x)]$ involves two variables, x and y.

Through propositions, predicates, logical operations and quantifiers, schemas can be formed. These schemas are called *monadic* if the predicates contain only one variable:

$\forall x\,[F(x) \to \exists y\,[G(y) \wedge H(x)]$ $F(x) \to \exists y\,[G(y) \wedge H(x)]$ $\exists y\,[G(y) \wedge H(x)]$ $G(y) \wedge H(x)$ are monadic schemas.

In a schema, a variable is *free* if no quantifier applies to it; otherwise, it is *bound*. A schema is *open* if it contains at least one free variable; otherwise, it is *closed*. Thus, the schema $\forall x\,[F(x) \wedge \exists y\,[G(y) \wedge H(x)]]$ is closed, as all variables are linked to a quantifier. Conversely, the schemas $F(x) \to \exists y\,[G(y) \wedge H(x)]$, $\exists y\,[G(y) \wedge H(x)]$, $G(y) \wedge H(x)$ are open, since the variable x is free and not bound to a quantifier.

In a schema, quantifiers can be manipulated in two opposing ways:

– Broadening the scope of a quantifier leads to a *prenex form*. For example, the proposition "$\exists x\,\forall y\,(P(x,y))$" can be transformed into "$\forall y\,\exists x\,(P(x,y))$".

– Restricting the scope of a quantifier leads to a *pure form*. For example, the proposition "$\forall x\,\exists y\,(P(x,y))$" can be transformed into "$\exists y\,\forall x\,(P(x,y))$".

In both scenarios, the previously defined passing rules can be applied. However, it is important to note that there are no passing rules for the biconditional operator ($\leftrightarrow$). Thus, it becomes necessary to "eliminate" biconditionals by substituting them with equivalent expressions. Let us consider the following examples to illustrate the application of the aforementioned rules:

EXAMPLE 1.20.– Consider the schema $P \leftrightarrow \forall x\, [F(x) \rightarrow \exists y\, [F(y) \wedge G(x)]]$, which is equivalent to $\{P \rightarrow \forall x\, [F(x) \rightarrow \exists y\, [F(y) \wedge G(x)]\} \wedge \{\forall x\, [F(x) \rightarrow \exists y\, [F(y) \wedge G(x)]] \rightarrow P\}$. To prevent ambiguity, this schema is rewritten as follows:

$$\{P \rightarrow \forall x\, [F(x) \rightarrow \exists y\, [F(y) \wedge G(x)]\} \wedge \{\forall z\, [F(z) \rightarrow \exists t\, [F(t) \wedge G(zx)]] \rightarrow P\}.$$

This diagram will be transformed into the prenex form. Using the passage rules, the successive steps are as follows:

$$\{\forall x\, [P \rightarrow \exists y\, [F(x) \rightarrow F(y) \wedge G(x)]]\} \wedge \{\exists z\, [\exists t\, [F(z) \rightarrow F(t) \wedge G(z)]] \rightarrow P\}$$

$$\{\forall x\, \exists y\, [P \text{ -> } (F(x) \rightarrow F(y)) \wedge G(x)]\} \wedge \{\exists z\, \forall t\, [F(z) \rightarrow F(t) \wedge G(z)] \rightarrow P]\}$$

$$\forall x\, \exists y\, \exists z\, \forall t\, [P \rightarrow (F(x) \rightarrow F(y) \wedge G(x))] \wedge [(F(z) \rightarrow F(t) \wedge G(z)) \rightarrow P]$$

The diagram is thus converted into the prenex form.

EXAMPLE 1.21.– Consider the schema: $\{\forall x\, [(\exists y\, (F(x) \leftrightarrow G(y))) \vee (\exists y\, F(y))]\} \wedge [\forall x\, (F(x) \vee G(x))]$. After modifying the variable letters for safety, it becomes:

$$\{\forall x\, [(\exists y\, (F(x) \leftrightarrow G(y))) \vee (\exists z\, F(z))]\} \wedge [\forall t\, (F(t) \vee G(t))].$$

First, the diagram will be converted in the prenex form:

$$\exists y\, (F(x) \leftrightarrow G(y)) \vee (\exists z\, F(z)) \equiv \exists y\, \exists z\, (F(z) \leftrightarrow G(y)) \vee F(z);$$
$$\text{hence, } \forall x\, \exists y\, \exists z\, \forall t\, [(F(x) \leftrightarrow G(y)) \vee F(z)] \wedge [F(t) \vee G(t)]$$

This can be summarized in the following prenex form:

$$\forall x\, \exists y\, [(F(x) \leftrightarrow G(y)) \vee F(z)] \wedge [F(t) \vee G(x)]$$

Now, purify the induced schema by:

$$\exists y\, (F(x) \leftrightarrow G(y)) \equiv \exists y\, \{(F(x) \rightarrow G(y)) \wedge (G(y) \rightarrow F(x))]$$

$$\equiv \exists y\, \{\neg F(x) \vee G(y)) \wedge (\neg G(y) \vee F(x))\}$$

$$\equiv \exists y\ \{(F(x)\wedge G(y))\vee(\neg F(x)\wedge\neg G(y))\vee(\neg F(x)\wedge F(x))\vee(G(y)\wedge\neg G(y))\}$$

The last two terms evaluate to "False", leading to the following expression:

$\exists y\ (F(x) \leftrightarrow G(y)) \equiv (\exists y\ (F(x)\wedge G(y)))\vee(\exists y\ (\neg F(x)\wedge\neg G(y))) \equiv (F(x)\wedge(\exists y\ G(y))\vee(\neg F(x)\wedge(\exists y \neg G(y)))$

Therefore, the expression is: $\forall x\ \{(F(x)\wedge(\exists y\ G(y)()\vee(\neg F(x)\wedge(\exists y\ \neg G(y)()\vee(\exists z\ F(z))\}\wedge[\forall t\ F(t)\vee G(t)]$

However,

$$(F(x)\wedge(\exists y\ G(y)))\vee(\neg F(x)\wedge(\exists y\ \neg G(y)))$$

$\equiv (F(x)\vee(\neg F(x)))\wedge((\exists y\ G(y))\vee(\exists y\neg G(y)))\wedge(F(x)\vee(\exists y\ \neg G(y)))\wedge((\exists y\ (y))\vee\neg F(x))$ which induces

$$\forall x\ [(F(x)\wedge(\exists y\ G(y)))\vee(\neg F(x)\wedge(\exists y\ \neg G(y)))\vee(\exists z\ F(z))]$$

$$\equiv \forall x\ [(F(x)\vee(\exists y\ \neg G(y)))\wedge((\exists y\ G(y))\vee\neg F(x))]$$

$$\equiv \forall x\ (F(x)\vee(\exists y\ \neg G(y)))\wedge \forall x\ (((\exists y\ G(y)))\vee\neg F(x))$$

$$\equiv (\forall x\ F(x)\vee(\exists y\ \neg G(y)))\wedge((\exists y\ G(y))\vee\forall x\ \neg F(x))$$

Finally, the pure form is achieved:

$$((\forall x\ F(x))\vee(\exists y\ \neg G(y)))\wedge((\forall x\ \neg F(x))\vee(\exists y\ G(y)))\vee(\exists z\ F(z))\wedge(\forall t\ (F(t)\vee G(t)))$$

2

Sets and Relationships

CONCEPTS COVERED IN THIS CHAPTER.–

The assumption is that the basics of set theory are known to the reader. However, it is deemed appropriate to provide a few reminders, if only to clarify the vocabulary.

After the reminders relating to sets and their algebra, the discussion will approach binary relationships, which actually play a major role in the universe of 0s and 1s.

References: [VEL 19, FAU 73, FAU 79, LIP 83, CON 68].

2.1. Generalities

The primitive terms of set theory are the words *element*, *membership* and *set*, which are related by the expression

$x \in E$ or "element x belongs to set E".

Such an expression is a proposition that takes a True or False value: an element either belongs or does not belong to a given set (as shown in Figure 2.1).

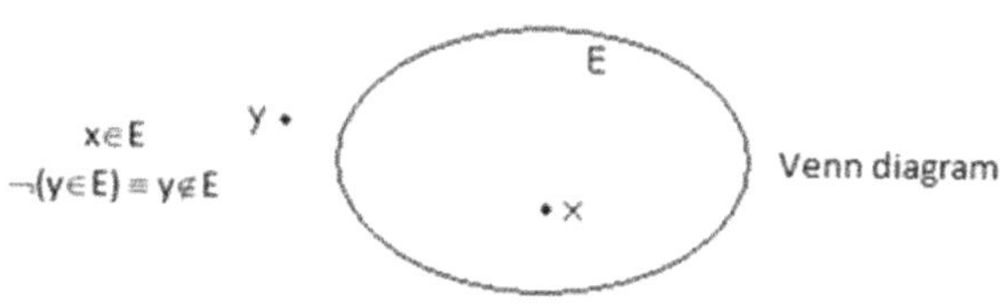

Figure 2.1. *Element, membership, set*

Mathematics for Digital Science 1,
by Gérard-Michel COCHARD and Mhand HIFI.

A set can be characterized in two different ways:

– In *extension*: by listing all the elements of the set using the braces {}: for example, $E = \{0, 1, 2, 3, 4\}$ means that set E consists of the elements 0, 1, 2, 3 and 4.

– In *comprehension*: by defining the set using a property that its elements must satisfy. The notation uses braces {} with a vertical bar | to separate the variable from the property. For example, $E = \{x \in N \mid x < 5\}$ means that set E consists of the natural integers x such that x is strictly less than 5, i.e. $E = \{0, 1, 2, 3, 4\}$.

Note the use of braces {…} to define a set. The elements inside the braces can be placed in any order.

Two sets are *equal* if and only if they contain the same elements: every element of one set is an element of the other and vice versa ($A = B$ if and only if for every element x, x belongs to A if and only if x belongs to B):

$$A = B \Leftrightarrow (x \in B \Leftrightarrow x \in A)$$

A set A is a *part* or *subset* of a set B if every element of A is also an element of B (see Figure 2.2). The relationship is denoted as $A \subseteq B$. If A and B are different, then $A \subset B$.

$$A \subset B \Leftrightarrow (\forall x \in A \Rightarrow x \in B)$$

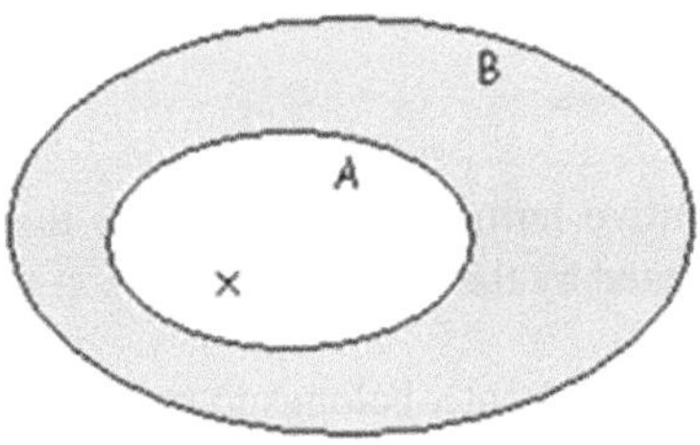

Figure 2.2. *Inclusion*

The equality of two sets A and B can be expressed by:

$$(A = B) \equiv (A \subset B) \wedge (B \subset A)$$

Let E *designate* the *universal* set, containing all the elements under consideration, and Ø the *empty* set, which by definition contains no elements. The following properties can then be easily demonstrated:

– $\forall A$, $\emptyset \subseteq A \subseteq E$: any set A contains the empty set and is a subset of the universal set.

– $\forall A, A \subseteq A$: any set A is a subset of itself.

– $(A \subseteq B) \wedge (B \subseteq C) \Rightarrow A \subseteq C$: if A is a subset of B, and B is a subset of C, then A is a subset of C.

EXAMPLE 2.1.– Let us illustrate this first wave of definitions with the following example borrowed from Lewis Caroll. Consider the following four propositions:

P1: my pans are the only objects I own that are made of pewter.

P2: all your gifts are very useful.

P3: not one of my pans has any use whatsoever.

P: the gifts you gave me are not made of pewter

Let us show that P1 $\wedge$ P2 $\wedge$ P3 $\rightarrow$ P is a tautology.

Figure 2.3 expresses the situation.

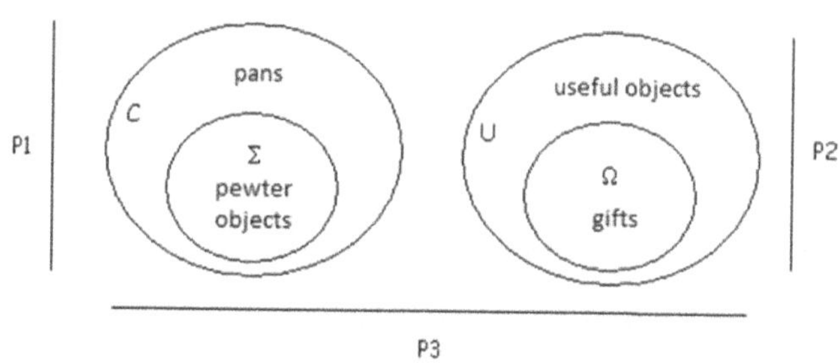

Figure 2.3. *Representation of propositions P1, P2, P3*

Proposition P1 expresses that the set of pans C contains the set of pewter objects Σ: $\Sigma \subset C$.

Proposition P2 expresses that the set of useful objects U contains the set of gifts Ω: $\Omega \subset U$.

Finally, proposition P3 expresses that the sets C and U have no common elements; it follows that Σ and Ω neither. So

$$x \in \Omega \Rightarrow x \notin \Sigma$$

which is proposition P.

In example 2.1, the sets Σ and Ω are *disjoint*. In general, the *intersection* of two sets A and B is the set of elements common to A and B (see Figure 2.4):

$$A \cap B = \{x \,/\, (x \in A) \wedge (x \in B)\}$$

Two sets are disjoint if their intersection is the empty set, denoted as ∅.

The *union* of two sets A and B is the set of elements belonging to A or B (see Figure 2.4)

$$A \cup B = \{x \,/\, ((x \in \mathrm{A}) \vee (x \in B)\}$$

The *complementary* $\bar{A}$ of a set A is defined as the set of all elements that do not belong to A; it is also denoted as C_A or $C_E A$ *or* $E\text{-}A$ (see Figure 2.4). Formally, it is written as:

$$\bar{A} = \{x \mid x \notin A\}.$$

The notation $\bar{A}$ means that $\bar{A}$ is the *set* of all x such that x does not belong to the set A.

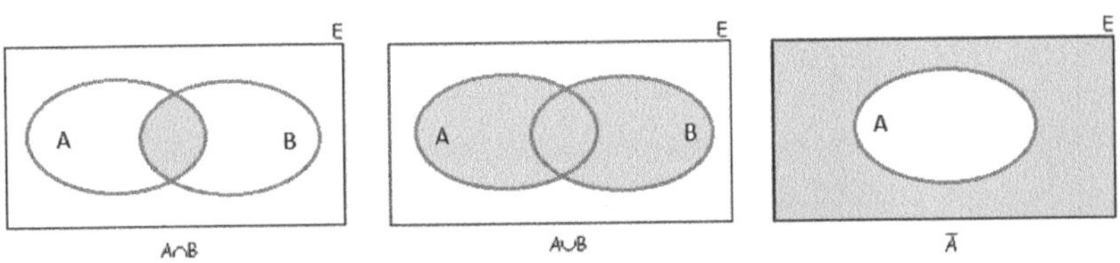

Figure 2.4. *Intersection, union, complementarity*

The cardinal of set A, denoted as #A, represents the number of elements contained in A. If $\#A$ is finite, set A is considered finite. Additionally, the cardinality of the empty set (sometimes denoted by the symbol "∅" or "{}") is equal to zero, i.e. $\#\emptyset = 0$.

The following theorems can be proven:

– If A and B are finite sets such that $A \cap B = \emptyset$, then their union $A \cup B$ is also finite, and $\#(A \cup B) = \#A + \#B$.

– If A and B are finite sets, then their union $A \cup B$ is finite, their intersection $A \cap B$ is finite, and $\#(A \cup B) = \#A + \#B - \#(A \cap B)$.

2.2. Algebra of sets

The operations of union (∪), intersection (∩) and complementarity (–) of sets have a certain number of properties, defined either by definition or by deduction. Below are listed these properties:

– Idempotence: $A \cup A = A$ and $A \cap A = A$.

– Associativity: $(A \cup B) \cup C = A \cup (B \cup C)$

$(A \cap B) \cap C = A \cap (B \cap C)$.

– Commutativity: $A \cup B = B \cup A$ and $A \cap B = B \cap A$.

– Distributivity: $A \cup (B \cap C) = (A \cup B) \cap (A \cup C)$

$A \cap (B \cup C) = (A \cap B) \cup (A \cap C)$.

– Identity: $A \cup \emptyset = A \quad A \cup E = E \quad A \cap \emptyset = \emptyset \quad A \cap E = A$.

– Involution: $\bar{\bar{A}} = A$.

– Complementarity: $A \cup \bar{A} = E \quad \bar{E} = \emptyset \quad \bar{\emptyset} = E \quad A \cap \bar{A} = \emptyset$.

– Morgan's laws: $\overline{A \cup B} = \bar{A} \cap \bar{B} \quad \overline{A \cap B} = \bar{A} \cup \bar{B}$.

The following example will be used to illustrate the demonstrations of some of these operations:

Example 2.2.– Recalling that the demonstration of Morgan's first law consists of showing that for any set of two or more elements A, B, C, etc., the union of their complements is equivalent to the complement of their intersection. Let us therefore prove Morgan's first law on two sets A and B, where the aim is to show that $x \in \overline{A \cup B} \Leftrightarrow x \in \bar{A} \cap \bar{B}$. Proceeding with the following steps:

1) $x \in \overline{A \cup B} \Leftrightarrow x \notin A \cup B$.

2) $x \notin A \cup B \Leftrightarrow \neg(x \in A \cup B)$.

3) $\neg(x \in A \cup B) \Leftrightarrow \neg((x \in A) \vee (x \in B))$.

4) $\neg((x \in A) \vee (x \in B)) \Leftrightarrow \neg(x \in A) \wedge \neg(x \in B)$

5) $\neg(x \in A) \wedge \neg(x \in B) \Leftrightarrow (x \notin A) \wedge (x \notin B)$.

6) $(x \notin A) \wedge (x \notin B) \Leftrightarrow (x \in \bar{\bar{A}}) \wedge (x \in \bar{\bar{B}})$.

– $(x \in \bar{A}) \wedge (x \in \bar{B}) \Leftrightarrow x \in \bar{A} \cap \bar{B}$.

2.3. Parts and partitions

Given a set E, we denote by $\mathcal{P}_E$ the set of its parts. First of all, it should be noted that $\mathcal{P}_E$ is not empty, because it contains at least the empty set ($\varnothing$) and itself (E).

EXAMPLE 2.3.– Consider the set $E = \{0, 1, 2, 3\}$, whose cardinality is $\#E = 4$.

The parts of E are the sets that contain one or more elements of E. Thus, the parts of E can be classified according to their number of elements:

– Parts with 0 elements: $\emptyset$.

– Parts with 1 element: $\{0\}, \{1\}, \{2\}, \{3\}$.

– Parts with 2 elements: $\{0, 1\}, \{0, 2\}, \{0, 3\}, \{1, 2\}, \{1, 3\}, \{2, 3\}$.

– Parts with 3 elements: $\{0, 1, 2\}, \{1, 2, 3\}, \{0, 2, 3\}, \{0, 1, 3\}$.

– The parts with 4 elements: $\{0, 1, 2, 3\}$, which is set E itself.

Thus, there are

$\mathcal{P}_E = \{\emptyset, \{0\}, \{1\}, \{2\}, \{3\}, \{0, 1\}, \{0, 2\}, \{0, 3\}, \{1, 2\}, \{1, 3\}, \{2, 3\}, \{0, 1, 2\}, \{1, 2, 3\}, \{0, 2, 3\}, \{0, 1, 3\}, E = \{0,1,2,3\}\}$.

The number of parts of set E, denoted as $\#\mathcal{P}(E)$, is equal to 2 raised to the power of the cardinal of E, denoted as $\#E$. In this specific case,

$$\#\mathcal{P}(E) = 16 = 2^4 = 2^{\#E}.$$

The result can be generalized to any finite set (and will be demonstrated later):

$$\#\mathcal{P}(E) = 2^{\#E}$$

A family of parts of set E, denoted as $p(E)$, is a *partition* of E if the three following conditions are verified:

– No element of $p(E)$ is empty: $\forall A \in p(E), A \neq \varnothing$.

– All the elements of $p(E)$ are pairwise disjoint:

$$\forall A \in p(E), \forall B \in p(E), A \neq B \quad A \cap B = \varnothing.$$

– The union of all the elements of p(E) is equal to E: the list of all the elements of p(E) contains exactly all the elements of E, and no others.

The definition of partition is important in mathematics because it allows a whole to be broken-down into non-overlapping parts that cover the entire whole.

EXAMPLE 2.4.– Figures 2.5–2.8 show some partitions of a given set E.

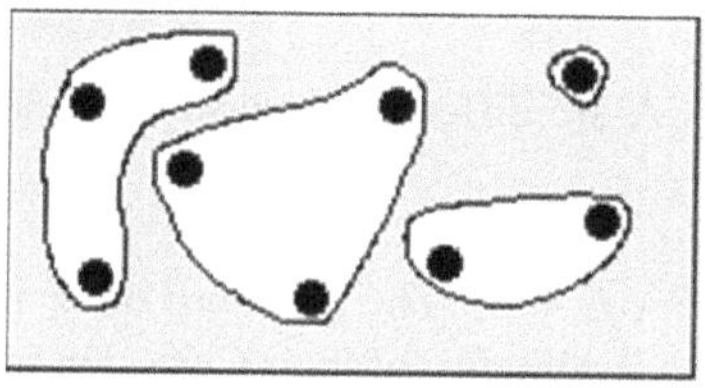

Figure 2.5. *Partition 1*

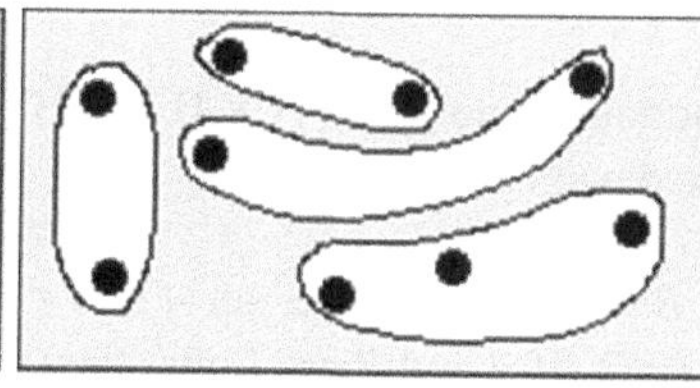

Figure 2.6. *Partition 2*

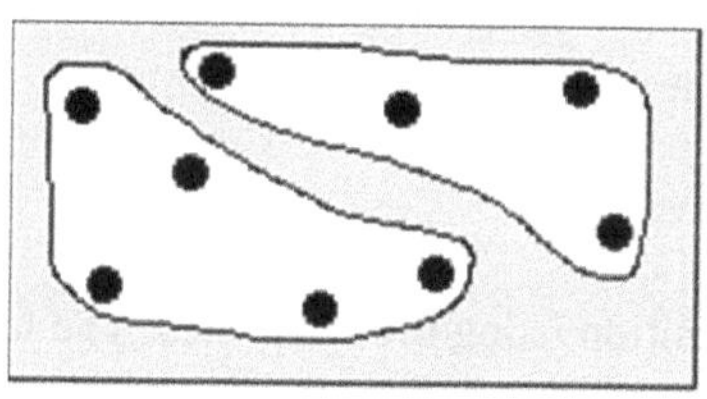

Figure 2.7. *Partition 3*

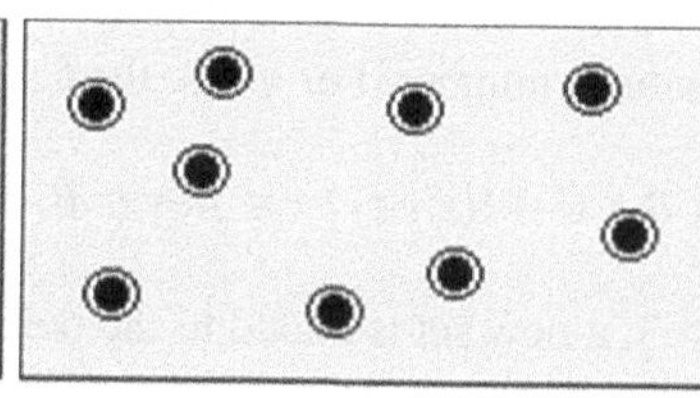

Figure 2.8. *Partition 4*

2.4. Cartesian product

A set of two elements $\{a, b\}$ can be associated with *two distinct ordered pairs*: (a, b) and (b, a). In an ordered pair (a, b), the first element is a and the second element is b. The ordered pairs (a, b) and (b, a) are different, and the equality of two ordered pairs is defined as follows:

$$(a, b) = (\mathrm{c}, \mathrm{d}) \Leftrightarrow (a = c) \wedge (b = d).$$

In general, given two sets A and B, the *Cartesian product* $A \times B$ *is defined* as the set of all ordered pairs (x, y) such that the first element x belongs to A and the second element y belongs to B:

$$A \times B = \{(x, y) \,/\, x \in A, y \in B\}.$$

The Cartesian product is an important tool in mathematics for constructing new sets from existing sets.

EXAMPLE 2.5.– If A = {0, 1} and B = {+, −, ×, ÷}, then we have:

$$A \times B = \{(0, +), (0, -), (0, \times), (0, \div), (1, +), (1, -), (1, \times), (1, \div)\}$$

$$B \times A = \{(+, 0), (+, 1), (-, 0), (-, 1), (\times, 0), (\times, 1), (\div, 0), (\div, 1)\}$$

It is important to note that A × B and B × A are different because their elements are ordered pairs.

If $A = B = E$, in this case, set A x $B = E$ x $E = E^2$. The *diagonal* of E^2 is the set $\{(x, x) / x \in E\}$. If A and B are finite, then the cardinality of their Cartesian product A x B is equal to the product of their respective cardinalities: $\#(A \text{ x } B) = \#A \text{ x } \#B$.

Finally, the notion of the Cartesian product generalizes to n sets, where n is a non-zero natural number. For $n = 3$, the Cartesian product is:

$$A \text{ x } B \text{ x } C = \{(x,y,z) / x \in A, y \in B, z \in C\}$$

For $n > 3$, a new set is added to the definition using the same idea. The Cartesian product of n sets $A_1, A_2, \ldots, A_n$ is defined as the set of all tuples $(x_1, x_2, \ldots, x_n)$ such that x_i belongs to A_i for all i from 1 to n:

$$A_1 \text{ x } A_2 \text{ x } \ldots \text{ x } A_n = \{(x_1, x_2, \ldots, x_n) / x_1 \in A_1, x_2 \in A_2, \ldots, x_n \in A_n\}$$

Note that, if all the sets $A_1, A_2, \ldots, A_n$ are finite, then the cardinality of the Cartesian product is given by:

$$\#(A_1 \text{ x } A_2 \text{ x } \ldots \text{ x } A_n) = (\#A_1)(\#A_2) \ldots (\#A_n)$$

2.5. Binary relationships

Consider two sets A and B and their Cartesian product A x B. Given a pair $(a,b) \in A$ x B, a relationship $\mathfrak{R}$ between the elements of A and the elements of B can be defined as follows:

– If the proposition $a\mathfrak{R}b$ is true, then a is related to b through $\mathfrak{R}$.

– If the proposition $a\mathfrak{R}b$ is false, then a is not related to b by $\mathfrak{R}$.

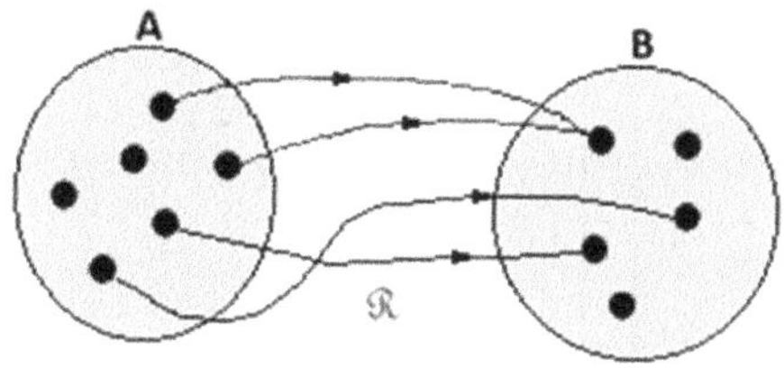

Figure 2.9. *ℜ relationship graph*

In all cases, ℜ is a *binary relationship* between *A* and *B*. Representing the relationships between the elements of *A* and *B* by arrows (see Figure 2.9) yields a graph of the relationship.

EXAMPLE 2.6.– Some illustrations.

Inclusion expresses a relationship between parts of the same set: for any set *E*, if *A* and *B* are subsets of *E*, then:

$$\forall A \in \mathcal{P}(E) \qquad \forall B \in \mathcal{P}(E) \qquad A \subset B \text{ or } A \not\subset B.$$

The expression "is married to" throughout a human community denotes a binary relationship. Within a human community, if *h* and *f* represent two individuals, a binary relationship ℜ ("is married to") between them can be defined such that

$h \Re f$ if and only if *h* is married to *f*.

In the set of real numbers, ℜ, the notations < or ≥ correspond to a binary relationship:

$$\forall x, y \in \text{R} \;\; x < y \text{ or } x \geq y$$

In the set N of natural numbers, the relationship ℜ expressing the integer division denoted by "|" expresses a binary relationship:

$\forall x \in \text{N}, \;\forall y \in \text{N} \;\; x|y$ is true or false ("|" means divide).

The graph of ℜ is defined as a subset *R* of the Cartesian product *A* x *B*. This subset perfectly defines the relationship ℜ, hence the common practice of using the same name for both.

EXAMPLE 2.7.– The relationship ℜ is defined by the expression $\Re \subseteq \text{A x B}$, which indicates that it is a subset of the Cartesian product A x B, as shown in Figure 2.10.

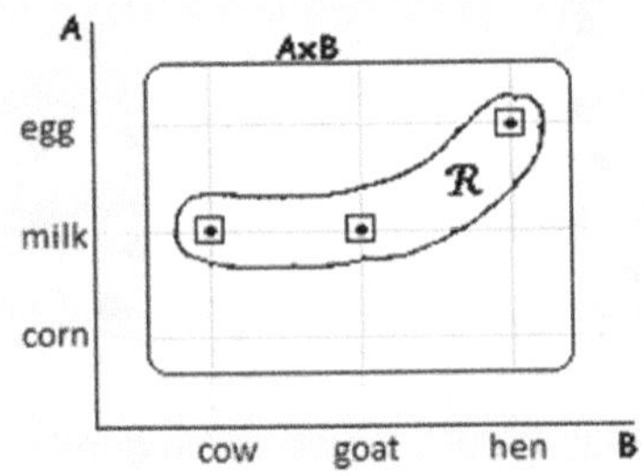

Figure 2.10. *ℜ relationship graph*

For a given binary relationship ℜ, its inverse relationship $\mathfrak{R}^{-1}$ is defined by reversing the direction of the arrows in the graph of the relationship ℜ, as shown on the right in Figure 2.11.

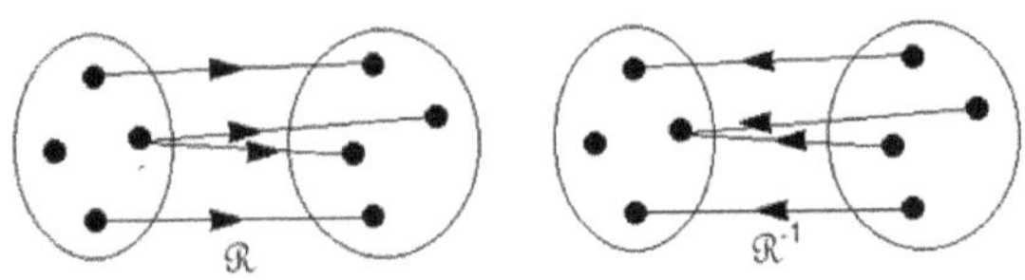

Figure 2.11. *Reverse of R*

We note $\mathfrak{R}^{-1} = \{(b,a) / (a,b) \in \mathfrak{R}\}$.

Consider two relationships ℜ and 𝔖, which are subsets of E x F and F x G, respectively (see Figure 2.12). An element x of E and an element z of G are linked by a relationship 𝔗 if there exists an element y of F such that $x\mathfrak{R}y$ and y𝔖z.

𝔗 is denoted as 𝔖oℜ (note the order of writing), and 𝔗 is referred to as the composite of ℜ and 𝔖.

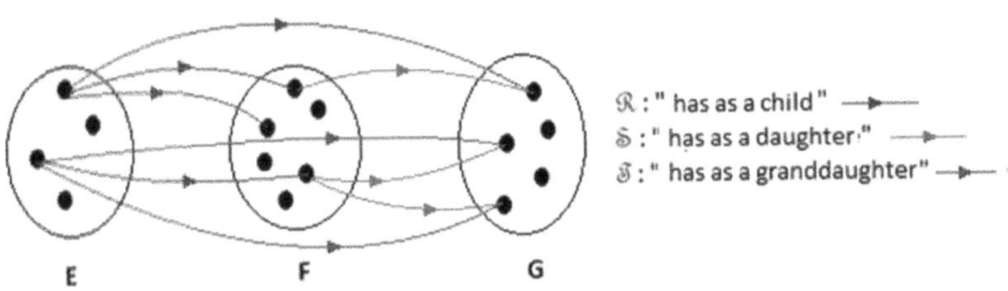

Figure 2.12. *Composition of relationships. For a color version of this figure, see www.iste.co.uk/cochard/mathematics1.zip*

2.6. Properties of relationships

Three properties can characterize binary relationships: reflexivity, symmetry and transitivity.

Reflexivity: for a binary relationship $\mathcal{R}$ in E^2, $\mathcal{R}$ is reflexive if and only if $\forall x \in E,\ x\mathcal{R}x$.

It follows that if $\mathcal{R}$ is reflexive, it must contain the diagonal of E^2, as shown in Figure 2.13.

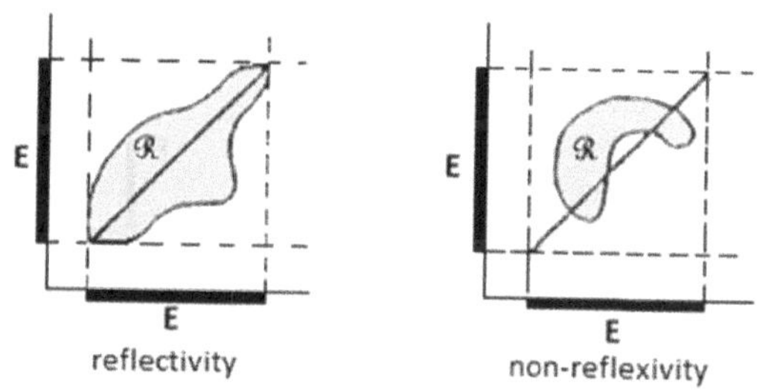

Figure 2.13. *Reflexivity and non-reflexivity*

EXAMPLE 2.8.– Let $\mathcal{R}$ be the binary relationship on E = N defined by: "having the same parity as". Thus, the pairs (0,0), (1,1), (2,2), … are part of $\mathcal{R}$. Additionally, $\mathcal{R}$ is a reflexive relationship.

EXAMPLE 2.9.– Let $\mathcal{R}$ be the binary relationship on E = N defined by: "have as square". So, (0,0) and (1,1) are part of $\mathcal{R}$, but (2,2) is not. It can be deduced that $\mathcal{R}$ is not a reflexive relationship.

It is easy to determine whether a relationship $\mathcal{R}$ on a set E is reflexive by examining its graph. Indeed, if $\mathcal{R}$ is reflexive, there will be a loop on each point of the graph. Figure 2.14 displays the graph of a reflexive relationship.

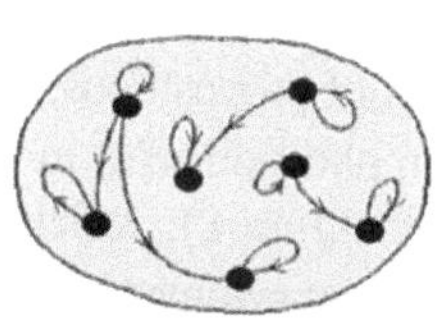

Figure 2.14. *Reflexivity*

Symmetry: a relationship $\Re$ on E^2 is symmetric if and only if

$$\forall x, y \in E, x \neq y, \quad x\Re y \Rightarrow y\Re x$$

Antisymmetry: a relationship $\Re$ on E^2 is antisymmetric if and only if

$$\forall x, y \in E, x \neq y, \quad x\Re y \nRightarrow y\Re x$$

Figure 2.15 shows examples of binary relationships on a set E and their symmetry and antisymmetry properties.

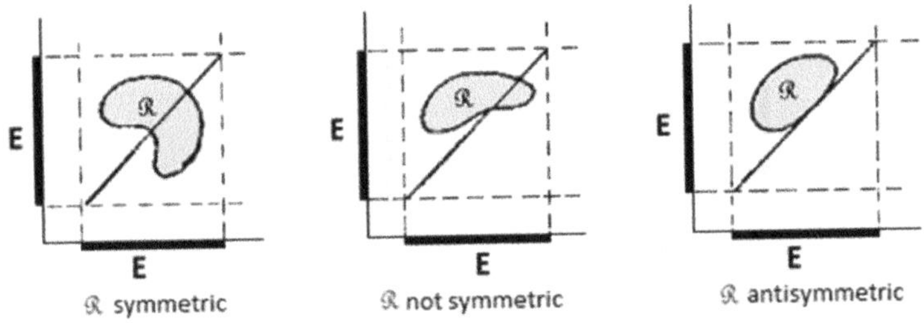

Figure 2.15. *Symmetry and antisymmetry*

EXAMPLE 2.10.– $\Re$: "have the same parity as", E = N, $\Re$ is symmetric. Thus (2,8), (8,2)$\in \Re$, (3,7), (7,3) $\in \Re$.

EXAMPLE 2.11.– $\Re$: "have as square", E = N, $\Re$ is antisymmetric as can be noted on all elements $x \neq y$. Thus (2,4)$\in \Re$ but (4,2) $\in \Re$.

EXAMPLE 2.12.– $\Re$: "less than or equal", the symbol $\Re$ = "$\leq$ ", E = N, $\Re$ is antisymmetric. Thus, (2,4) $\in \Re$, but (4,2) $\notin \Re$, since we can write $2 \leq 4$, but not $4 \leq 2$.

EXAMPLE 2.13.– $\Re$: "$\Rightarrow$", E = set of propositions. In general, if P and Q belong to the set of propositions E, the implication of P to Q ($P \Rightarrow Q$) does not logically imply the implication of Q to P ($Q \Rightarrow P$), which shows that the relationship $\Re$ represented by the symbol "$\Rightarrow$" is not symmetric.

Graphs allow us to determine whether a relationship is symmetric, asymmetric (not symmetric) or antisymmetric (see Figure 2.16).

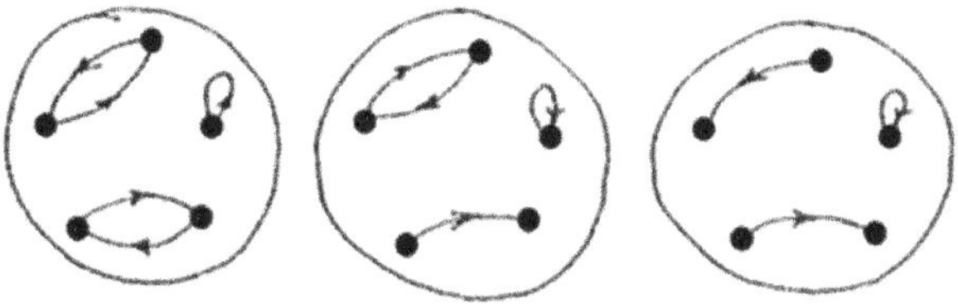

Figure 2.16. *Symmetry, antisymmetry*

Transitivity: the transitivity of a relationship $\mathfrak{R}$ in set E^2 is characterized by the following property: $\mathfrak{R}$ is transitive if and only if, for any triplet (x, y, z) of elements of E^2 such that x is in relationship $\mathfrak{R}$ with y and y is in relationship $\mathfrak{R}$ with z, then x is in relationship $\mathfrak{R}$ with z. In other words, the following logical implication is verified:

$$(x\mathfrak{R}y) \wedge (y\mathfrak{R}z) \Rightarrow x\mathfrak{R}z$$

Figure 2.17 shows two examples of relationships, one being transitive and the other not.

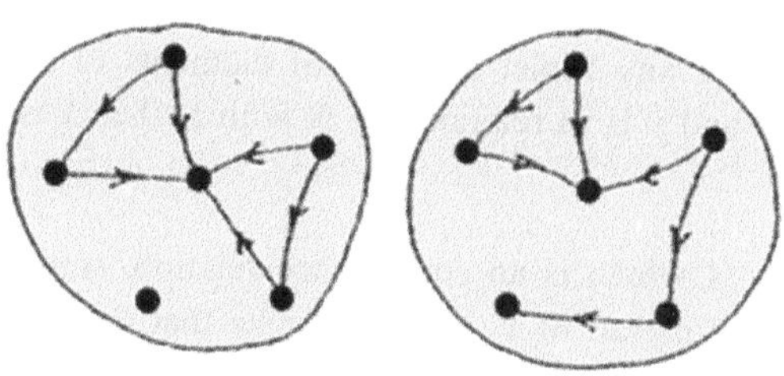

Figure 2.17. *Transitivity*

EXAMPLE 2.14.– Let $\mathfrak{R}$ be a relationship defined on set E of integers N, representing the relationship: "have the same parity as".

The relationship $\mathfrak{R}$ is transitive, meaning that for any triplet of integers (x, y, z) such that x is in relationship $\mathfrak{R}$ with y and y is in relationship $\mathfrak{R}$ with z, then x is in relationship $\mathfrak{R}$ with z. Thus, given that $(2, 8) \in \mathfrak{R}$, $(8, 12) \in \mathfrak{R}$ and $(2, 12) \in \mathfrak{R}$, it follows that the relationship $\mathfrak{R}$ is indeed transitive.

EXAMPLE 2.15.– Non-transitivity. Let $\mathfrak{R}$ be a relationship defined on set E of integers N, representing the relationship: "have as square". Although (2, 4) and (4, 16) belong to $\mathfrak{R}$, it is evident that (2, 16) does not belong to $\mathfrak{R}$, indicating that the relationship $\mathfrak{R}$ is not transitive.

EXAMPLE 2.16.– Transitivity. Let $\mathfrak{R}$ be a relation defined on set E of integers N, representing the relationship $\mathfrak{R}$: "to be greater than or equal to" (denoted by "≥"). It is known that if x is in relationship $\mathfrak{R}$ with y ($x \geq y$) and y is in relationship $\mathfrak{R}$ with z ($x \geq z$), then x is in relationship $\mathfrak{R}$ with z ($x \geq z$). Thus, the relationship $\mathfrak{R}$ is transitive.

A relationship $\mathfrak{R}$ defined on a set E is an *equivalence relationship* if it satisfies the following three properties:

– $\mathfrak{R}$ is reflexive: for any element x of E, x is in relationship $\mathfrak{R}$ with itself (noted $x \mathfrak{R} x$).

– $\mathfrak{R}$ is symmetric: for any element x and y of E, if x is in relationship $\mathfrak{R}$ with y ($x \mathfrak{R} y$), then y is in relationship $\mathfrak{R}$ with x ($y \mathfrak{R} x$).

– $\mathfrak{R}$ is transitive: for any triplet (x, y, z) of elements of E such that x is in relationship $\mathfrak{R}$ with y and y is in relationship $\mathfrak{R}$ with z, then x is in relationship $\mathfrak{R}$ with z.

For a relationship $\mathfrak{R}$ which is an equivalence relation, sometimes the notation $x \equiv y \pmod{\mathfrak{R}}$ is used instead of $x \mathfrak{R} y$ to indicate that the elements x and y are equivalent according to the relationship $\mathfrak{R}$.

EXAMPLE 2.17.– Equivalence relationship. Let $\mathfrak{R}$ be defined on set E of real numbers R, representing the equality relationship (denoted by "="). The following properties hold: $\mathfrak{R}$ is reflexive because every real number x is equal to itself ($x = x$). $\mathfrak{R}$ is symmetric because for any pair of real numbers x and y, if x is equal to y ($x = y$), then y is equal to x ($y = x$). $\mathfrak{R}$ is transitive because for any triple of real numbers x, y and z, if x is equal to y ($x = y$) and y is equal to z ($y = z$), then x is equal to z ($x = z$). Thus, the relationship $\mathfrak{R}$ is an equivalence relationship.

EXAMPLE 2.18.– Equivalence relationship. Let the relationship $\mathfrak{R}$ be defined on set E of lines in space, representing the parallelism relationship (denoted by "//"). The following properties hold: $\mathfrak{R}$ is reflexive because every line D is parallel to itself (D // D). $\mathfrak{R}$ is symmetric because for any pair of lines D and D', if D is parallel to D' (D // D'), then D' is also parallel to D (D' // D). $\mathfrak{R}$ is transitive because for any triplet of lines D, D' and D", if D is parallel to D' (D // D') and D' is parallel to D" (D' // D"), then D is also parallel to D" (D // D"). Thus, $\mathfrak{R}$ is an equivalence relationship.

Given an equivalence relationship $\mathfrak{R}$ on set E, the *equivalence class* of the element a of E, denoted as $\mathring{a}$, is the set of all the elements of E which are in relationship with a according to $\mathfrak{R}$. In other words, the equivalence class $\mathring{a}$ is defined as follows:

$$\mathring{a} = \{x \in E \mid x \,\mathfrak{R}\, a\}$$

Using modular notation, it can also be written as:

$$\mathring{a} = \{x \in E \mid x \equiv a \ (\mathrm{mod}\ \mathfrak{R})\},$$

which is the equivalence relationship defined by $\mathfrak{R}$.

The equivalence classes are parts of set E, constituting a *partition* of E. This is evident because

– For any element a of E, the corresponding equivalence class, denoted as $\mathring{a}$, is non-empty because it contains at least the element a ($a \in \mathring{a}$).

– $E = \cup_a \mathring{a}$ (obvious).

– The equivalence classes are disjoint: if x belongs to two different equivalence classes $\mathring{a}$ and $\mathring{a}'$, then $\mathring{a} = \mathring{a}'$ (if $x \in \mathring{a}$ and $x \in \mathring{a}'$, then $\mathring{a} = \mathring{a}'$).

Conversely, any partition of E corresponds to an equivalence relationship. The relationship $\mathfrak{R}$: "belongs to the same class as" is reflexive, symmetric and transitive, thereby defining an equivalence relationship. The set of equivalence classes of E for a relationship $\mathfrak{R}$ is called the quotient set $E/\mathfrak{R}$. Such a set is itself a set whose elements are the equivalence classes of E. For example, if E is the set of relative integers and $\mathfrak{R}$ is the equivalence relationship defined by "x and y have the same remainder in division by 3", then the quotient set $E/\mathfrak{R}$ contains three elements:

– the class of integers which have a remainder of 0 in division by 3;

– the class of integers which have a remainder of 1;

– the class of integers which have a remainder of 2.

Figure 2.18 shows the set of equivalence classes of E for a relationship $\mathfrak{R}$ (partition of E), called the quotient set $E/\mathfrak{R}$:

$$E/\mathfrak{R} = \{A, B, C\}$$

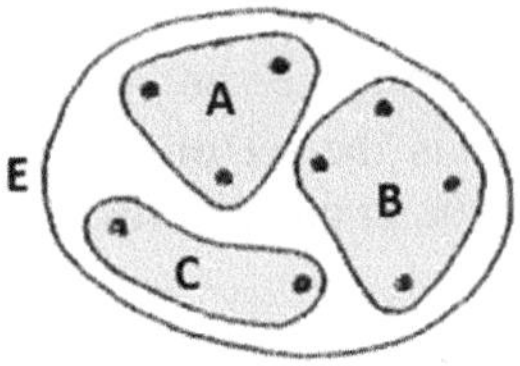

Figure 2.18. *Equivalence classes*

A relationship $\mathfrak{R}$ defined on a set E is a *(partial) order relation* if the following conditions hold:

– $\mathfrak{R}$ is reflexive: $\forall x \in E, x \,\mathfrak{R}\, x$.

– $\mathfrak{R}$ is antisymmetric: $\forall x, y \in E$, if $x \,\mathfrak{R}\, y$ and $y \,\mathfrak{R}\, x$, then $x = y$.

– $\mathfrak{R}$ is transitive: $\forall$x, y, z $\in$ E, if x $\mathfrak{R}$ y and y $\mathfrak{R}$ z, then x $\mathfrak{R}$ z.

A *total order relationship*, in addition to these conditions, must also satisfy the property $\mathfrak{R}$ is total: $\forall x, y \in E, x \,\mathfrak{R}\, y$ or $y \,\mathfrak{R}\, x$.

A partial order relationship defines a preorder structure on E, while a total order relationship defines a total order structure.

EXAMPLE 2.19.– Let $E = \{1, 2, 3, 4, 5, 6, 7, 8, 9, 10, 11, 12\}$ and the relationship $\mathfrak{R}$ defined on E by "divides" or "|". $\mathfrak{R}$ is an order relationship on E because it satisfies the following three properties: $\mathfrak{R}$ is reflexive because every element of E divides itself, i.e. for all x in E, $x|x$; $\mathfrak{R}$ is antisymmetric because if $x|y$ and $y|x$, then $x = y$, i.e. if x and y are two divisors of each other, they must be equal; $\mathfrak{R}$ is transitive because if $x|y$ and $y|z$, then $x|z$, i.e. if x divides y and y divides z, then x divides z.

Figure 2.19 shows the relationship $\mathfrak{R}$: each point represents the division relationship between two numbers (the point (3, 6) means 3|6). It is evident that $\mathfrak{R}$ satisfies the three properties of the order relationship, and therefore it is indeed an order relationship on E.

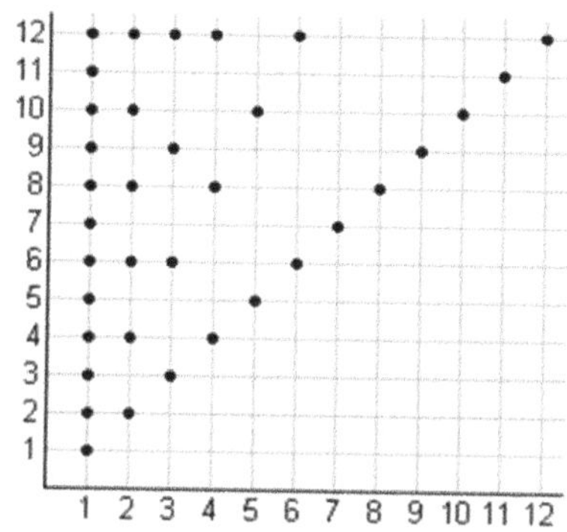

Figure 2.19. *Order relationship*

EXAMPLE 2.20.– Let E be the set of towns located on a given river and $\mathfrak{R}$ be the relationship "is downstream of". $\mathfrak{R}$ is an order relationship on E, because it satisfies the following properties: $\mathfrak{R}$ is reflexive: each city is downstream of itself; $\mathfrak{R}$ is antisymmetric: if a city A is downstream of B, then B is not downstream of A; $\mathfrak{R}$ is transitive: if a city A is downstream from a city B, and B is downstream from a city C, then A is downstream from C.

An order can be total *if* there exists a comparison relationship between all the elements of E. Otherwise, the order is said to be *partial* (see Figure 2.20).

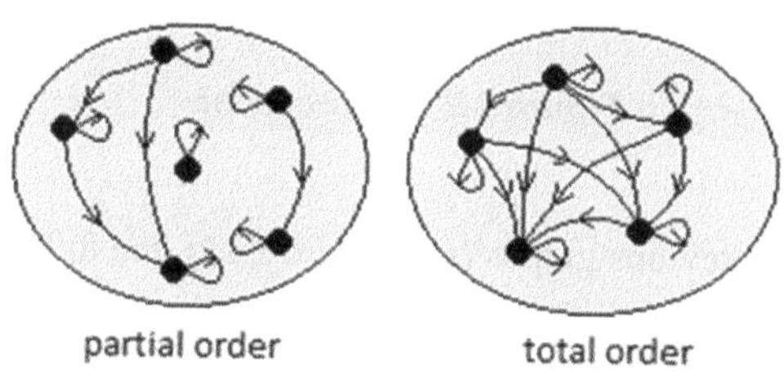

Figure 2.20. *Partial order, total order*

EXAMPLE 2.21.– "$\subset$" defines a partial order on a set of sets. If E is the set of finite sets of natural integers, the relationship "inclusion" or "$\subset$" is a partial order on E, because "$\subset$" is reflexive: every set is included in itself, "$\subset$" is antisymmetric: if A is included in B and B is included in A, then A = B, "$\subset$" is transitive: if A is included in B and B is included in C, then A is included in C. However, this order is not total, because there are sets that are not comparable with each other. For example, $\{1, 2\}$ and $\{2, 3\}$ are neither included in each other nor equal.

EXAMPLE 2.22.– "$\geq$" defines a total order on a set of numbers. For example, if we consider the numbers 3, 5 and 7, the following total order can be established: $7 \geq 5 \geq 3$. This total order facilitates the comparison of numbers and allows them to be classified hierarchically.

2.7. Applications

Consider a binary relationship f from a set E to a set F. The relationship f is an *application* (or *function*) if for every element x of E, there exists a unique element y of F such that x is associated with y by the relationship f, which is generally noted $y = f(x)$ instead of $x\,f\,y$. Then, y is considered the *image* of x by the function f, or x is considered an *antecedent* of y by the function f. This notation facilitates the understanding and manipulation of functions by associating each element of the starting set E with a unique element of the arrival set F.

Consider the graphs in Figure 2.21.

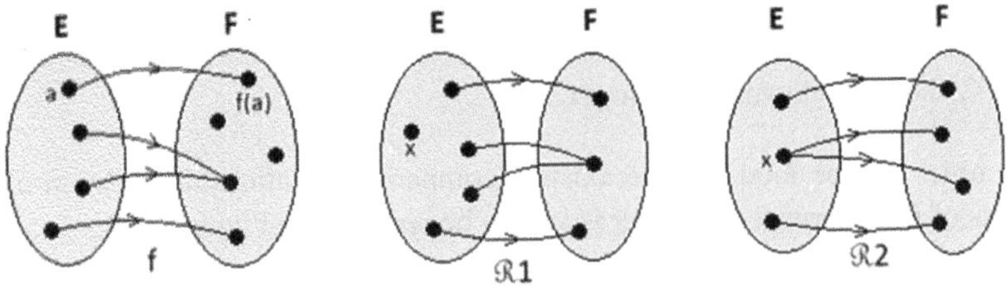

Figure 2.21. *Applications*

By examining the previous graphs (see Figure 2.21), the following is evident:

– The binary relationship f is an application, because each element of the starting set is associated with a unique element of the ending set.

– The binary relationship $\Re 1$ is not an application, because the element x of the starting set is not associated with any element of the target set, meaning it does not have an image.

– The binary relationship $\Re 2$ is not an application, because the element x of the starting set is associated with two different elements of the target set, which contradicts the definition of a function that must associate a unique element of the arrival set to each element of the starting set.

Set E is the *domain of definition* of the function f, representing the set of values for which the function is defined and can be evaluated. It is generally noted as:

$$E = \text{Dom(f) or } E \xrightarrow{f} \text{F}$$

Such a notation explicitly specifies the starting set of the function and facilitates the understanding and manipulation of the function.

A function f from set E to set F is:

– *surjective* if any element of the target set F is the image of at least one element of the starting set E;

– *injective* if any element of the target set F has at most one image in the starting set E;

– bijective if the function f is both surjective and injective, meaning every element of the target set F has exactly one image in the starting set E.

To better understand these definitions, Figure 2.22 graphically shows the concepts of surjectivity, injectivity and bijectivity of a function.

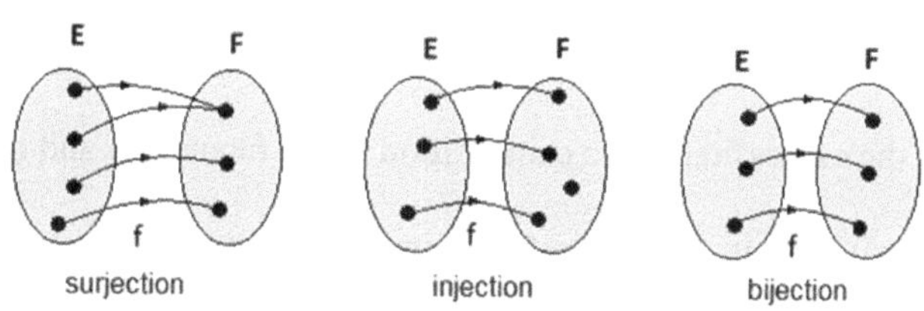

Figure 2.22. *Surjection, injection and bijection*

It is important to note that if f is a bijection from set E to set F, then the sets E and F have the same cardinality. It means that there is a one-to-one correspondence between the elements of set E and those of set F, implying that these two sets contain the same number of elements. Such a property is often used to establish equivalences between different sets and facilitate calculations in set theory.

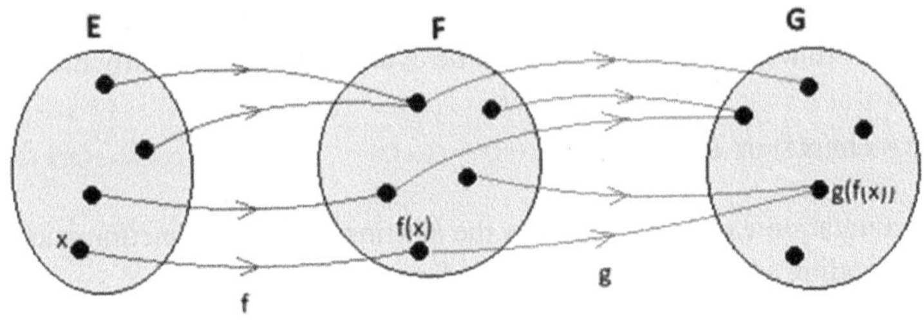

Figure 2.23. *Composition of two applications. For a color version of this figure, see www.iste.co.uk/cochard/mathematics1.zip*

Consider three sets E, F and G, as well as two applications f from E to F and g from F to G, as shown in Figure 2.23.

Note that the composition of two functions f and g defines a new function h, called the composite of f and g. For any element x of set E, the function $h(x)$ is defined as the image of x by the composition of the functions f and g, obtained by successively applying f and then g to x. Thus, any element in set E will have an image in set G via the composite function h. This operation is noted as:

$$h = gof$$

and simplifies the expression of the composition of the functions f and g.

If f is a bijection from set E to set F, it implies the existence of a reciprocal or inverse application of f, denoted as f^{-1}, which enables the transition from F to E by reversing the arrows of the graph of f. The function f^{-1} is also a bijection, indicating that each element y of F corresponds to a unique element x of E such that $y = f(x)$ and that each element x of E corresponds to a unique element y of F such that $y = f(x)$. In other words, if $x = f^{-1}(y)$, then $y = f(x)$ and vice versa. The reciprocal function f^{-1} is therefore defined by $x = f^{-1}(y) \Leftrightarrow y = f(x)$.

If I designates the identical application which is defined on a set E by $\forall x \in E\ I(x) = x$, we have $fof^{-1} = f^{-1}of = I$.

Now, let us examine the number of distinct applications from the finite set A to the finite set B. To answer this question, we can reason from the diagram in Figure 2.24, where each element of set A is connected to all the elements of set B.

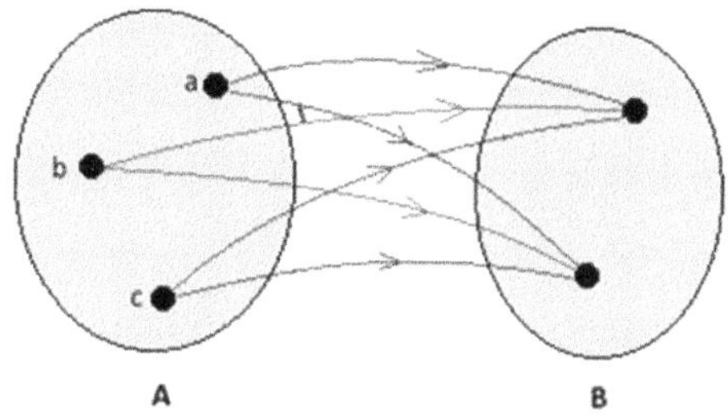

Figure 2.24. *Number of applications from A to B. For a color version of this figure, see www.iste.co.uk/cochard/mathematics1.zip*

To construct a given application f from A to B, we must draw an arrow starting from a ($\#B$ possible choices), then an arrow starting from b ($\#B$ possible choices), then an arrow starting from c ($\#B$ possible choices), etc. In total, there are $\#A$ times the product $\#B.\#B....\#B$. The number of possible choices is therefore $\#B^{\#A}$. Consequently, there are $\#B^{\#A}$ different applications from A to B.

Now, consider a set E and a part A of E. Let us define the map φ_A from E to $B = \{0,1\}$ by:

– $\varphi_A(x) = 1$ if $x \in A$;

– $\varphi_A(x) = 0$ if $x \notin A$.

as shown in Figure 2.25.

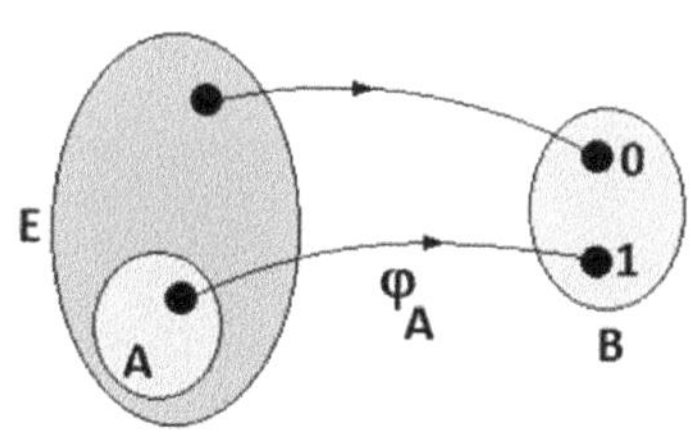

Figure 2.25. *Characteristic function*

φ_A is called the *characteristic function* of A. There are as many distinct characteristic functions as there are elements in $\mathcal{P}$(E). Now, the number of characteristic functions (functions from E to B) is $\#B^{\#E}$, or $2^{\#E}$. Consequently, the number of parts of a set E is $2^{\#E}$, a result previously announced previously in section 2.3.

3

Counting and Combinatorial Analysis

CONCEPTS COVERED IN THIS CHAPTER.–

Counting problems are extremely difficult to solve manually. Computers have proven highly effective in tackling these fundamental problems, which quickly surpass human calculation capabilities. The ability to identify and count object pairings falls within the domain of combinatorial analysis.

Therefore, the examination will successively cover the counting of permutations, arrangements and combinations. In the latter case, the properties of binomial coefficients will be detailed. All these counting formulas involve calculating the factorial function of an integer n. To address cases where n is very large, an approximate formula, known as Stirling's formula, will be presented along with a detailed demonstration.

References: [VEL 19, ENG 76, COC 19].

3.1. Set cardinals

The counting problem involves finding the number of elements in a finite set. The number is called the *cardinal* of the set and is denoted by $\#E$ for a set E.

Consider the fundamental operations of set algebra (see Figure 3.1). The following propositions can be easily proved using the basic definitions and properties of set algebra:

Mathematics for Digital Science 1,
by Gérard-Michel COCHARD and Mhand HIFI.

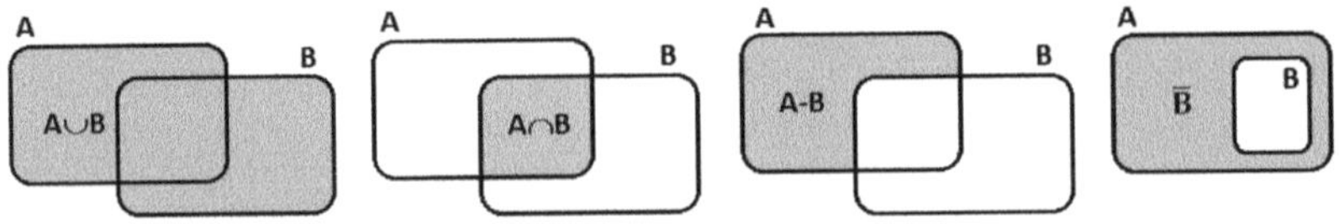

Figure 3.1. *Fundamental operations*

– For any sets A and B: $\#(A \cup B) = \#A + \#B - \#(A \cap B)$

Special case: if A and B are disjoint, then $\#(A \cup B) = \#A + \#B$.

– For arbitrary sets A and B: $\#(A - B) = \#A - \#(A \cap B)$.

– If B is included in A ($B \subseteq A$): $\#(A - B) = \#A - \#B$.

– If $A_1, A_2, \ldots, A_n$ is a partition of A: $\#A = \#A_1 + \#A_2 + \ldots + \#A_n$.

– For finite sets A and B: $\#(A \times B) = \#A.\#B$.

– For finite sets $A_1, A_2, \ldots, A_n$ $\#(A_1 \times A_2 \times \ldots \times A_n) = \#A_1.\#A_2.\ldots.\#A_n$.

– In particular, if A is a finite set: $\#(A^n) = (\#A)^n$.

The following results were also presented in a previous chapter:

– For a set A of cardinal n, we have: $\#A = n$. Let $\mathcal{P}(A)$ be the set of parts of A; then $\#(\mathcal{P}(A)) = 2^n$.

– For a set A of cardinal n, the number of parts of k elements of A is given by:

$$\frac{n.(n-1).(n-2)\ldots.(n-k+1)}{k.(k-1)(k-2)\ldots.1} = \frac{n!}{(n-k)!\,k!}$$

Throughout this chapter, unless specified otherwise, only finite sets will be considered.

3.2. Permutations, arrangements, combinations

Let E be a set of cardinal n (composed of n elements).

DEFINITION.– A *permutation* of E is an ordered list of its elements: the first, the second, …, the n-th.

Consider the set $E = \{1,2,3,4\}$. The list $(4,2,3,1)$ is a permutation of E. To find the number of permutations P_n of a set with n elements, the method is as follows:

– Choose the first element: there are n possibilities.

– Choose the second element: there are $n-1$ possibilities since one element has already been chosen.

– Choose the third element: there are $n-2$ possibilities since two elements have already been chosen.

– Continue with such process until the choice of the n-th and last element, for which there remains only one possibility.

The total number of permutations of the set E is therefore given by the product of the numbers of possibilities at each step, i.e.:

$$n.(n-1).(n-2).\dots.(2).(1) = n!$$

Therefore, we have

$$P_n = n!$$

EXAMPLE 3.1.– Consider a deck of 52 cards, all distinct. The number of possible permutations of the cards is 52!, an extremely large number of the order of 8.10^{67}. On the contrary, if only the 4 aces of the deck of cards are considered, the number of permutations is smaller: it is equal to $4! = 24$.

DEFINITION.– An ordered list of k distinct elements chosen from a set of elements n is called an *arrangement*.

Note A_n^k as the number of arrangements of k elements taken in n elements.

An element taken by itself is an arrangement of 1 element among n. There are obviously n arrangements of this type, therefore $A_n^1 = n$.

To obtain the arrangements of 2 elements among n elements, simply take each arrangement with 1 element and add one of the $n-1$ remaining elements to it. Thus, the number of arrangements of 2 elements among n elements is $A_n^2 = n.(n-1)$. Continuing this reasoning, the number of arrangements of k elements among n elements is given by the formula $A_n^k = n.(n-1).(n-2)\dots.(n-k+1) = n!/(n-k)!$.

$$A_n^k = \frac{n!}{(n-k)!}$$

EXAMPLE 3.2.– In a horse race with $n = 20$ horses, the number of trifectas in order (i.e. the number of ways to rank three horses in order first, second and third) is given by the number of arrangements of $k = 3$ elements among 20, i.e. $A_{20}^3 = 20!/(20-3)! = 20!/17! = 20.19.18 = 6840$.

Returning to the problem of choosing k elements from n elements, but this time without taking into account the order.

DEFINITION.– A *combination* of k elements taken from n without repetition is a subset of k elements of the set of n elements.

Note C_n^k as the number of combinations of k elements taken from n elements. In other words, a combination is a selection of k elements from n elements where the order does not matter. The number of combinations C_n^k can be calculated using the number of arrangements A_n^k because each combination of k elements corresponds to $k!$ different arrangements. Therefore,

$$C_n^k = \frac{A_n^k}{k!} = \frac{n!}{(n-k)!\,k!}$$

EXAMPLE 3.3.– Returning to the example of horse racing with $n = 20$ horses, the number of combinations of $k = 3$ horses that can be formed is given by the formula:

$$C_{20}^3 = \frac{20!}{(20-3)!3!} = \frac{20!}{17!3!} = 1140$$

or

$$C_{20}^3 = \frac{A_{20}^3}{3!} = \frac{20!}{(20-3)!} = \frac{20!}{17!} \times \frac{1}{3!} = \frac{20!}{17!3!} = 1140$$

Therefore, a trifecta out of order brings in less than a trifecta in order because the number of combinations of three horses out of orders is greater than the number of permutations of three horses in order (typically six times more).

From the previous relationships, we can establish the following relationship:

$$C_n^k = \frac{A_n^k}{P_k}$$

The coefficients C_n^k are called *binomial coefficients* because of their connection to the expansion of Newton's binomial. These coefficients are also noted as:

$$\binom{n}{k}$$

3.3. Properties of binomial coefficients

Given the relationship that defines it, the binomial coefficient C_n^k has the following immediate properties.

> PROPERTY 3.1.– The binomial coefficient C_n^k has the immediate property of symmetry: $C_n^k = C_n^{n-k}$.
>
> Special cases: $C_n^0 = C_n^n = 1$ and $C_n^1 = C_n^{n-1} = n$.

The following property holds:

$$C_n^k = \frac{n!}{(n-k)!\,k!} = \frac{n!}{k!\,(n-k)!} = C_n^{n-k}$$

> PROPERTY 3.2.– The binomial coefficient C_n^k has the following property, known as Pascal's:
> $C_n^k + C_n^{k+1} = C_{n+1}^{k+1}$

Indeed, expanding the sum $C_n^k + C_n^{k+1}$ by replacing each binomial coefficient with its expression:

$$\begin{aligned} C_n^k + C_n^{k+1} &= \frac{n!}{(n-k)!\,k!} + \frac{n!}{(n-k-1)!\,(k+1)!} \\ &= \frac{n!}{(n-k-1)!\,k!}\left(\frac{1}{n-k} + \frac{1}{k+1}\right) \\ &= \frac{n!}{(n-k-1)!\,k!} \cdot \frac{n+1}{(n-k)(k+1)} = \frac{(n+1)!}{(n-k)!\,(k+1)!} \\ &= \frac{(n+1)!}{((n+1)-(k+1))!\,(k+1)!} = C_{n+1}^{k+1} \end{aligned}$$

The above result forms the basis of the famous Pascal's triangle (see Figure 3.2).

		k								
		0	1	2	3	4	5	6	7	8
	0	1								
	1	1	1							
	2	1	2	1						
	3	1	3	3	1					
n	4	1	4	6	4	1				
	5	1	5	10	10	5	1			
	6	1	6	15	20	15	6	1		
	7	1	7	21	35	35	21	7	1	
	8	1	8	28	56	70	56	28	8	1

Figure 3.2. *Pascal's triangle*

The name of the binomial coefficient comes from Newton's binomial, a formula given by Isaac Newton which is a remarkable identity.

NEWTON'S BINOMIAL FORMULA.–

$$(x+y)^n = \sum_{k=0}^{n} C_n^k x^{n-k} y^k$$

The result can be proven by induction. First of all,

$$(x+y)^0 = 1 = C_0^0 = C_0^0 x^0 y^0$$

$$(x+y)^1 = x + y = C_1^0 x^1 y^0 + C_1^1 x^0 y^1$$

$$(x+y)^2 = x^2 + 2xy + y^2 = C_2^0 x^2 y^0 + C_2^1 x^1 y^1 + C_2^2 x^0 y^2$$

The formula is true for $n = 0, 1, 2$. Suppose it is true for $n > 2$. So, for $n + 1$

$$(x+y)^{n+1} = (x+y)\sum_{k=0}^{n} C_n^k x^{n-k} y^k = x\sum_{k=0}^{n} C_n^k x^{n-k} y^k + y\sum_{k=0}^{n} C_n^k x^{n-k} y^k$$

$$= x[C_n^0 x^n + \sum_{k=1}^{n} C_n^k x^{n-k} y^k] + y[\sum_{k=0}^{n-1} C_n^k x^{n-k} y^k + C_n^n y^n]$$

$$= x^{n+1} + \sum_{k=1}^{n} C_n^k x^{n-k+1} y^k + \sum_{k=0}^{n-1} C_n^k x^{n-k} y^{k+1} + y^{n+1}$$

Evaluate each of the two sums:

$$\sum_{k=1}^{n} C_n^k \, x^{n-k+1} y^k = C_n^1 x^n y^1 + C_n^2 x^{n-1} y^2 + \cdots + C_n^n x^1 y^n$$

$$\sum_{k=0}^{n-1} C_n^k \, x^{n-k} y^{k+1} = C_n^0 x^n y^1 + C_n^1 x^{n-1} y^2 + \cdots + C_n^{n-1} x^1 y^n$$

Therefore,

$$\sum_{k=1}^{n} C_n^k \, x^{n-k+1} y^k + \sum_{k=0}^{n-1} C_n^k \, x^{n-k} y^{k+1}$$

$$= (C_n^0 + C_n^1) x^n y^1 + (C_n^1 + C_n^2) x^{n-1} y^2 + \cdots + (C_n^{n-1} + C_n^n) x^1 y^n$$

$$= \sum_{k=0}^{n-1} (C_n^k + C_n^{k+1}) x^{n-k} y^{k+1} = \sum_{k=0}^{n-1} C_{n+1}^{k+1} x^{n-k} y^{k+1}$$

According to property 3.2, the following result holds:

$$(x+y)^{n+1} = x^{n+1} + \sum_{k=0}^{n-1} C_{n+1}^{k+1} x^{n-k} y^{k+1} + y^{n+1}$$

$$= C_{n+1}^0 x^{n+1} + C_{n+1}^1 x^n y^1 + C_{n+1}^2 x^{n-1} y^2 + \cdots + C_{n+1}^n x^1 y^n + C_{n+1}^{n+1} y^{n+1}$$

$$= \sum_{k=0}^{n+1} C_{n+1}^k x^{n+1-k} y^k$$

which proves the development of Newton's binomial.

An immediate application of Newton's binomial formula leads to the following property.

PROPERTY 3.3.–

$$\sum_{k=0}^{n} C_n^k = 2^n$$

Indeed, by replacing $x = y = 1$ in the above formula (property 3.3), the following holds:

$$(1+1)^n = \sum_{k=0}^{n} C_n^k 1^{n-k} 1^k \quad \text{either} \quad 2^n = \sum_{k=0}^{n} C_n^k$$

This property is often called *Vandermonde identity*. Note that this property is not new because the sum of the binomial coefficients represents the number of parts of a set with n elements, which is equal to 2^n. Furthermore, this relationship can be "read" directly on Pascal's triangle. Indeed, taking for example the line for $n = 5$ (see Figure 3.1), the sum of the binomial coefficients is $1 + 5 + 10 + 10 + 5 + 1 = 32 = 2^5$.

3.4. Stirling's formula

As seen, combinatorial calculation primarily involves the factorial $n!$. The value of this factorial increases drastically with n, and it is convenient to use an approximate expression that allows a simpler calculation. This expression is given by Stirling's formula (also due to Moivre).

STIRLING'S FORMULA For $n \to \infty$ $\quad n! \sim \sqrt{2\pi} n^{n+1/2} e^{-n}$

Although the formula is valid for $n \to \infty$, it provides an approximate value of $n!$ for large values of n, as shown in the table in Figure 3.3.

n	n!	Stirling
1	1	0.92213809
10	3628800	3598699.83
100	9.333E+157	9.325E+157

Figure 3.3. *Comparison between n! and the value of Stirling*

To demonstrate this magnificent formula, some intermediate results are needed. The calculation is a bit long, but it offers a good analytical exercise.

First of all, consider the sequence defined by

$$u_n = ln \left[\frac{n!}{n^{n+\frac{1}{2}} e^{-n}} \right] \qquad n \in N$$

The expression in square brackets is derived from the formula to be demonstrated. For $n \to \infty$, we should find $ln(\sqrt{2\pi})$. Demonstrating that this sequence is bounded both lower and upper involves considering the integral calculation based on the trapezoid method. Figure 3.4 shows the function $y = ln(x)$ between the values $x = k$ and $x = k + 1$.

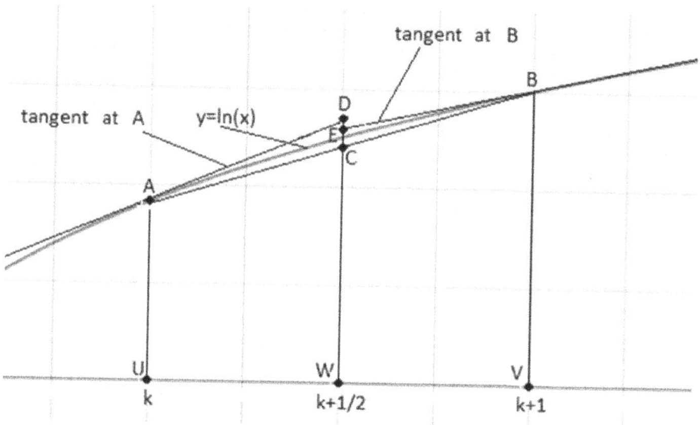

Figure 3.4. *Trapezoid method*

Also, consider the abscissa point $(k + \frac{1}{2})$, along with the tangents to the curve at A (abscissa $x = k$) and B (abscissa $x = k + 1$). The equations of the two tangents are:

$$y = \frac{x - k}{k} + ln\,(k) \quad \text{tangent at A}$$

$$y = \frac{x - k - 1}{k + 1} + ln\,(k + 1) \quad \text{tangent at B}$$

with C being the midpoint of AB, the vertical line WC intersects the tangent at A at D and the tangent at B at E. Since the curve $y = ln(x)$ is concave, points D and E lie above C. The coordinates of D and E are:

$$x_D = x_E = k + \frac{1}{2} \qquad y_D = \frac{1}{2k} + \ln(k) \qquad y_E = -\frac{1}{2(k+1)} + ln(k+1)$$

The area bounded by the curve $y = ln(x)$, the abscissa axis, and the straight lines $x = k$ and $x = k+1$, is smaller than the sum of the areas of the trapezoids ADWU and EBVW, on the one hand, and greater than the area of the trapezoid ABVU, on the other hand, i.e.:

$$area\ ABVU < \int_k^{k+1} \ln(x)\, dx < area\ ADWU + area\ EBVW$$

However,

$$area\ ABVU = \frac{1}{2}[\ln(k) + \ln(k+1)]$$

$$area\ ADWU = \frac{1}{4}[2\ln(k) + \frac{1}{2k}]$$

$$area\ EBVW = \frac{1}{4}[2\ln(k+1) - \frac{1}{2(k+1)}]$$

Therefore,

$$\frac{1}{2}[\ln(k) + \ln(k+1)] < \int_k^{k+1} \ln(x)\, dx < \frac{1}{2}[\ln(k) + \ln(k+1)] + \frac{1}{8}(\frac{1}{k} - \frac{1}{k+1})$$

Now, integrate $ln(x)$ from 1 to n, which corresponds over k from 1 to $n-1$ on both sides:

$$\frac{1}{2}\sum_{k=1}^{n-1}[\ln(k) + \ln(k+1) < \int_1^n \ln(x)\, dx < \frac{1}{2}\sum_{k=1}^{n-1}[\ln(k) + \ln(k+1)] + \frac{1}{8}\sum_{k=1}^{n-1}(\frac{1}{k} - \frac{1}{k+1})$$

Therefore,

$$\frac{1}{2}\sum_{k=1}^{n-1}[\ln(k) + \ln(k+1]$$
$$= \frac{1}{2}[l(2) + l(2) + \ln(3) + l(3) + l(4) + \cdots + l(n-1)$$
$$+ l(n)] = l(n!) - \frac{1}{2}l(n)$$

$$\frac{1}{8}\sum_{k=1}^{n-1}[\frac{1}{k} - \frac{1}{k+1}] = \frac{1}{8}\left[1 - \frac{1}{2} + \frac{1}{2} - \frac{1}{3} + \frac{1}{3} - \frac{1}{4} + \cdots \frac{1}{n-1} - \frac{1}{n}\right] = \frac{1}{8}(1 - \frac{1}{n})$$

$$\int_1^n \ln(x)\,dx = [xln(x)]_1^n - \int_1^n dx = nln(n) - n + 1$$

From this, we can deduce

$$\ln(n!) - \frac{1}{2}\ln(n) < nln(n) - n + 1 < \ln(n!) - \frac{1}{2}\ln(n) + \frac{1}{8}(1 - \frac{1}{n})$$

Hence, by rearranging the terms

$$\left(n + \frac{1}{2}\right)\ln(n) - n + \frac{7}{8} + \frac{1}{8n} < \ln(n!) < \left(n + \frac{1}{2}\right)\ln(n) - n + 1$$

By noting that

$$\left(n + \frac{1}{2}\right)\ln(n) - n = \ln\left(n^{n+\frac{1}{2}}\right) - nln(e\,) = \ln\left(n^{n+\frac{1}{2}}\right) + \ln(e^{-n})$$
$$= \ln[n^{n+\frac{1}{2}}e^{-n}]$$

The double inequality is written as $\frac{7}{8} + \frac{1}{8n} < \ln(n!) - \ln\left[n^{n+\frac{1}{2}}e^{-n}\right] < 1$ for $n \to \infty, \frac{7}{8} < \ln\left[\frac{n!}{n^{n+\frac{1}{2}}e^{-n}}\right] < 1.$

The sequence u_n is therefore limited. But this sequence is decreasing. In fact, calculate $u_{n+1} - u_n$. A fairly easy calculation shows that:

$$u_{n+1} - u_n = -\left(n + \frac{1}{2}\right)f(n) \quad \text{with} \quad f(n) = ln\left(\frac{n+1}{n}\right) - \frac{1}{n + \frac{1}{2}}$$

The function $f(x)$ for $x \in [1, \infty[$ is decreasing because its derivative is negative:

$$f'(x) = \frac{1}{x+1} - \frac{1}{x} + \frac{1}{\left(x+\frac{1}{2}\right)^2} = -\frac{1}{4x(x+1)(x+\frac{1}{2})^2}$$

Since $f(1) = ln(2) - \frac{2}{3} > 0$ and $f(x) \to 0$ for $x \to \infty$, $f(x) > 0$ on $[1, \infty[$.

So $u_{n+1} - u_n < 0$. The sequence u_n is decreasing and bounded by 7/8. It therefore tends towards a limit $\lambda \in]7/8, 1[$when $n \to \infty$, and we therefore have at the limit:

$$ln\left[\frac{n!}{n^{n+\frac{1}{2}}e^{-n}}\right] \sim \lambda \ either\ \frac{n!}{n^{n+\frac{1}{2}}e^{-n}} \sim e^{\lambda} = K \text{ and therefore } n! \sim Kn^{n+\frac{1}{2}}e^{-n}$$

To obtain Stirling's formula, we now need to calculate *K*. For this, the following Wallis integral is used:

$$I_n = \int_0^{\pi/2} cos^n(x)dx \qquad n \in \mathbb{N}$$

The first Wallis integrals are (easy calculation) $I_0 = \frac{\pi}{2}$ and $I_1 = 1$. All Wallis integrals can be calculated from I_0 and I_1. Indeed,

$$I_n = \int_0^{\pi/2} cos^{n-1}(x)\cos(x)\,dx = [\sin(x)cos^{n-1}(x)]_0^{\pi/2} + (n-1)\int_0^{\pi/2} cos^{n-2}(x)sin^2(x)dx$$

$$= (n-1)\left\{\int_0^{\frac{\pi}{2}} cos^{n-2}\,(x)dx - \int_0^{\frac{\pi}{2}} cos^{n}\,(x)dx\right\}\} = (n-1)(I_{n-2} - I_n)$$

Hence, the recurrence relationship is

$$I_n = \frac{n-1}{n} I_{n-2}$$

To continue the calculation, it is necessary to distinguish between n even and n being odd. So,

$$I_{2p} = \frac{2p-1}{2p} I_{2p-2} = \frac{(2p-1)(2p-3)\dots .1}{(2p)(2p-2)(2p-4)\dots .2} I_0$$
$$= \frac{(2p-1)(2p-3)\dots .1}{2^p p!} I_0$$

However,

$$(2p-1)(2p-3)\dots .1 = \frac{(2p)(2p-1)(2p-2)(2p-3)(2p-4)\dots\dots .1}{(2p)(2p-2)(2p-4)(2p-6)\dots\dots 2}$$
$$= \frac{(2p)!}{2^p p!}$$

so that:

$$I_{2p} = \frac{(2p)!}{2^{2p}(p!)^2} I_0$$

An analogous calculation, not detailed here, yields for n odd,

$$I_{2p+1} = \frac{2^{2p}(p!)^2}{(2p+1)!} I_1 \quad \text{and therefore} \quad I_{2p+1} = \frac{\pi}{2(2p+1)I_{2p}}$$

We will use these relations to obtain K. When $n \to \infty$,

$$p! \sim K p^{p+1/2} e^{-p} \qquad (2p)! \sim K(2p)^{2p+1/2} e^{-p} \quad \text{therefore} \quad I_{2p} \sim \frac{\pi\sqrt{2}}{2K\sqrt{p}}$$

It can be deduced that:

$$I_{2p+1} \sim \frac{K}{2\sqrt{2}\sqrt{p}}$$

The sequence I_n is decreasing. Indeed,

$$I_{n+1} - I_n = \int_0^{\pi/2} cos^n(x)[\cos(x) - 1]dx$$

However, for $x \in [0, \pi/2]$, $cos(x) - 1 \leq 0$ and $cos^n(x) \geq 0$. Therefore, $cos^n(x)[cos(x) - 1] \leq 0$.

The integral will therefore be negative, which demonstrates that I_n is a decreasing sequence. Thus,

$$I_{n+2} \leq I_{n+1} \leq I_n \text{ and } \frac{n+1}{n+2} I_n \leq I_{n+1} \leq I_n$$

using the recurrence relation between I_n and I_{n+2}.

For $n \to \infty$, the previous relationship leads to $I_n \leq I_{n+1} \leq I_n$, which signifies that $I_{n+1} \sim I_n$ and, in particular, $I_{2p+1} \sim I_{2p}$. Taking into account the previous result giving I_{2p+1} and I_{2p} as a function of K, it follows that

$$\frac{K}{2\sqrt{2}\sqrt{p}} \sim \frac{\pi\sqrt{2}}{2K\sqrt{p}} \quad \text{hence } K^2 \sim 2\pi \quad \text{and } K \sim \sqrt{2\pi}$$

Therefore, the result that was missing to obtain Stirling's formula is now obtained.

4

Boolean Algebra and Boolean Functions

CONCEPTS COVERED IN THIS CHAPTER.–

The computer universe is Boolean: it only handles 0s or 1s (at least in current computers). The construction of computers is based on logic circuits, which will be explained in the next chapter. Before this, it is appropriate to recall the mathematics relating to Boolean functions.

The notion of ordered sets with their particular elements will be approached, leading to lattices and Boolean algebras. Next, Boolean functions will be explained in detail.

References: [VEL 19, FAU 79, LIP 83].

4.1. Special elements of an ordered set

Let E be a set on which an order relation is defined, denoted by $\angle$. E is termed an *ordered set*. From the definition, the relation $\angle$ has the properties of reflexivity, antisymmetry and transitivity.

Mathematics for Digital Science 1,
by Gérard-Michel COCHARD and Mhand HIFI.

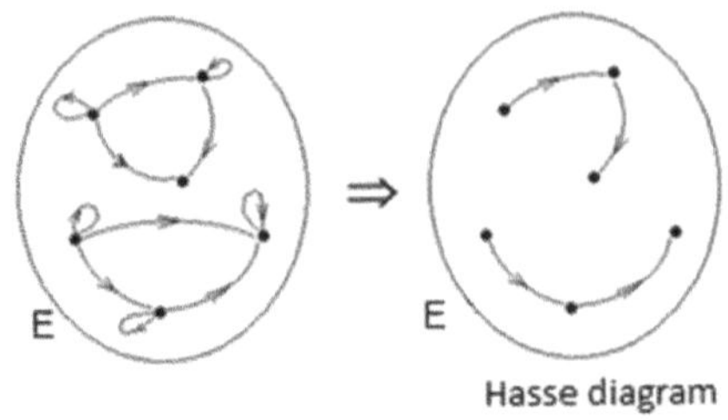

Figure 4.1. *Hasse diagram*

Practically, when representing a relationship by its graph, if it is an order relation, simplification involves removing:

– loops (for reflexivity);

– arrows representing transitivities.

Then a *Hasse diagram* is obtained, as shown in Figure 4.1. Note that a Hasse diagram displays at most a single arrow between two elements.

EXAMPLE 4.1.– Consider the set $E = \{0, 1, 2, 3, 4, 5, 6\}$ and the relationship "≥". Figure 4.2 shows the corresponding Hasse diagram.

Figure 4.2. *Hasse diagram*

EXAMPLE 4.2.– Consider the set E of positive integer divisors of 30, i.e. $E = \{1, 2, 3, 5, 6, 10, 15, 30\}$. The relationship is "divided". The Hasse diagram for this relationship is shown in Figure 4.3.

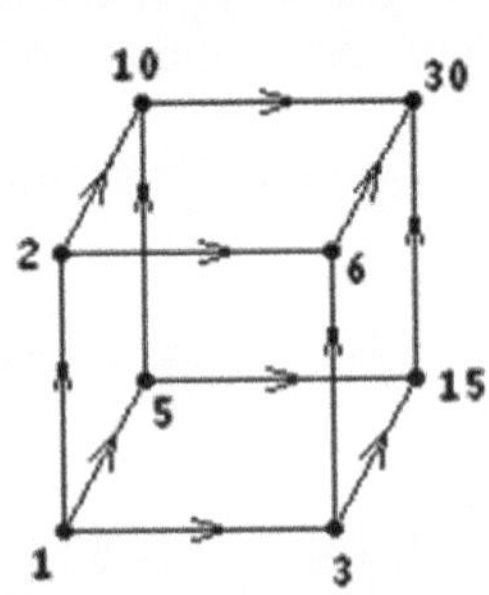

Figure 4.3. *Hasse diagram*

Now, consider a subset P of E. Here are some definitions relating to special elements of E.

DEFINITION 4.1.– *Upper bound and lower bound*

An *upper bound* of P is any element a of E such that $x \in P$ precedes a: $\forall x \in P \ x \angle a$.

A *lower bound* of P is any element b of E such that for every x in P, b precedes x: $\forall x \in \mathrm{P} \ b \angle x$.

EXAMPLE 4.3.– Consider the set $E = \{1, 2, 3, 5, 6, 10, 15, 30\}$ and the relationship "divides". If $P = \{2, 3, 5, 10\}$, there exists a single upper bound, 30, and a single lower bound, 1 (as shown in Figure 4.4).

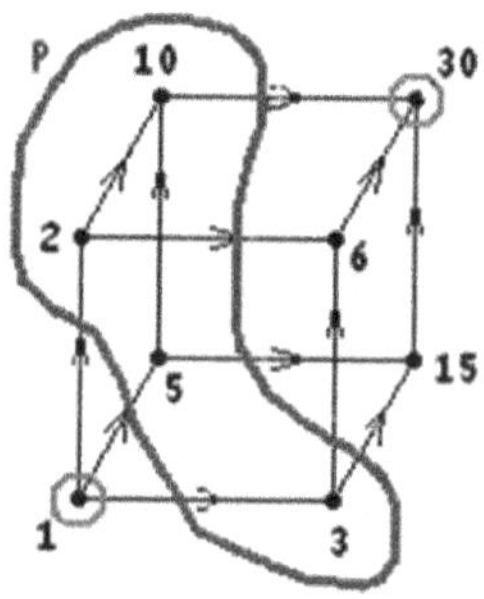

Figure 4.4. *P = {2, 3, 5, 10}. For a color version of this figure, see www.iste.co.uk/cochard/mathematics1.zip*

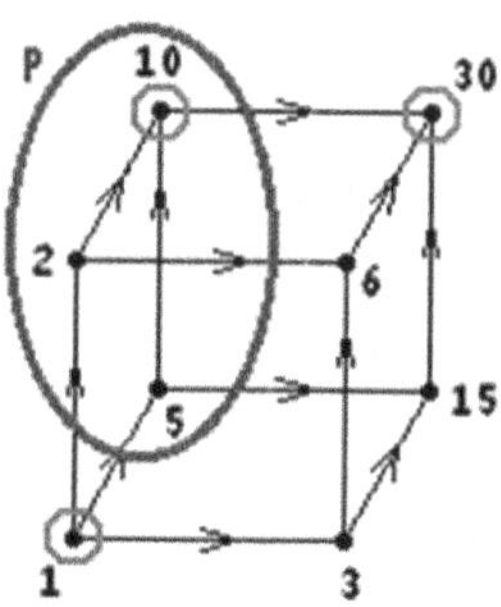

Figure 4.5. *P = {2, 5, 10}. For a color version of this figure, see www.iste.co.uk/cochard/mathematics1.zip*

If $P = \{2, 5, 10\}$, there are two upper bounds, 10 and 30, and a single lower bound, 1 (as shown in Figure 4.5).

DEFINITION 4.2.– *Maximal element and minimal element*

An element a of P is a *maximal element* of P if there is no element of P strictly greater than a: $a \in P, \forall x \in P, \neg(a \angle x)$.

An element b of P is a *minimal element* of P if there is no element of P strictly less than b: $b \in P, \forall x \in P, \neg(x \angle b)$.

EXAMPLE 4.4.– Let $E = \{1,2,3,5,6,10,15,30\}$ and consider the relationship "divides".

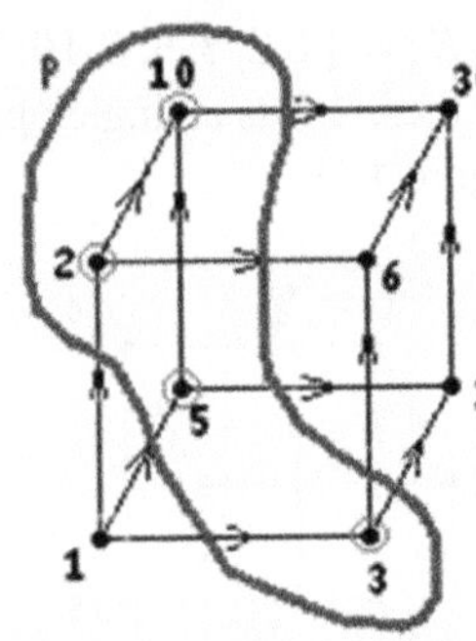

Figure 4.6. *P = {2, 3, 5, 10}. For a color version of this figure, see www.iste.co.uk/cochard/mathematics1.zip*

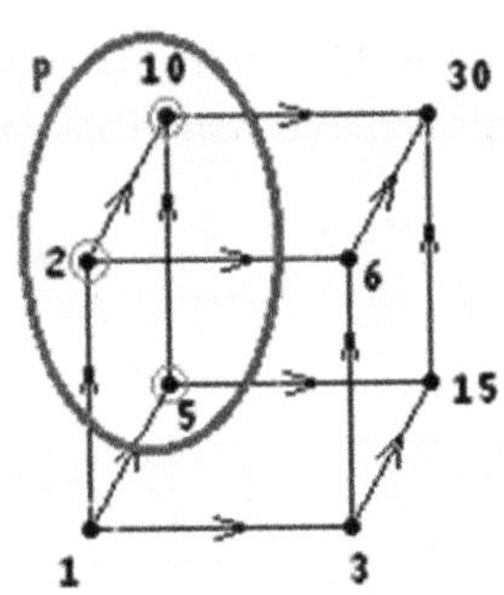

Figure 4.7. *P = {2, 5, 10}. For a color version of this figure, see www.iste.co.uk/cochard/mathematics1.zip*

If $P = \{2,3,5,10\}$, there is a single maximal element, 10, and three minimal elements, 2, 3 and 5 (as shown in Figure 4.6).

If $P = \{2,5,10\}$, there is a single maximal element, 10, and two minimal elements, 2 and 5 (as shown in Figure 4.7).

DEFINITION 4.3.– *Largest element/smallest element*
The *largest element* (*le*) of P is the element a of P such that $a \in P, \forall x\ P, x \angle a$.
The *smallest element* (*se*) of P is the element b of P such that $b \in P, \forall x \in P$, $b \angle x$.

EXAMPLE 4.5.– Let $E = \{1, 2, 3, 5, 6, 10, 15, 30\}$ and consider the relationship "divides".

If $P = \{2, 3, 5, 10\}$, there is neither *le* nor *se* in this set.

If $P = \{2, 5, 10\}$, the *le* of P is 10, and there is no *se* in this set.

DEFINITION 4.4.– Least upper bound/greatest lower bound
The *least upper bound*, $LUB(P)$, of P is the smallest element among the upper bounds of P, if it exists. In other words, $LUB(P)$ is the *se* of the set of upper bounds of P.
The *greatest lower bound* $GLB(P)$ of P is the largest element among the lower bounds of P, if it exists. In other words, $GLB(P)$ is the *le* of the set of lower bounds of P.

EXAMPLE 4.6.– Let $E = \{1, 2, 3, 5, 6, 10, 15, 30\}$ and consider the relationship "divides".

If $P = \{1, 3, 5, 30\}$, then the set of upper bounds of P is $\{30\}$ and the set of lower bounds of P is $\{1\}$. Thus, the least upper bound of P is $LUB(P) = 30$ and the greatest lower bound of P is $GLB(P) = 1$.

If $P = \{2, 5, 10\}$, then the set of upper bounds of P is 10 and the set of lower bounds of P is $\{1\}$. Thus, the least upper bound of P is $LUB(P) = 10$ and the greatest lower bound of P is $GLB(P) = 1$.

NOTE.– The least upper bound $LUB(P)$ (respectively greatest lower bound, $GLB(P)$) does not always exist because it is possible that the set of upper bounds (respectively lower bounds) of P does not have a (se) (respectively (le)).

DEFINITION 4.5.– The *universal element* of E is the largest element (le) of E; the *null element* of E is the smallest element (se) of E.

EXAMPLE 4.4.– Consider the set $E = \{1, 2, 3, 5, 6, 10, 15, 30\}$ with the relationship "divides". Then, 30 is the universal element of E, meaning that any element of E divides 30. Similarly, 1 is the null element of E, as it is divisible by all the elements of E.

4.2. Lattice

A *lattice* is an ordered set E such that for any pair of elements x and y of E, there exists a least upper bound ($LUB(x, y)$) and a greatest lower bound ($GLB(x, y)$) of x and y in E. That is, $LUB(x, y)$ is the smallest element of E which is greater than x and y, and $GLB(x, y)$ is the largest element of E which is smaller than x and y. Using the Hasse diagram (shown in Figure 4.1), we can easily visualize the order relationships between the elements of E and determine whether E is a lattice or not. The Hasse diagram is a graphical means of representing the partial order between the elements of an ordered set: each element is represented by a point and each arrow connects the elements that are related by the order relationship. If the Hasse diagram satisfies the properties of the definition of a lattice, then the ordered set is a lattice.

EXAMPLE 4.8.– Figure 4.8 shows Hasse diagrams of ordered sets.

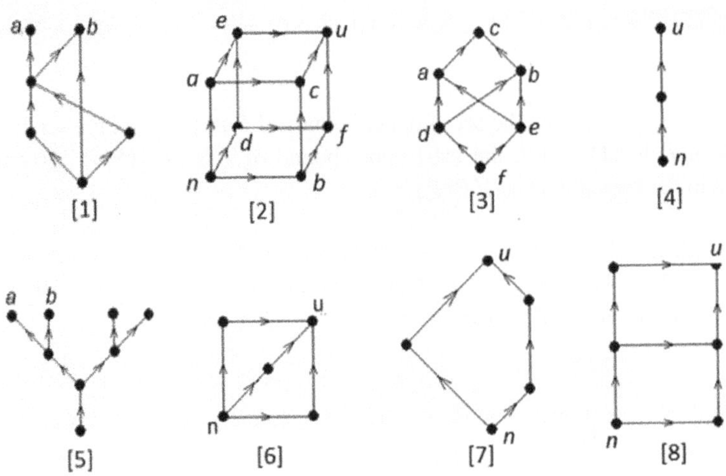

Figure 4.8. *Where are the lattices?*

[1] is not a lattice because the subset $\{a, b\}$ has no upper bound. [3] is not a lattice because the subset $\{a, b\}$ has only one upper bound, c (it is the (se), therefore $LUB\{a, b\} = c$), but it has three lower bounds d, e and f, and $\{d, e, f\}$ has no

(le). [5] is not a lattice because the part $\{a, b\}$ has no upper bound. On the contrary, [2], [4], [6], [7], [8] are lattices. In these cases, the null element is n and the universal element is u.

NOTE.– If the order is total, E is a lattice. Indeed, in this case, all the elements are comparable two by two, and therefore for any pair of elements, there exists an LUB and a GLB. We can therefore find the le and se necessary to determine whether E is a lattice.

A lattice can be redefined algebraically by setting:

$$x\nabla y = LUB\{x, y\}$$

$$x\Delta y = GLB\{x, y\}$$

The least upper and greatest lower bounds are thus defined as the result of operations that verify the following properties:

– idempotence: $x\nabla x = x, \ x\Delta x = x$;

– associativity: $x\nabla(y\nabla z) = (x\nabla y)\nabla z, \ x\Delta(y\Delta z) = (x\Delta y)\Delta z$;

– commutativity: $x\nabla y = y\nabla x, \ x\Delta y = y\Delta x$;

– identity: $x\nabla u = u\nabla x = u, \ x\Delta n = n\Delta x = n$, where n and u denote, respectively, the null element and the universal element.

The whole E can be considered structured by the laws ∇ and Δ, obtaining the algebraic definition of a lattice while disregarding the order relationship. The preceding properties are then accepted as axioms, and the two definitions are equivalent.

A lattice can have additional properties:

– A lattice E is *complemented* if for every element x of E, there exists a *complement* $\bar{x}$ such that $x\nabla\bar{x} = u$ and $x\Delta\bar{x} = n$:

$$\forall x \in E, \exists\bar{x} : x\nabla\bar{x} = u \text{ and } x\Delta\bar{x} = n.$$

There can be several complements for the same element x.

– A lattice is *distributive* if for any triplet of elements x, y and z of E, the following holds:

$$x\Delta(y\nabla z) = (x\Delta y)\nabla(x\Delta z), \ x\nabla(y\Delta z) = (x\nabla y)\Delta(x\nabla z).$$

These additional properties are significant in the study of lattices as they help characterize certain types of lattices and establish results on their structure. For example, any distributive lattice is complemented, but the converse is not always true. In other words, there exist complemented lattices that are not distributive.

EXAMPLE 4.9.– The lattice [2] shown in Figure 4.8 is both distributive and complemented.

Indeed, for any element x of E, there exists a complement $\bar{x}$ such that $x\Delta\bar{x} = n$ and $x\nabla\bar{x} = u$. For example, for the element a, its complement is $\bar{a} = f$, because $a\Delta f = n$ and $a\nabla f = u$.

Furthermore, it can be verified that this lattice is distributive using the operations ∇ and Δ defined on the elements of E. For example, for the elements a, b and c of E, the following holds:

$$a\Delta(b\nabla c) = a\Delta c = a \qquad (a\Delta b)\nabla(a\Delta c) = n\nabla a = a$$

and for the elements a, b and d of E, the following holds:

$$a\nabla(b\Delta d) = a\nabla n = a \qquad (a\nabla b)\Delta(a\nabla d) = c\Delta a = a$$

These two equalities verify the distribution condition, which confirms that the lattice [2] is indeed distributive.

4.3. Boolean algebra

DEFINITION 4.6.– A *Boolean algebra E* is a distributive and complemented lattice.

IMMEDIATE PROPERTY.– In a Boolean algebra, for each element x, there exists a unique complement denoted by $\bar{x}$.

Proof: let y_1 and y_2 be two complements of x, any element of E. By definition, $x\nabla y_1 = u$, $x\Delta y_1 = n$, $x\nabla y_2 = u$, $x\Delta y_2 = n$. Now $(x\nabla y_1)\Delta y_2 = u\Delta y_2 = y_2$ and $(x\Delta y_1)\Delta y_2 = (x\Delta y_2)\nabla(y_1\Delta y_2) = u\nabla(y_1\Delta y_2) = y_1\Delta y_2$ by distributivity. It implies that $y_2 = y_1\Delta y_2$ and by symmetry $y_1 = y_2\Delta y_1$. According to commutativity, $y_1 = y_2$, showing that x has a unique complement.

Finally, the algebraic properties of a Boolean algebra are given below.

Idempotence: $x\nabla x = x \quad x\Delta x = x$.
Associativity: $x\nabla(y\nabla z) = (x\nabla y)\nabla z \quad x\Delta(y\Delta z) = (x\Delta y)\Delta z$.
Commutativity: $x\nabla y = y\nabla x \quad x\Delta y = y\Delta x$.
Identity: $x\nabla n = x \quad x\nabla u = u \quad x\Delta n = n \quad x\Delta u = x$.
Distributivity: $x\nabla(y\Delta z) = (x\nabla y)\Delta(x\nabla z) \quad x\Delta(y\nabla z) = (x\Delta y)\nabla(x\Delta z)$.
Complementarity: $x\nabla\bar{x} = u \quad x\Delta\bar{x} = n \quad \bar{n} = u \quad \bar{u} = n$.
Involution: $\bar{\bar{x}} = x$.
Morgan's laws: $\overline{x\nabla y} = \bar{x}\Delta\bar{y} \quad \overline{x\Delta y} = \bar{x}\nabla\bar{y}$.

These properties can be used to define Boolean algebra formally. It is worth noting that these properties have already been encountered twice:

– In propositional calculus, where propositional variables can be interpreted as elements of a Boolean algebra.

– In the algebra of subsets of a set, where union, intersection and complement operations are examples of binary operations on a set that satisfy the properties of Boolean algebras.

E Boolean algebra	V set of propositions	$\mathcal{P}_E$ set of the parts of E
∇	$\vee$	$\cup$
Δ	$\wedge$	$\cap$
$\bar{x}$	$\neg P$	$\bar{A}$
u	v	E
n	f	$\varnothing$
$\angle$	$\rightarrow$	$\subset$

Figure 4.9. *Correspondence between E, V, $\mathcal{P}_{(E)}$*

Generalizing these examples, Boolean algebras can be seen as abstract algebraic systems that obey these properties. This approach enables the formal definition of Boolean algebras, without referencing particular examples. Figure 4.9 shows the correspondences between the algebraic properties of a Boolean algebra and the operations on the sets V of propositions or on the parts $\mathcal{P}(E)$ of a set E. Thus, it can be stated that the set V and the set $\mathcal{P}(E)$ have a Boolean algebra structure.

Stone's theorem (1934) expresses several interesting results about finite Boolean algebras. Admitting this theorem, it can be formulated in three steps:

STONE'S THEOREM.– The Hasse diagram of a finite Boolean algebra is always an n-cube for a certain n (suitably chosen).

– Two finite Boolean algebras having the same number of elements have isomorphic (identical) Hasse diagrams.

– The number of elements of a finite Boolean algebra can only be 2^n for a certain integer n.

EXAMPLE 4.10.– Consider the set E of divisors of 110, given by $E = \{1, 2, 5, 10, 11, 22, 55, 110\}$. This set contains eight items.

Two binary operations are defined on E:

$x \nabla y = LCM(x, y)$ (the LCM of x and y is the lowest common multiple of x and y) and $x \Delta y = GCD(x, y)$ (the GCD of x and y is the greatest common divisor of x and y).

These laws correspond to the "divides" order relation. The truth tables of these two laws are presented in the table in Figure 4.10.

∇	1	2	5	10	11	22	55	110
1	1	2	5	10	11	22	55	110
2	2	2	10	10	22	22	110	110
5	5	10	5	10	55	110	55	110
10	10	10	10	10	110	110	110	110
11	11	22	55	110	11	22	55	110
22	22	22	110	110	22	22	110	110
55	55	110	55	110	55	110	55	110
110	110	110	110	110	110	110	110	110

Δ	1	2	5	10	11	22	55	110
1	1	1	1	1	1	1	1	1
2	1	2	1	2	1	2	1	2
5	1	1	5	5	1	1	5	5
10	1	2	5	10	1	2	5	10
11	1	1	1	1	11	11	11	11
22	1	2	1	2	11	22	11	22
55	1	1	5	5	11	11	55	55
110	1	2	5	10	11	22	55	110

x	$\bar{x}$
1	110
2	55
5	22
10	11
11	10
22	5
55	2
110	1

Figure 4.10. *Truth tables*

Thus, $n = 1$ and $u = 110$, implying that the Hasse diagram is the 3-cube shown in Figure 4.11.

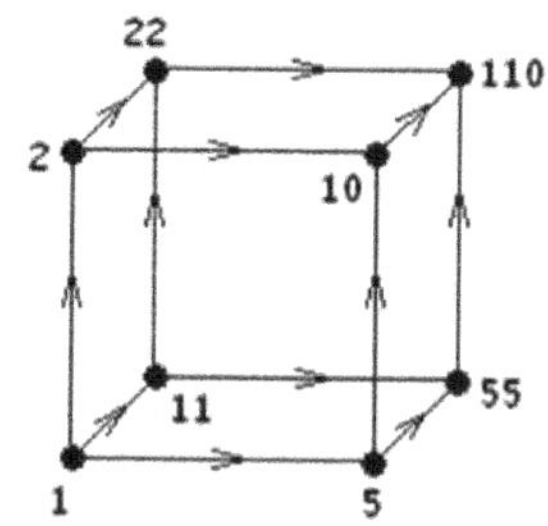

Figure 4.11. *Hasse diagram*

EXAMPLE 4.11.– Consider the set $\Sigma = \{a, b, c\}$. All parts of Σ is denoted by $\mathcal{P}(E) = \{\emptyset, a, b, c, \{a, b\}, \{a, c\}, \{b, c\}, \Sigma\}$, where $\emptyset$ is the empty set, and $\{a\}$, $\{b\}$, $\{c\}$, $\{a, b\}$, $\{a, c\}$, $\{b, c\}$ and Σ are the sets containing, respectively, a single element of Σ, two elements of Σ and Σ itself (see Figure 4.12). The set $\mathcal{P}(E)$ is ordered by the relationship $\subseteq$, which expresses the inclusion between the sets.

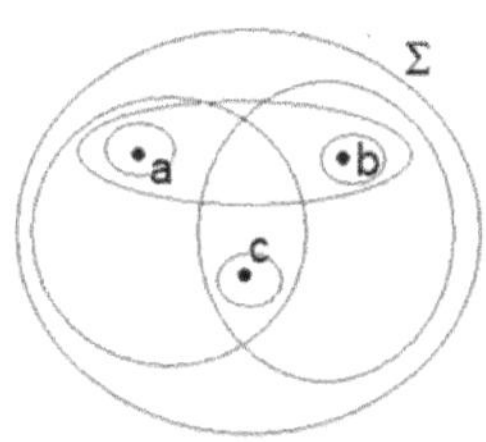

Figure 4.12. *Inclusions*

The truth tables of the laws $\cup$, $\cap$ and "inverse" are shown in Figure 4.13. With $n = \emptyset$ and $u = \Sigma$, the Hasse diagram becomes a 3-cube, as shown in Figure 4.14.

∪	Ø	A	B	C	D	E	F	Σ
Ø	Ø	A	B	C	D	E	F	Σ
A	A	A	C	C	E	E	Σ	Σ
B	B	C	B	C	F	Σ	F	Σ
C	C	C	C	C	Σ	Σ	Σ	Σ
D	D	E	F	Σ	D	E	F	Σ
E	E	E	Σ	Σ	E	E	Σ	Σ
F	F	Σ	F	Σ	F	Σ	F	Σ
Σ	Σ	Σ	Σ	Σ	Σ	Σ	Σ	Σ

∩	Ø	A	B	C	D	E	F	Σ
Ø	Ø	Ø	Ø	Ø	Ø	Ø	Ø	Ø
A	Ø	A	Ø	A	Ø	A	Ø	A
B	Ø	Ø	B	B	Ø	Ø	B	B
C	Ø	A	B	C	Ø	A	B	C
D	Ø	Ø	Ø	Ø	D	D	D	D
E	Ø	A	Ø	A	D	E	D	E
F	Ø	Ø	B	B	D	D	F	F
Σ	Ø	A	B	C	D	E	F	Σ

X	$\overline{X}$
Ø	Σ
A	F
B	E
C	D
D	C
E	B
F	A
Σ	Ø

Figure 4.13. *Truth tables*

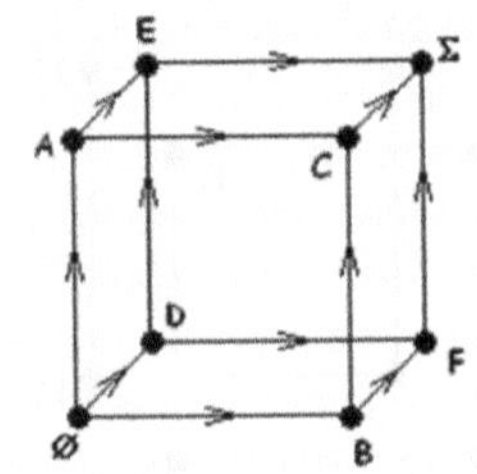

Figure 4.14. *Hasse diagram*

4.4. Boolean functions

Consider the set $B = \{0, 1\}$, provided with the order relationship $\leq$. B has a Boolean algebra structure, where the operations of union and intersection are, respectively, defined by $a \nabla b = max(a, b)$ and $a \Delta b = min(a, b)$.

The truth tables of laws ∇ and Δ and "inverse" as well as the Hasse diagram are given in Figure 4.15.

∇	0	1
0	0	1
1	1	1

Δ	0	1
0	0	0
1	0	1

x	$\bar{x}$
0	1
1	0

n = 0

u = 1

Figure 4.15. *Truth tables and Hasse diagram*

In this particular case, ∇ represents the logical "or" and Δ the logical "and".

Let us denote by B_n the set of n-tuples $(a_1, a_2, \dots, a_n)$, where $a_i \in B$. This set contains 2^n elements, because each term of an n-tuple can take two values (0 or 1).

In fact, $B_n = B \times B \times B \times \dots \times B$ (n times), which means that B_n is the Cartesian product of n copies of the set B :$B_n = B^n$.

EXAMPLE 4.12.– Consider the example where $n = 3$. In this case, B^3 is the set of triples (a, b, c), where a, b and c belong to B. Thus, it can be expressed as:

$$B^3 = \{(a, b, c) | a \in B, b \in B, c \in B\}.$$

Expressing it further, B^3 is equivalent to $B \times B \times B$, implying that it contains $2^3 = 8$ elements in total. Using the notation $\{0, 1\}$ for set B, it can be illustrated as:

$$B^3 = \{(0,0,0), (0,0,1), (0,1,0), (1,0,0), (0,1,1), (1,0,1), (1,1,0), (1,1,1)\}$$

A *Boolean function* is an application B_n *to* B.

EXAMPLE 4.13.– Consider $f: B^3 \rightarrow B$ a Boolean function defined by the truth table presented in Figure 4.16.

a	b	c	f(a,b,c)
0	0	0	0
0	0	1	0
0	1	0	0
0	1	1	1
1	0	0	1
1	0	1	0
1	1	0	0
1	1	1	1

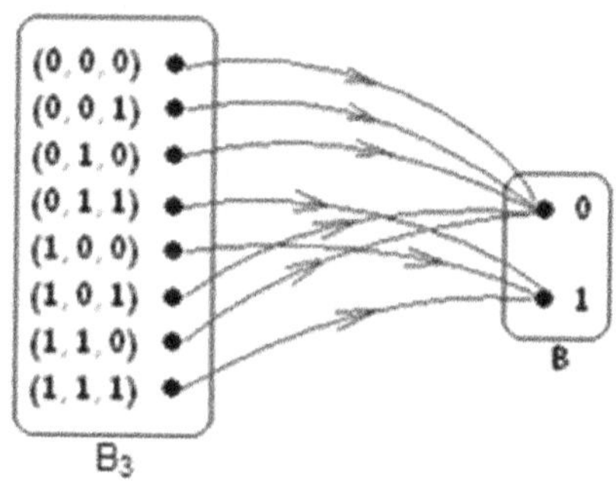

Figure 4.16. *Example of a Boolean function*

The number of Boolean functions $B^n \rightarrow B$ can be calculated using the formula $\#B^{\#A}$. In the case where $A = B^n$, $\#A = 2^n$ and $\#B = 2$. Therefore, there are $2^{(2^n)}$different Boolean functions from B^n to B:

– $n = 1$: 4 Boolean functions;

– $n = 2$: 16 Boolean functions;

– $n = 3$: 256 Boolean functions;

– $n = 6$: 18,446,744,073,709,551,616 Boolean functions!

Let F_n denote the set of all Boolean functions from B_n to B. On this set, an order relation can be defined as follows:

$$f \in F_n, g \in F_n,$$
$$f \angle g \Leftrightarrow f(a_1, a_2, a_3, \ldots, a_n) \leq g(a_1, a_2, a_3, \ldots, a_n), \forall (a_1, a_2, a_3, \ldots, a_n) \in B_n$$

With this order relation, the set F_n forms a Boolean algebra.

EXAMPLE 4.14.– F_1, denoted as $B_1 = B$, includes only two elements (singletons) 0 and 1 (see Figure 4.17).

$a \in B_1$	f_1	f_2	f_3	f_4
0	0	0	1	1
1	0	1	0	1

$f_1 \angle f_2$ $f_1 \angle f_3$ $f_1 \angle f_4$
$f_2 \angle f_4$ $f_3 \angle f_4$

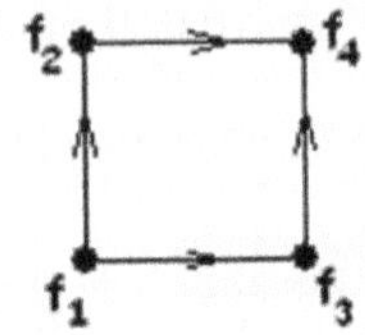

Figure 4.17. *Boolean algebra B_1*

EXAMPLE 4.15.– $F_2 : B_2 = \{(0,0), (0,1), (1,0), (1,1)\}$. The truth table and Hasse diagram are shown in Figures 4.18 and 4.19.

$(a,b) \in B_2$	f_1	f_2	f_3	f_4	f_5	f_6	f_7	f_8	f_9	f_{10}	f_{11}	f_{12}	f_{13}	f_{14}	f_{15}	f_{16}
0 0	0	1	0	0	0	1	1	1	0	0	0	1	1	1	0	1
0 1	0	0	1	0	0	1	0	0	1	1	0	1	0	1	1	1
1 0	0	0	0	1	0	0	1	0	1	0	1	1	1	0	1	1
1 1	0	0	0	0	1	0	0	1	0	1	1	0	1	1	1	1

Figure 4.18. *Boolean algebra B_2*

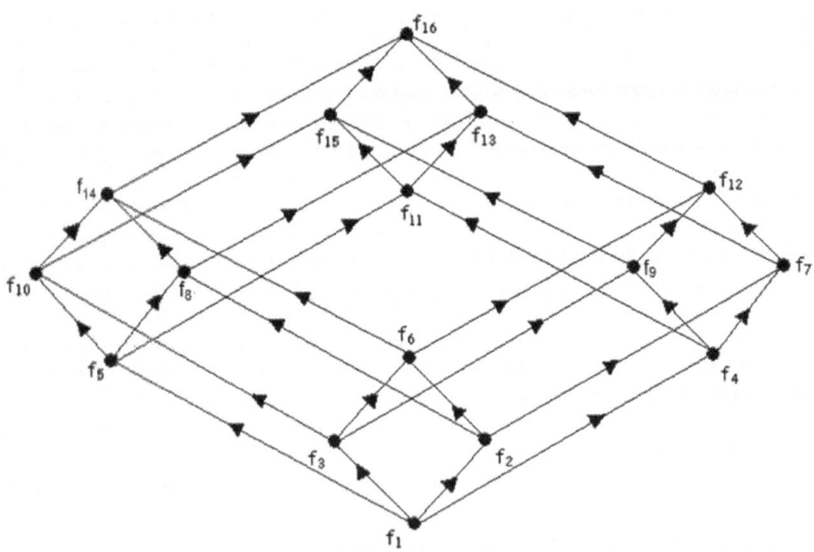

Figure 4.19. *Hasse diagram*

5

Logic Circuits

CONCEPTS COVERED IN THIS CHAPTER.–

Even without the intention of building a computer, gaining an understanding of how it operates through electronic circuits can be valuable.

Building on the previous chapter, this chapter continues the study of Boolean functions, focusing on simplifying them using the Karnaugh array method.

A computer's logical functions are built from assemblies of logic circuits, which are physically represented by electronic circuits. An aside on computer electronics will precede a detailed explanation of constructing logical functions using the example of a binary adder.

References: [WOO 80, VEL 19, HUR 79].

5.1. Canonical forms of a Boolean function

This chapter focuses on the set $E = \{0,1\}$. The two elements of this set represent binary information, or bits, which are the fundamental components of computer operations.

In the set E, the two laws of internal composition, "+" and ".", can be defined as equivalents of the *general* laws ∇ and Δ discussed in Chapter 4.

Mathematics for Digital Science 1,
by Gérard-Michel COCHARD and Mhand HIFI.

Thus, there will be:

$\bar{0} = 1$	$0 + 0 = 0$	$0\,0 = 0$
	$0 + 1 = 1 + 0 = 1$	$0.1 = 1.0 = 0$
$\bar{1} = 0$	$1 + 1 = 1$	$1.1 = 1$

The "+" law, often referred to as logical addition, shares similarities with arithmetic addition, except that in this context, 1 + 1 equals 1 instead of 2. The "." law, known as logical multiplication, parallels its arithmetic counterpart without variation. In this structure, elements 0 and 1 are complements of each other. As discussed in the preceding chapter, E illustrates a Boolean algebraic structure.

Previously, it was shown that Boolean functions $f_n: E_n \rightarrow E$ can be defined, where $E_n \equiv E^n$ denotes the set of n-tuples $(x_1, x_2, \dots, x_n)$ with each x_i belonging to E. However, for large n, representing the truth tables of 2^{2^n} Boolean functions f_n becomes impractical. Fortunately, there is a method to express these functions in the form of formulas that incorporate elements of E and the operations "+", "." and "inversion". This method is based on the fact that any Boolean function can be represented by a Boolean expression, which is a combination of elements of E and the operations "+", "." and "inversion". The method uses De Morgan's laws to express the negations of each variable x_i in terms of conjunctions and disjunctions, followed by simplification using distributive and associative laws of "+", "." and "inversion". Finally, this yields a Boolean formula that describes the Boolean function f_n.

Consider F_n as the set of all Boolean functions $E^n \longrightarrow E$, where E^n represents the set of n-tuples $(x_1, x_2, \dots, x_n)$, with each x_i belonging to E. An order relation can be defined on F_n such that:

$$f_i \angle f_j \Leftrightarrow f_i(x_1, x_2, \dots, x_n) \angle f_j(x_1, x_2, \dots, x_n), \forall x_1, x_2, \dots, x_n \in E.$$

Indeed, F_n represents a Boolean algebra structure. Within F_n, there are "special" functions known as *minterms*, which take the value 1 exactly once. These *minterms* are the immediate successors of the null element, which always takes the value 0. They are sometimes referred to as *atoms*. It should be noted that *minterms* are important in logic circuit theory. They are used in the synthesis of Boolean functions in combinatorics, as discussed later.

EXAMPLE 5.1.– Consider F_1 the set of Boolean functions $E \longrightarrow E$. There are four possible functions represented schematically in Figure 5.1.

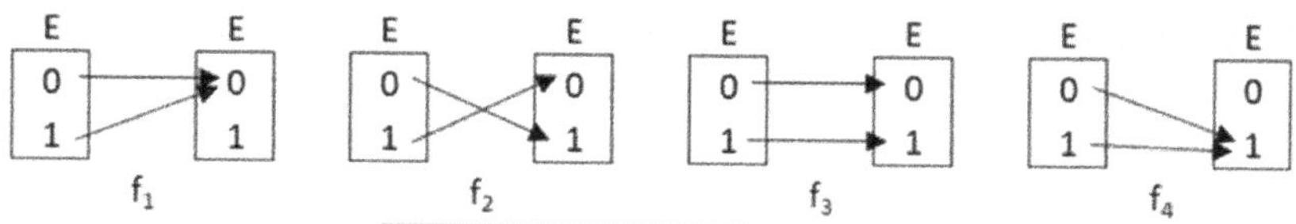

x	f_1	f_2	f_3	f_4
0	0	1	0	1
1	0	0	1	1

truth table

Figure 5.1. *The four possible functions*

The null element in the set F_2 is the function f_1, which always returns 0 for all possible inputs.

The functions f_2 and f_3 are *minterms*, i.e. they return the value 1 for a single combination of inputs, while other combinations return 0.

In contrast, the function f_4 always returns 1 for all possible inputs and is therefore not a minterm.

EXAMPLE 5.2.– The set F_2 is composed of all Boolean functions from E^2 to E. This set contains 16 distinct functions, shown schematically in Figure 5.2.

x,y	f_1	f_2	f_3	f_4	f_5	f_6	f_7	f_8	f_9	f_{10}	f_{11}	f_{12}	f_{13}	f_{14}	f_{15}	f_{16}
00	0	1	0	0	0	1	1	1	0	0	0	1	1	1	0	1
01	0	0	1	0	0	1	0	0	1	1	0	1	0	1	1	1
10	0	0	0	1	0	0	1	0	1	0	1	1	1	0	1	1
11	0	0	0	0	1	0	0	1	0	1	1	0	1	1	1	1

Figure 5.2. *The 16 possible functions of* F_2

The null element is f_1, the minterms are f_2, f_3, f_4 and f_5.

Since truth tables have 2^n rows, F_n contains 2^n minterms. Regarding the process, the following result is assumed.

THEOREM. 5.1.– Any Boolean function F_n can be expressed from the minterms, which are less than it: $f(a_1, a_2, \dots, a_n) = m_1 + m_2 + \cdots \dots + m_k, \forall m_j \angle f$.

EXAMPLE 5.3.– In the set F_1: $f_4 = f_2 + f_3$.

EXAMPLE 5.4.– In the set F_2: $f_8 = f_5 + f_2$. The observation from the truth tables depend on the variables involved in functions f_2 and f_5. Understanding these truth tables is crucial for assessing the validity of the expression of f_8.

Moreover, each minterm can be represented by the conjunctive combinations of a_i and $\bar{a}_j$, as stated in the following theorem, which will also be admitted.

THEOREM 5.2.– The minterms of F_n are expressed in the form $m = a_1.a_2.\dots.a_p$, where each a_i is either x_i or $\bar{x}_i$.

EXAMPLE 5.5.– In the set F_1, where x represents a variable:

f_2 is a minterm of F_1 where $f_2 = \bar{x}$

f_3 is a minterm of F_1 where $f_3 = x$

EXAMPLE 5.6.– In the set F_2:

f_3 is a minterm of F_2, where $f_3 = \bar{x}.y$

f_4 is a minterm of F_2, where $f_4 = x.\bar{y}$

f_5 is a minterm of F_2, where $f_5 = x.y$

f_2 is a minterm of F_2, where $f_2 = \bar{x}.\bar{y}$

It is evident that using the two previous theorems allows for the expression of a Boolean function as a disjunction of conjunctions, known as the *disjunctive canonical form*. This form is unique for each Boolean function. The practical method for finding the *minterms* from the truth table is straightforward:

– Identify the combinations of values of x_i which give a value 1 for the function $f(x_1, x_2, \dots, x_n)$. These are the minterms.

– Form the corresponding minterm using the values of $x_1, x_2, \dots, x_n$ by setting $m = a_1.a_2 \dots. a_n$ where $a_i = x_i$ if $x_i = 1$ or $a_i = \bar{x}_i$ if $x_i = 0$ for each minterm m identified.

The Boolean function f can then be expressed simply as the sum of the previously identified minterms: $f = m_1 + m_2 + \cdots.. + m_p$, where m_1, m_2, ... m_p are the minterms.

EXAMPLE 5.7.– In the set F_2, consider the function f_{12}, whose minterms are provided in Figure 5.3.

$$f_{12} = m_1 + m_2 + m_3 = \bar{x}.\bar{y} + \bar{x}.y + x.\bar{y}$$

x,y	f_{12}	minterms
00	1	$m_1 = \bar{x}.\bar{y}$
01	1	$m_2 = \bar{x}.y$
10	1	$m_3 = x.\bar{y}$
11	0	

Figure 5.3. *Identification of minterms*

EXAMPLE 5.8.– In the set F_3, consider the function f. The minterms are defined in Figure 5.4.

$$f = m_1 + m_2 + m_3 + m_4 = \bar{x}.\bar{y}.\bar{z} + \bar{x}.y.z + x.\bar{y}.z + x.y.z$$

x, y, z	f	minterms
000	1	$m_1 = \bar{x}.\bar{y}.\bar{z}$
001	0	
010	0	
011	1	$m_2 = \bar{x}.y.z$
100	0	
101	1	$m_3 = x.\bar{y}.z$
110	0	
111	1	$m_4 = x.y.z$

Figure 5.4. *Identification of minterms*

The laws "+" and "." being dual, it is clear that there exists another canonical form known as the *conjunctive canonical form*, which is expressed as a conjunction of disjunctions.

The notion of *maxterm*, which is the function of F_n that evaluates to 0 exactly once, is the dual concept of minterm. Let us illustrate this with the following examples.

EXAMPLE 5.9.– Consider the set F_2. Consider the function f_{12} for which the maxterm M corresponds to the "0"s of the function in Figure 5.5.

$$f_{12} = M = \bar{x} + \bar{y}$$

x,y	f_{12}	maxterms
00	1	
01	1	
10	1	
11	0	$M = \bar{x} + \bar{y}$

Figure 5.5. *Identification of the maxterm*

EXAMPLE 5.10.– Consider the set F_3. Consider f, one of the Boolean functions in this set. The maxterms corresponding to the "0"s of this function as shown in Figure 5.6.

$$f = M_1.M_2.M_3.M_4 = (x + y + \bar{z}).(x + \bar{y} + z).(\bar{x} + y + z).(\bar{x} + \bar{y} + z)$$

x, y, z	f	maxterms
000	1	
001	0	$M_1 = x + y + \bar{z}$
010	0	$M_2 = x + \bar{y} + z$
011	1	
100	0	$M_3 = \bar{x} + y + z$
101	1	
110	0	$M_4 = \bar{x} + \bar{y} + z$
111	1	

Figure 5.6. *Identification of maxterms*

5.2. Reduction of a Boolean function

The canonical disjunctive and conjunctive forms are valuable because they are unique and facilitate comparison of two Boolean expressions to determine their equivalence or disparity. However, they can be redundant, and there are instances where simplification is feasible.

EXAMPLE 5.11.– Reduction of F_2 functions (see Figure 5.7).

canonical disjunctive form	reduced form
$f_1 = 0$	$f_1 = 0$
$f_2 = \bar{x}.\bar{y}$	$f_2 = \bar{x}.\bar{y}$
$f_3 = \bar{x}.y$	$f_3 = \bar{x}.y$
$f_4 = x.\bar{y}$	$f_4 = x.\bar{y}$
$f_5 = x.y$	$f_5 = x.y$
$f_6 = (\bar{x}.\bar{y}) + (\bar{x}.y)$	$f_6 = \bar{x}$
$f_7 = (\bar{x}.\bar{y}) + (x.\bar{y})$	$f_7 = \bar{y}$
$f_8 = (\bar{x}.\bar{y}) + (x.y)$	$f_8 = (\bar{x}.\bar{y}) + (x.y)$
$f_9 = (\bar{x}.y) + (x.\bar{y})$	$f_9 = (\bar{x}.y) + (x.\bar{y})$
$f_{10} = (\bar{x}.y) + (x.y)$	$f_{10} = y$
$f_{11} = (x.\bar{y}) + (x.y)$	$f_{11} = x$
$f_{12} = (\bar{x}.\bar{y}) + (\bar{x}.y) + (x.\bar{y})$	$f_{12} = \bar{x} + \bar{y}$
$f_{13} = (\bar{x}.\bar{y}) + (x.\bar{y}) + (x.y)$	$f_{13} = x + \bar{y}$
$f_{14} = (\bar{x}.\bar{y}) + (\bar{x}.y) + (x.y)$	$f_{14} = \bar{x} + y$
$f_{15} = (\bar{x}.y) + (x.\bar{y}) + (x.y)$	$f_{15} = x + y$
$f_{16} = (\bar{x}.\bar{y}) + (\bar{x}.y) + (x.\bar{y}) + (x.y)$	$f_{16} = 1$

Figure 5.7. *Reduction of F_2 functions*

In Boolean algebra, a prime monomial in an expression A is defined as a monomial M such that $M + A = A$, and M does not contain a simpler monomial with the same property.

EXAMPLE 5.12.– Let us show that $(x.\bar{z})$ is a prime monomial in $A = (x.\bar{y}) + (x.y.\bar{z}) + (\bar{x}.y.\bar{z})$.

Indeed, firstly, there is no simpler form of this monomial. For instance, $(x, \bar{y})$ can be expressed as $x.\bar{y} = x.\bar{y}.(z + \bar{z}) = (x.\bar{y}.z) + (x.\bar{y}.\bar{z})$. The canonical form of A then becomes:

$$A = (x.\bar{y}.z) + (x.\bar{y}.\bar{z}) + (x.y.\bar{z}) + (\bar{x}.y.\bar{z})$$

The canonical form of $(x.\bar{z})$ is $x.(y + \bar{y}).\bar{z} = (x.y.\bar{z}) + (x.\bar{y}.\bar{z})$.

The equation simplifies to $(x.\bar{z}) + A = (x.y.\bar{z}) + (x.\bar{y}.\bar{z}) + (x.\bar{y}.z) + (x.\bar{y}.\bar{z}) + (x.y.\bar{z}) + (\bar{x}.y.\bar{z}) = (x.\bar{y}.z) + (x.\bar{y}.\bar{z}) + (x.y.\bar{z}) + (\bar{x}.y.\bar{z}) = A$

Therefore, $(x.\bar{z})$ is indeed a prime monomial of A.

Prime monomials have a particular importance, as indicated by the following theorem (which is admitted).

THEOREM 5.3.– If a Boolean expression A is in a reduced disjunctive form, each of its monomials is a prime monomial of A.

This means that the reduced disjunctive form of A is a disjunction of the prime monomials of A (but not necessarily all). The practical method for finding all prime monomials of a Boolean expression A is based on the notion of *consensus*: given two monomials of forms $x.m_1$ and $\bar{x}.m_2$, the monomial $m_1.m_2$ is called the *consensus* of monomials $x.m_1$ and $\bar{x}.m_2$.

The method can be implemented in several stages:

1) Putting A to its polynomial form and creating the list L of its monomials.

2) Iterating over each variable x:

a) Splitting list L into three sublists: L1 (monomials containing x), L2 (monomials containing $\bar{x}$) and L3 (other monomials).

b) If L1 or L2 is empty, proceeding to the next variable; otherwise, forming the list L′ by computing consensuses from L1 and L2 (multiplying the monomials of each pair and eliminating duplicate variables). Adjusting L′ into L involves adding monomials not containing the variable x and removing those divisible by existing monomials in L.

3) The resulting list L obtained from step 2 contains all prime monomials of A.

EXAMPLE 5.13.– Given $A = (x.y.\bar{z}) + (\bar{x}.\bar{y}.z) + (y.z)$,

$\mathrm{L} = \{(x.y.\bar{z}), (\bar{x}.\bar{y}.z), (y.z)\}$,

$\mathrm{L1} = \mathrm{L}(x) = \{(x.y.\bar{z})\}$,

$\mathrm{L2} = \mathrm{L}(\bar{x}) = \{(\bar{x}.\bar{y}.z)\}$,

$\mathrm{L3} = \{(y.z)\}$.

Consensus of $(L1, L2) = y.\bar{z}.\bar{y}.z = 0$, so nothing is added to L.

For $\mathrm{L1} = \mathrm{L}(y) = \{(x.y.\bar{z}), (y.z)\}$,

$\mathrm{L2} = \mathrm{L}(\bar{y}) = \{(\bar{x}.\bar{y}.z)\}$,

$\mathrm{L3} = \emptyset$.

Consensus of (L1, L2) $= \{(x.\bar{z}.\bar{x}.z), (z.\bar{x}.z)\} = \{0, (\bar{x}.z)\}$. The non-zero monomial to add to L is $(\bar{x}.z)$.

Initially, $\mathrm{L} = \{(x.y.\bar{z}), (\bar{x}.\bar{y}.z), (y.z), (\bar{x}.z)\}$, and after refinement, the set becomes: $\mathrm{L} = \{(x.y.\bar{z}), (y.z), (\bar{x}.z)\}$.

For $\mathrm{L1} = \mathrm{L}(z) = \{(y.z), (\bar{x}.z)\}$, $\mathrm{L2} = \mathrm{L}(\bar{z}) = \{(x.y.\bar{z})\}$.

Consensus of $(\mathrm{L1}, \mathrm{L2}) = \{(\mathrm{y.x.y}), (\bar{\mathrm{x}}.\mathrm{x.y})\} = \{(x.y), 0\}$. The non-zero monomial to add to L is (x.y).

Initially, $\mathrm{L} = (x.y.\bar{z}), (y.z), (\bar{x}.z), (x.y) \Rightarrow \mathrm{L} = (y.z), (\bar{x}.z), (x.y)$.

However, $(y.z) + (x.y) + (\bar{x}.z)$ is not the reduced form of A. The reduced form is obtained by constructing a double-entry table (see Figure 5.8), listing the minterms of the disjunctive canonical form of A alongside the prime monomials:

$$A = (x.y.\bar{z}) + (\bar{x}.\bar{y}.z) + (y.z).(x + \bar{x}) = (x.y.\bar{z}) + (\bar{x}.\bar{y}.z) + (x.y.z) + (\bar{x}.y.z)$$

		minterms from the canonical form			
		$\mathrm{x.y.\bar{z}}$	$\mathrm{\bar{x}.\bar{y}.z}$	$\mathrm{x.y.z}$	$\mathrm{\bar{x}.y.z}$
prime monomials	x.y				
	y.z				
	$\bar{\mathrm{x}}$.z				

Figure 5.8. *Practical table*

$(x.y)$ generates $(x.y.z)$ and $(x.y.\bar{z})$ since $(x.y) = (x.y)(z + \bar{z}) = (x.y.z) + (x.y.\bar{z})$.

Also, $(y.z)$ generates $(x.y.z)$ and $(\bar{x}.y.z)$. Finally, $(\bar{x}.z)$ generates $(\bar{x}.y.z)$ and $(\bar{x}.\bar{y}.z)$.

Highlighting the corresponding boxes in the table (see Figure 5.8), it becomes (x.y) and $(\bar{x}.z)$ accounting for all the minterms, thus forming the reduced form: $A = (x.y) + (\bar{x}.z)$.

5.3. Karnaugh tables

A graphical method known as the Karnaugh method, developed by Veitch and Karnaugh, simplifies the process of obtaining the reduced form of a Boolean function $E_n \longrightarrow E$.

This method uses tables called Karnaugh tables, illustrated for $n = 1, 2, 3$ and 4 in Figure 5.9:

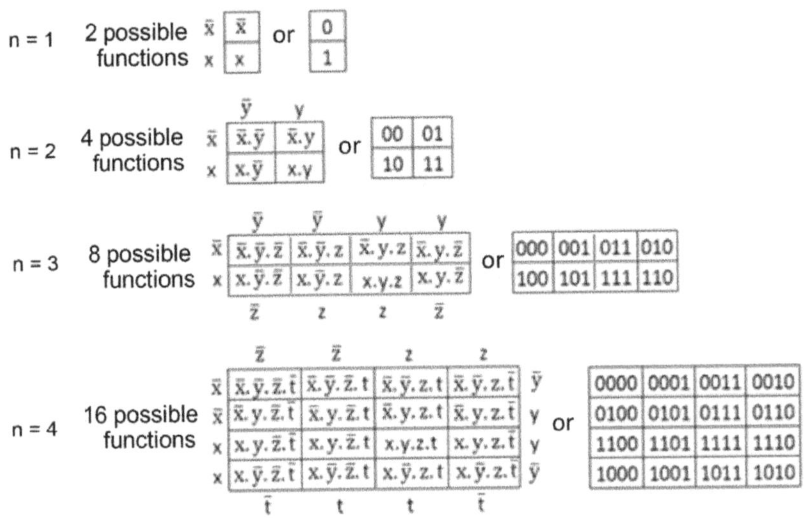

Figure 5.9. *Karnaugh tables*

– Each table contains 2^n boxes. In the first series of tables, these boxes represent the different combinations of variables.

– In the second series (sometimes referred to as a Veitch diagram), each variable is represented by 1 and its inverse (complement) by 0.

– By convention, in these tables, the boxes of the first row are adjacent to those in the last row, and similarly, the boxes in the left column are adjacent to those in the right column.

Here is how to use Karnaugh tables to simplify a three-variable Boolean function (x, y, z) with an output E. Figure 5.10 shows the corresponding table with 0s and 1s:

– Place 1s and 0s: fill in the Karnaugh table by placing a 1s in each box corresponding to the combinations of variables for which the function *E* equals 1. Place 0s in all other boxes.

– Identify groups of 1s: look for all possible groups of adjacent 1s in the table. Groups can be horizontal or vertical, but not diagonal. Each group can contain 1, 2, 4 or 8 boxes. Boxes must be adjacent in the table, and groups can wrap around the edges of the board (edges).

– Write expressions for groups: for each group of 1s, write an expression that includes all the variables in the table (*x, y,* and *z*) for which the boxes in the group contain 1s. If a group includes a box with a 0, ignore that box in the expression.

– Simplify expressions: apply Boolean laws to simplify each expression derived from the groups of 1s.

– Combine expressions: write the simplified Boolean function by combining all the simplified expressions using the "or" operator.

This yields a simplified function with fewer terms than the original function.

000	001	011	010
100	101	111	110

Figure 5.10. *Table with 0s and 1s*

Highlight a combination of variables by shading the boxes (see Figure 5.11).

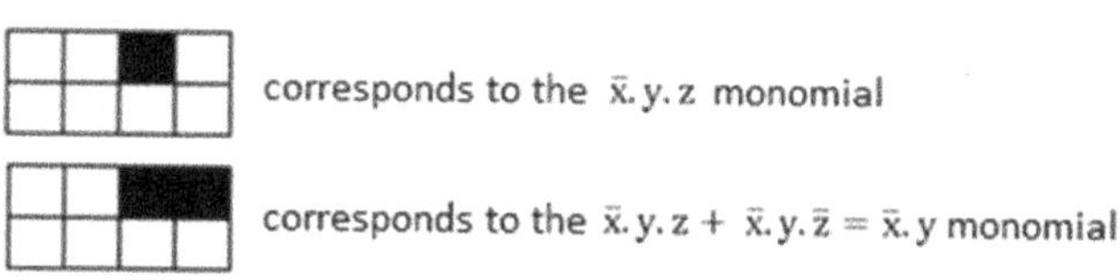

Figure 5.11. *Practical form of Karnaugh tables*

Note that adjacent boxes cause certain variables to disappear from the expression.

EXAMPLE 5.14.– Consider function *A* from the previous example and construct its truth table (see Figure 5.12).

$$A = (x.y.\bar{z}) + (\bar{x}.\bar{y}.z) + (x.y.z) + (\bar{x}.y.z)$$

The function is represented by the Karnaugh table in Figure 5.13 (the boxes corresponding to the value 1 are shaded). This diagram is decomposed as shown in Figure 5.14, using adjacent boxes to eliminate variables, resulting in $A = \bar{x}.z + x.y$. Thus, the previously obtained reduced form is derived.

abc	A
000	0
001	1
010	0
011	1
100	0
101	0
110	1
111	1

Figure 5.12. *Truth table*

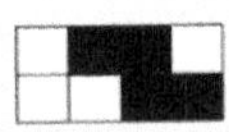

Figure 5.13. *Karnaugh table*

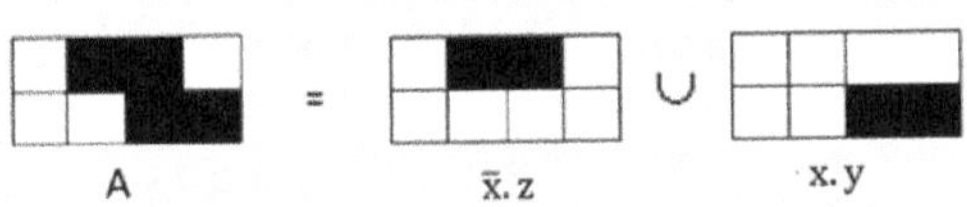

Figure 5.14. *Decomposition*

EXAMPLE 5.15.– Let us provide additional examples illustrating a Boolean function A of three variables. It is important to consider the adjacency of the peripheral boxes (see Figure 5.15(a)–(c))

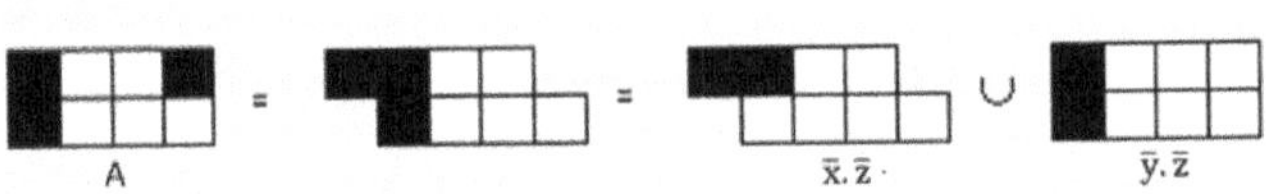

Figure 5.15(a). $A = \bar{x}.\bar{z} + \bar{y}.\bar{z}$

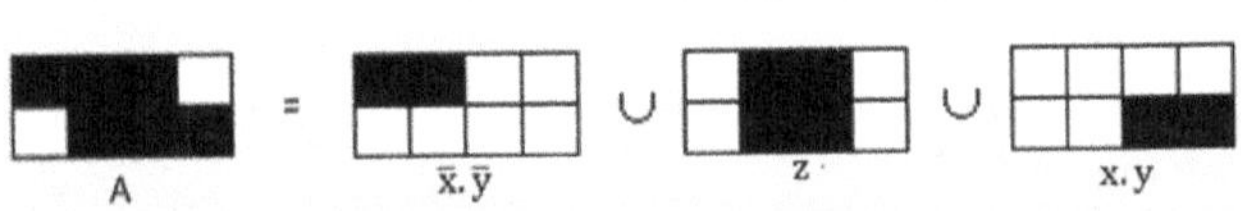

Figure 5.15(b). $A = \bar{x}.\bar{y} + z + x.y$

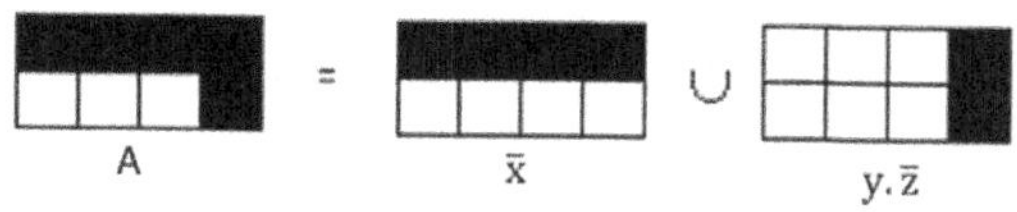

Figure 5.15(c). $A = \bar{x} + y.\bar{z}$

5.4. Elementary logic circuits

In logic, the following operations are studied on propositions: the negation (NOT) noted $\bar{x}$; the conjunction (AND) noted $\wedge$; the disjunction (INCLUSIVE OR) noted $\vee$; and the "exclusive OR" operator (XOR), each with their truth table where the values are T (true) or F (false). These operations are then applied to computer circuits. Modern computers process basic information known as bits, which have two possible values: 0 and 1. Logical values T (true) and F (false) correspond to bits 1 and 0, respectively. Thus, computer logic operates with a two-element set: $E = \{0, 1\}$. Common terminology and representations in electronics are used, with elementary logic functions including:

– *NOT* function (corresponding to the *NOT* operation): any element x of E corresponds to its opposite $\bar{x}$ (see Figure 5.16).

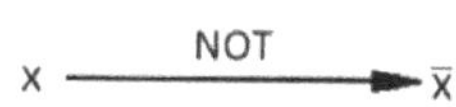

Figure 5.16. *NOT function*

The truth table for this logic function is straightforward, as shown in Figure 5.17. This logic function corresponds to an electronic circuit known as a *NOT gate*, typically represented as shown in Figure 5.18.

x	$\bar{x}$
0	1
1	0

Figure 5.17. *Truth table for NOT*

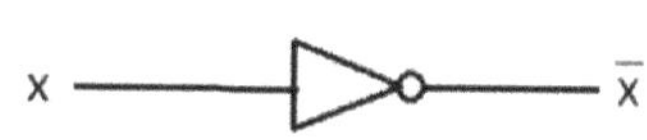

Figure 5.18. *NOT gate*

– The small loop in the previous diagram is important as it signifies the negation of the input bit. If omitted, the identity function is obtained, as shown in Figure 5.19.

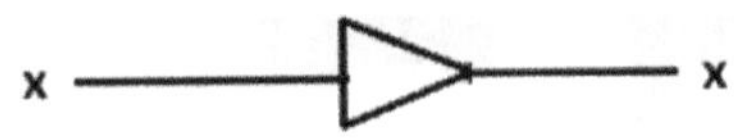

Figure 5.19. *YES circuit*

However, this representation is only used when other operations occur without altering the input "bit", such as signal amplification or temporary recording (buffer).

– The logical function corresponding to *AND* and its truth table (see Figure 5.20) is identical to that seen previously in propositional logic but noted in bits. The result of this AND function, denoted *as* $x.y$, is sometimes called "logical multiplication". It operates similarly to arithmetic multiplication. The corresponding logic circuit for this function is an *AND gate*, as shown in Figure 5.21.

x	y	x.y
0	0	0
0	1	0
1	0	0
1	1	1

Figure 5.20. *Truth table for AND*

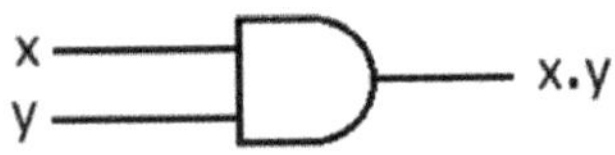

Figure 5.21. *AND gate*

– The *OR* function corresponds to the "inclusive OR" operation, whose truth table is shown in Figure 5.22.

x	y	x+y
0	0	0
0	1	1
1	0	1
1	1	1

Figure 5.22. *Truth table for OR*

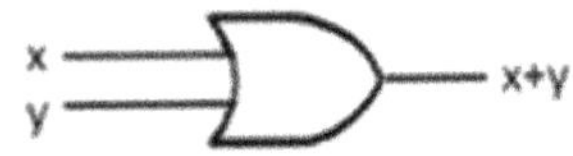

Figure 5.23. *OR gate*

The result is denoted by $x + y$ because this function is sometimes referred to as "logical addition". However, it differs from arithmetic addition because $1 + 1 = 1$ in binary (with only two bits). The logic circuit corresponding to this function is an *OR gate*, shown in Figure 5.23.

– The *XOR* function corresponds to exclusive OR. Its truth table is shown in Figure 5.24.

x	y	x⊕y
0	0	0
0	1	1
1	0	1
1	1	0

Figure 5.24. *Truth table for XOR*

Figure 5.25. *XOR gate*

The notation x⊕y is used to indicate the result. The distinction from the OR function is observed in the last row of the truth table. The logic circuit corresponding to this function is an *XOR gate,* represented by Figure 5.25.

– The *NAND* function is the negation of the AND function, as shown in Figure 5.26.

x	y	$\overline{x.y}$
0	0	1
0	1	1
1	0	1
1	1	0

Figure 5.26. *Truth table for NAND*

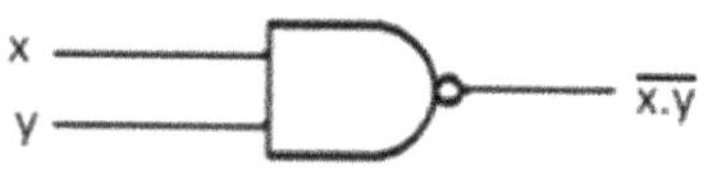

Figure 5.27. *NAND gate*

The corresponding logic circuit is the *NAND gate,* which is simply represented by adding a circle (negation) on the AND gate (see Figure 5.27). This circuit can also be represented as shown in Figure 5.28.

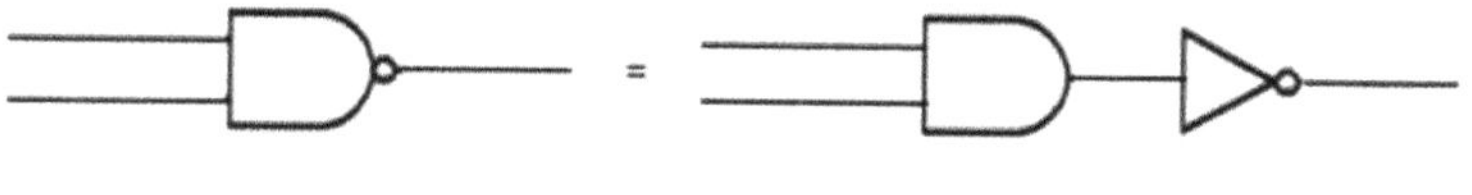

Figure 5.28. *Equivalence of circuits*

– In the same manner, the logical function *NOR* is the negation of the OR function, where the result is $\overline{x + y}$ (see Figure 5.29).

x	y	$\overline{x+y}$
0	0	1
0	1	0
1	0	0
1	1	0

Figure 5.29. *Truth table for XOR*

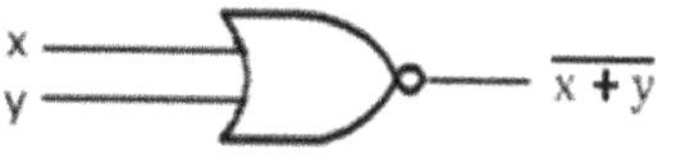

Figure 5.30. *XOR gate*

The corresponding logic circuit is the *NOR gate*, as shown in Figure 5.30. This circuit can be implemented using an OR gate and a NOT gate, as shown in Figure 5.31.

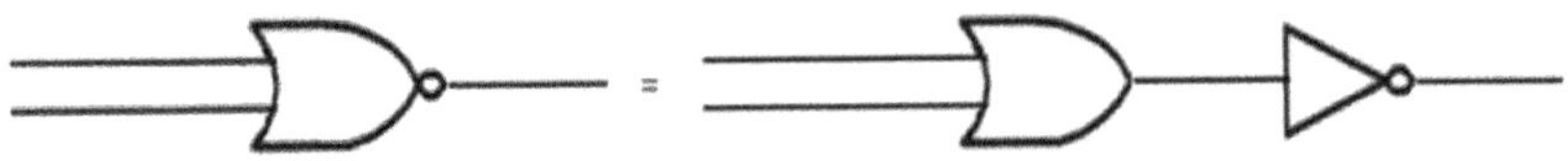

Figure 5.31. *Equivalence of circuits*

At this point, it is useful to recall some of the results previously presented that will be beneficial later:

1) Equipped with the operations NOT, OR, AND, the set $E = \{0, 1\}$ forms a Boolean algebra structure $(E, +, .)$ which can be axiomatized with the following postulates:

$\bar{0} = 1$	$0 + 0 = 0$	$0.0 = 0$
	$0 + 1 = 1 + 0 = 1$	$0.1 = 1.0 = 0$
$\bar{1} = 0$	$1 + 1 = 1$	$1.1 = 1$

2) The laws of addition ("+") and multiplication (".") are both commutative and associative. For example,

$$(x + y) + z = z + (y + x) = (z + y) + x = \ldots$$

3) The laws of addition ("+") and multiplication (".") are distributive with respect to each other. For example,

$$x + (y.z) = (x + y).(x + z)$$

$$x.(y + z) = (x.y) + (x.z)$$

4) Generalized Morgan's theorem (or Shannon's theorem): For any logical expression f, the following holds:

$$\overline{f(x, y, z, \ldots., +, .)} = f(\bar{x}, \bar{y}, \bar{z}, \ldots, ., +, .)$$

For example, $(x.y) + (z.t) = \overline{(\overline{x.y}).(\overline{z.t})}. = \overline{(\bar{x} + \bar{y}).(\bar{z} + \bar{t})}$ or

$$
\begin{aligned}
(x.\bar{y}) + (y.\bar{x}) &= [x + (y.\bar{x})].[\bar{y} + (y.\bar{x})] \\
&= [(x + y).(x + \bar{x})].[(\bar{y} + y).(\bar{y} + \bar{x})] \\
&= (x + y).(\bar{y} + \bar{x}) = (x + y).(\overline{x.y}) = \overline{(\bar{x}.\bar{y})}.(\overline{x.y}) \\
&= \overline{(\bar{x}.\bar{y}) + (x.y)}
\end{aligned}
$$

5) Decomposition theorem: any logical function of several Boolean variables can be expressed in two canonical forms:

$$f(x, y, z) = (x + y + z).(\bar{x} + y + z).(x + \bar{y} + z) \dots \dots .(\bar{x} + \bar{y} + \bar{z})$$

$$f(x, y, z) = (x.y.z) + (\bar{x}.y.z) + (x.\bar{y}.z) + \cdots .. + (\bar{x} + \bar{y} + \bar{z})$$

In the first form, the decomposition is a logical product of “basic” logical sums, and in the second form, the decomposition is a logical sum of “basic” logical products. The “basic” terms may not all appear in the decomposition and are obtained by combinations of variables and their complements.

So, $f(x, y) = \overline{(\bar{x}.\bar{y}) + (x.y)} = (x.\bar{y}) + (\bar{x}.y) = (x + y).(\bar{x} + \bar{y})$

The six logic circuits NOT, OR, AND, NOR, NAND, XOR are basic circuits because combinations of these circuits facilitate the creation of more complex circuits. A *complete logic set* is a minimal set of basic circuits that allows the construction of the six basic circuits mentioned earlier. It should be noted that NOT, AND and OR circuits can generate NAND, NOR and XOR circuits (see Figure 5.32).

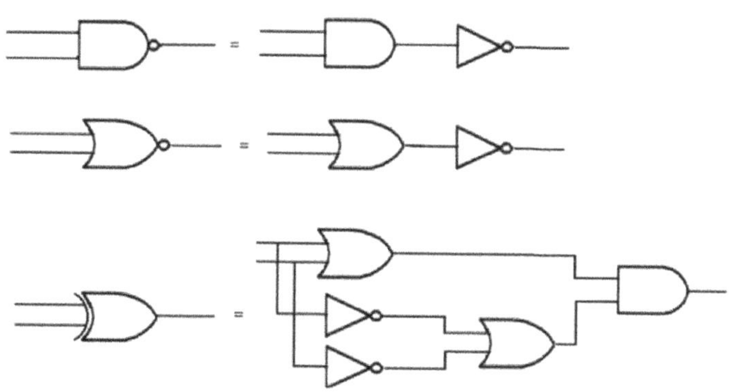

Figure 5.32. *Complete logic group {NAND, NOR, XOR}*

Furthermore, the decomposition theorem states that the three circuits NOT, OR and AND can be obtained from combinations of NOT with AND or NOT with OR. It results in two primary complete logical groups:

– The complete logical group {NOT, AND} (see Figure 5.33).

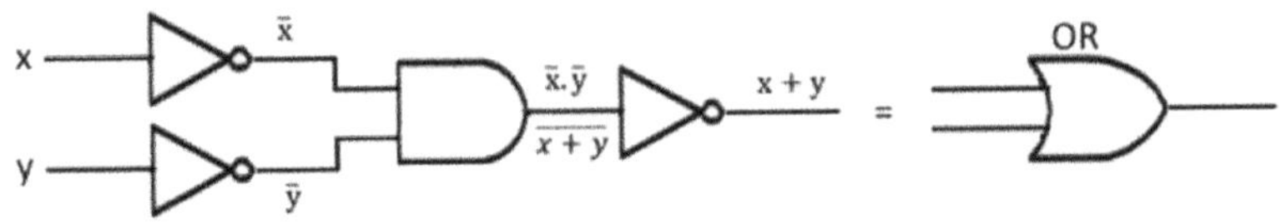

Figure 5.33. *Complete logical group {NOT, AND}*

– The complete logical group {NOT, OR} (see Figure 5.34).

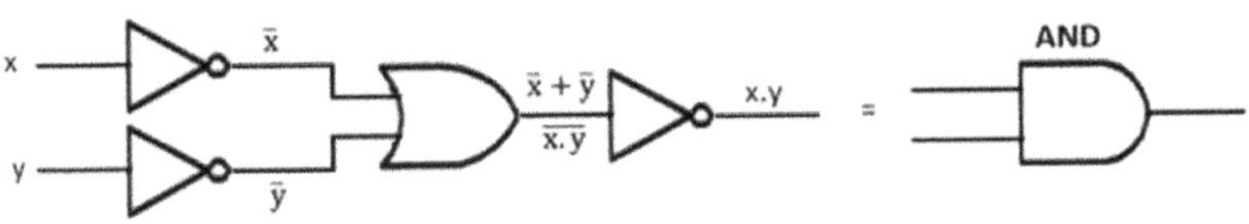

Figure 5.34. *Complete logical group {NOT, OR}*

However, there are even simpler complete logical groups:

– The complete logical group {NOR} (see Figure 5.35).

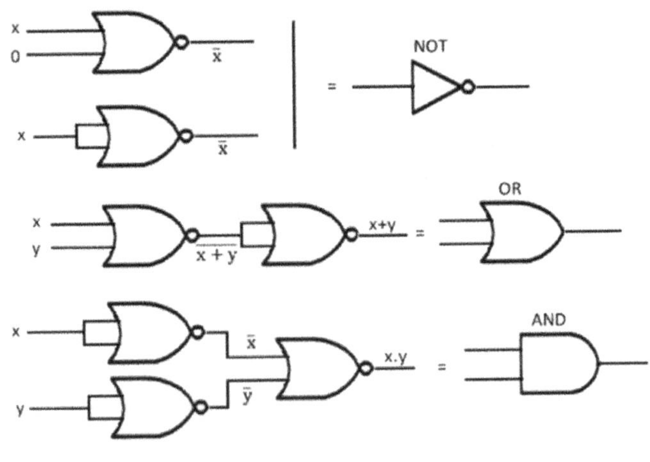

Figure 5.35. *Complete logic group {NOR}*

– The complete logic group {NAND} (see Figure 5.36).

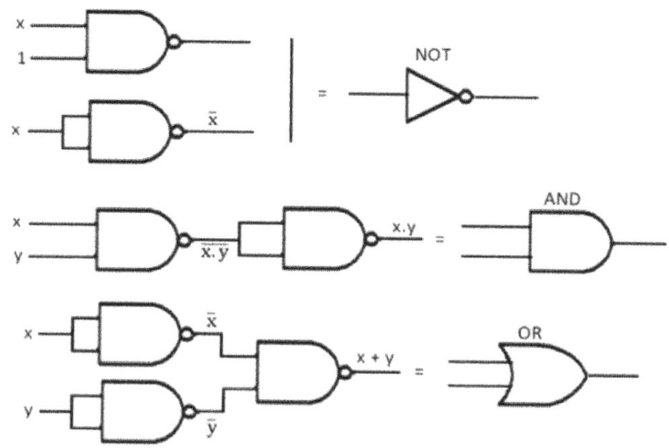

Figure 5.36. *Complete logic group {NAND}*

5.5. A little on electronics

Logic circuits are electrical circuits designed to transform one or more input signals into an output signal based on the logic function implemented by an electronic device.

Diodes and transistors are the basic and fundamental components of logic circuits, and their introduction into electronics has generated a technological revolution. Without extensively discussing the detailed principles of these components, let us explore their role in digital electronics.

An N-type semiconductor is a doped semiconductor (such as silicon or germanium) that contains impurities releasing electrons, leading to an excess of electrons and a deficiency of positive charges. These free electrons act as negative charge carriers. Conversely, a P-type semiconductor is doped with impurities that create electron void vacancies known as holes, which behave as positive charge carriers.

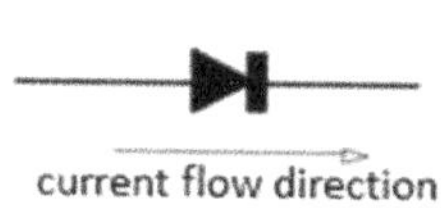

Figure 5.37. *Diode*

A diode is an electronic component that consists of a PN junction, where P represents the P-type region and N the N-type region. It allows the passage of electric current in only one direction, from the P-type region to the N-type region, while blocking current in the opposite direction. Its conventional representation is shown in Figure 5.37.

A transistor is essentially composed of two diodes sharing a common region. Consequently, two types of transistors can be obtained:

– The PNP transistor (see Figure 5.38) is shown schematically in Figure 5.39.

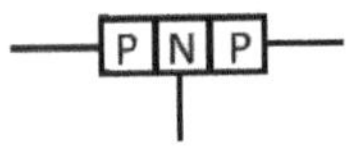

Figure 5.38. *PNP transistor*

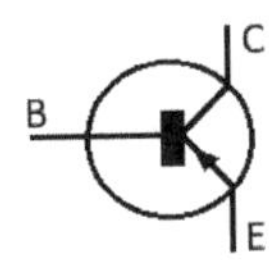

Figure 5.39. *PNP schematization*

– The NPN transistor (see Figure 5.40) is schematically shown in Figure 5.41.

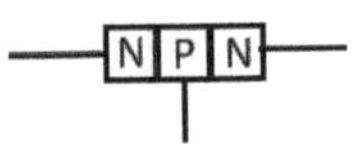

Figure 5.40. *NPN transistor*

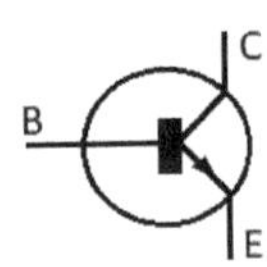

Figure 5.41. *NPN schematization*

A transistor consists of three electrodes: the base (B), the emitter € and the collector (C).

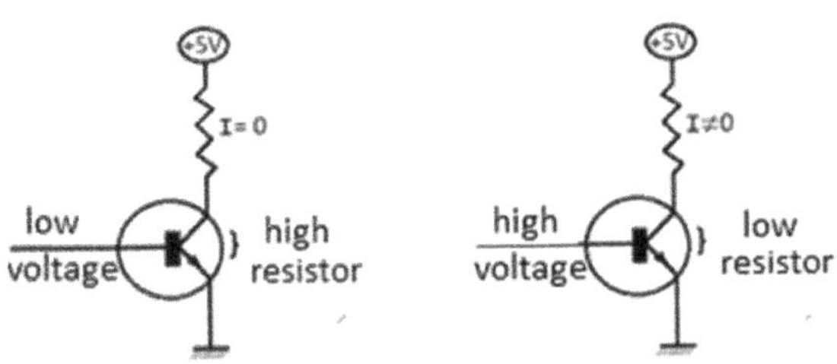

Figure 5.42. *Transistor operation*

In digital circuits, transistors are often used in switching, operating in two states: either they are in saturation mode (allowing current to pass) or in cutoff mode (blocking current). Depending on the voltage applied to the base, current can flow between the collector and the emitter (saturation mode: passing regime) or be blocked (cutoff mode: non-passing regime) (see Figure 5.42).

In bipolar transistors, which include those described above, conduction occurs due to the simultaneous movement of electric charge carriers of opposite signs: negative electrons and positive holes. Bipolar transistors are part of bipolar technology (TTL family and derivatives). There are also unipolar transistors (MOS family and derivatives) which require the movement of only one type of charge carrier, either electrons or holes. Unipolar transistors are of two types: P channel and N channel. These components tend to replace bipolar-type components.

Digital electronics is primarily based on the use of digital integrated circuits. These circuits, which vary in complexity, consist of numerous components manufactured simultaneously in a single process from semiconductors. A measure of this complexity is the number of transistors contained within each integrated circuit.

In 1976, with VLSI (Very-Large-Scale Integration) technology, integrated circuits contained between 10,000 and 100,000 transistors, also referred to as electronic chips due to their small size achieved through miniaturization. Today, each electronic chip typically contains several hundred million transistors. Current microprocessors, the most complex integrated circuits, incorporate approximately 3 billion transistors. The simplest commercial integrated circuits are packaged in boxes with pins for soldering onto a printed circuit board, which facilitates connections within an electronic system. A common package type is the DIL (Dual-In-Line) package, as shown in Figure 5.43, with a 14-pin (integrated circuit) DIL package.

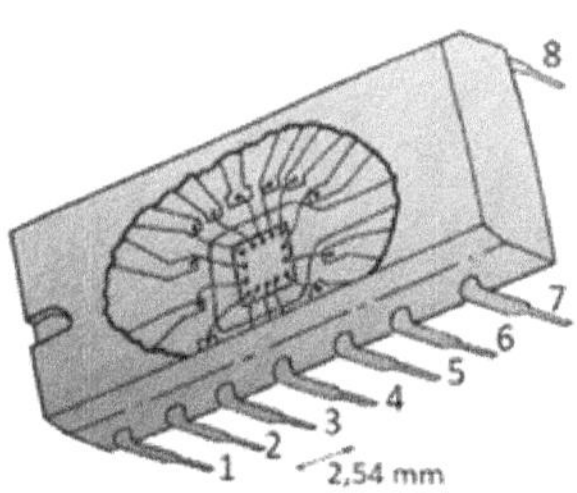

Figure 5.43. *A hardware circuit in a DIL package*

Figure 5.44. *Microprocessor*

A microprocessor, such as the one shown in Figure 5.44 (a square with a side of 3 cm), requires a significant number of pins.

To convert a logical circuit into its hardware implementation, it is necessary to establish a logical convention, which associates bits 0 and 1 with values of a physical quantity, such as an electrical voltage. In most cases, the voltage is fixed at 0 or 5 volts (for TTL or MOS circuits). The logical convention adopted is as follows: bit 0 → 0 volt and bit 1 → x volts.

However, in reality, due to inevitable voltage losses caused by internal resistors, bit 0 and bit 1 correspond to ranges of values. For TTL circuits, these ranges are shown in Figure 5.45.

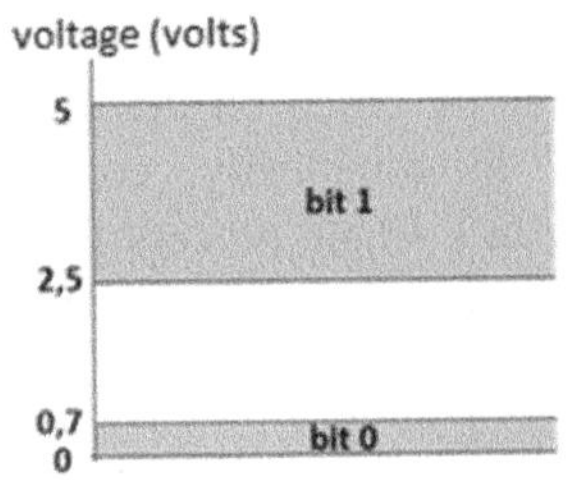

Figure 5.45. *Voltage definition of bits 0 and 1*

To illustrate the implementation of this logic convention, Figure 5.46 briefly shows how a NOT gate can be realized using an NPN transistor.

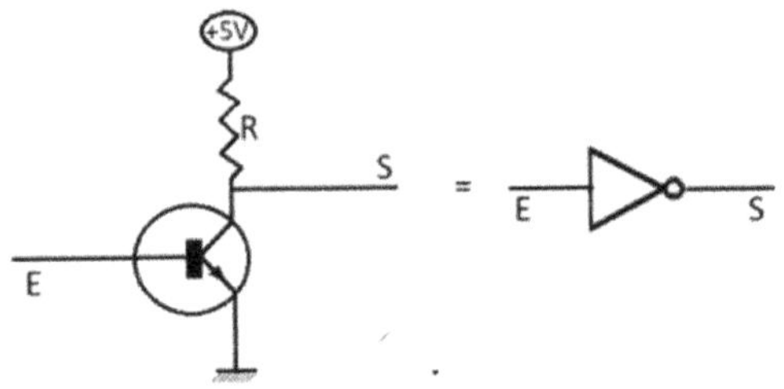

Figure 5.46. *NOT gate made by a transistor*

– If a voltage of +5V is applied to emitter E (bit 1), the NPN transistor enters saturation, causing the voltage on the collector C to be close to 0V (low level).

– If the emitter E is connected to ground (bit 0), the NPN transistor is blocked, and no current flows through the resistor R. The voltage on the collector then approaches +5V (high level).

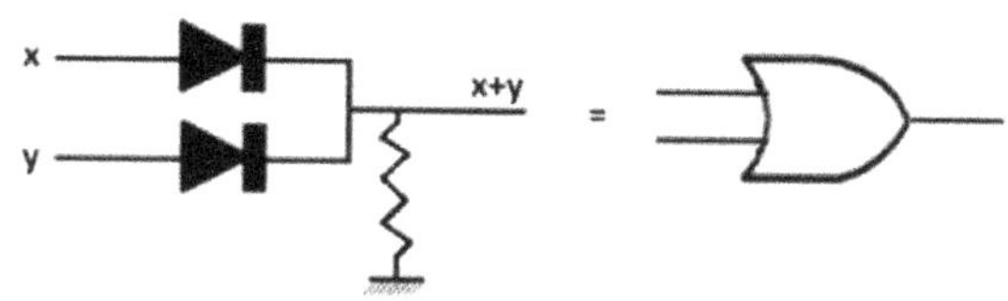

Figure 5.47. *OR gate made with two diodes*

Figure 5.47 shows the hardware implementation of an OR gate with two diodes.

5.6. Construction of logical functions

Karnaugh tables are valuable tools for constructing logic circuits corresponding to a specified Boolean function. Here are some examples to illustrate their utility.

EXAMPLE 5.16.– Let us design a logic circuit for the Boolean function $f(x, y, z, t) = x.y.z.t + \bar{x}.y.z.t + \bar{x}.y.\bar{z}.t + \bar{x}.\bar{y}.\bar{z}.t$. This function is already in canonical form, and its truth table is provided in Figure 5.48.

x	y	z	y	f	x	y	z	y	f
0	0	0	0	0	1	0	0	0	0
0	0	0	1	1	1	0	0	1	0
0	0	1	0	0	1	0	1	0	0
0	0	1	1	0	1	0	1	1	0
0	1	0	0	0	1	1	0	0	0
0	1	0	1	1	1	1	0	1	0
0	1	1	0	0	1	1	1	0	0
0	1	1	1	1	1	1	1	1	1

Figure 5.48. *Truth table of f(x, y, z)*

For four variables, the Karnaugh table representing this function is shown in Figure 5.49 and can be easily decomposed. Therefore, the reduced form of the function is $f = \bar{x}.\bar{z}.t + y.z.t = (\bar{x}.\bar{z} + y.z).t$. This logical form facilitates the simplification of the logic circuit corresponding to this function.

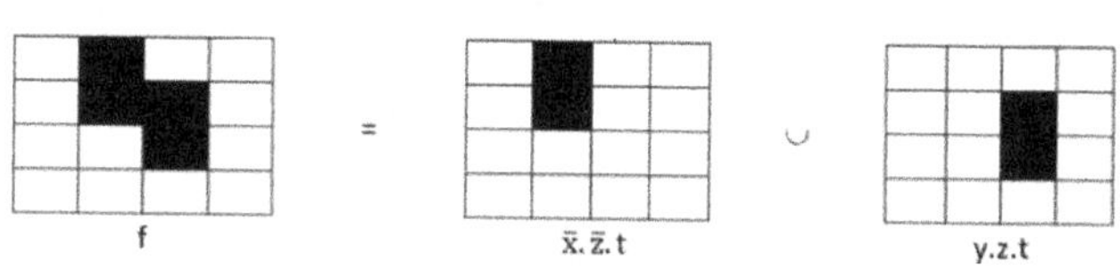

Figure 5.49. *Karnaugh tables*

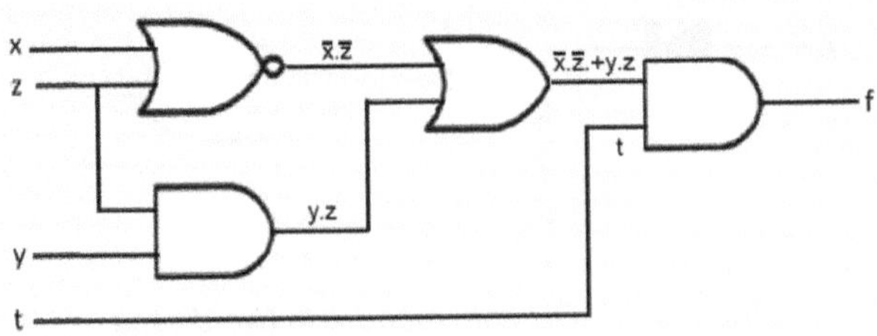

Figure 5.50. *Logic circuit proposal*

Noting that $\bar{x}.\bar{z} = \overline{x + z}$, a proposed logic circuit can be seen in Figure 5.50.

EXAMPLE 5.17.– Let us find the logic circuits for the functions of two variables $S = \bar{x}.y + x.\bar{y}$ and $R = x.y$. These functions are already in their reduced form, as confirmed by the Karnaugh tables (see Figure 5.51). Since there are no adjacent boxes, S and R are indeed in reduced form. The corresponding circuits are shown in Figure 5.52.

Figure 5.51. *Karnaugh tables*

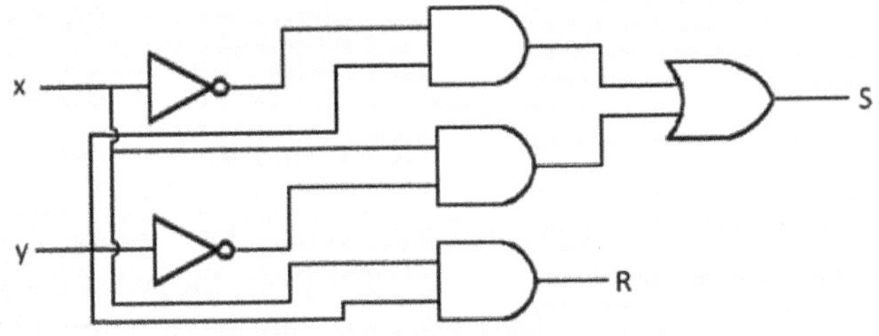

Figure 5.52. *First proposal for a logic circuit*

This logic circuit uses three types of logic gates. It can be verified that only two types of logic gates are needed, as shown in the equivalent circuit in Figure 5.53.

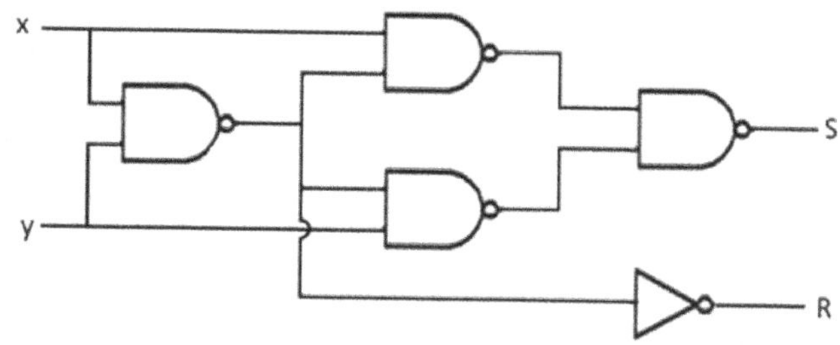

Figure 5.53. *Second logic circuit proposal*

Remember this logic circuit because it corresponds to a half-adder. There is an even simpler circuit, as will be shown shortly.

EXAMPLE 5.18.– Consider the Boolean function $f = \bar{x}.\bar{y}.\bar{z} + x.y.\bar{z} + \bar{x}.\bar{y}.z + \bar{x}.y.\bar{z} + x.\bar{y}.z$. Let us try to simplify this function using Karnaugh tables. The Karnaugh table of f can be broken-down as shown in Figure 5.54.

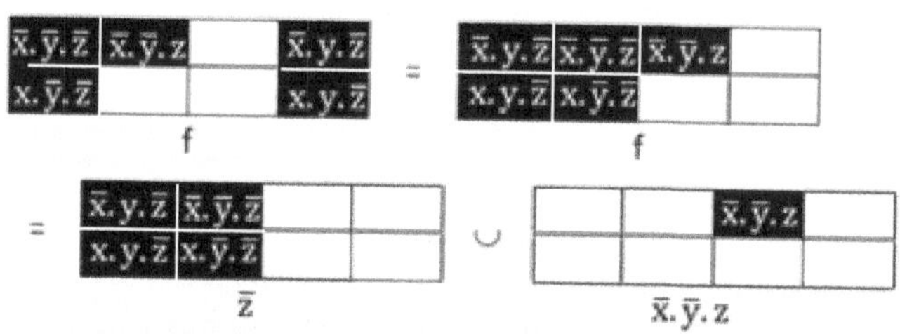

Figure 5.54. *Karnaugh tables*

Therefore, the reduced form is $f = \bar{z} + \bar{x}.\bar{y}.z$, and the corresponding logic circuit is given in Figure 5.55.

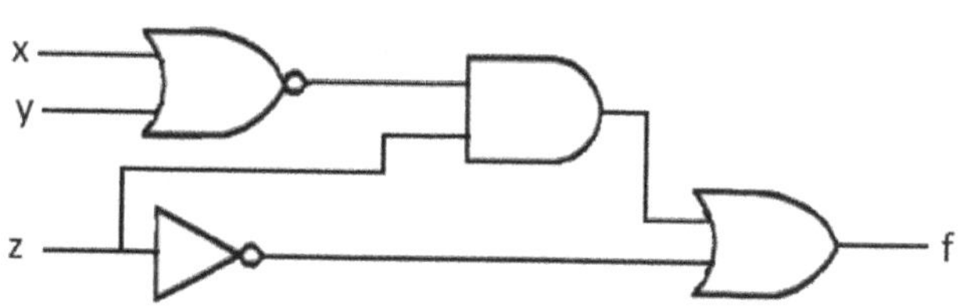

Figure 5.55. *Proposed logic circuit*

5.7. A basic circuit: the binary adder

Numbers are represented in the binary system for computer use. The computer, via its processing unit, performs calculations on these numbers. The simplest operation is addition. Addition in the binary system is carried out similarly to addition in the decimal system. For example, to perform the decimal addition 24 + 57 = 81, it can be translated into the binary system by adding 11000 (corresponding to 24 in binary) with 111001 (corresponding to 57 in binary), resulting in 1010001 (corresponding to 81 in binary), as shown in Figure 5.56.

In binary arithmetic, as in decimal arithmetic, the addition of two numbers can generate carries. In the binary system, there are only two possible digits, 0 and 1, so adding two digits can give a sum of 0, 1, 10 (equivalent to 2 in decimal) or 11 (equivalent to 3 in decimal) if a carry is generated.

```
        1
    1   1   1   0   0   0
+   1   1   1   0   0   1
-------------------------
1   0   1   0   0   0   1
```

Figure 5.56. *Binary addition*

In the given example, the sum of 24 and 57 in decimal is 81. In binary, this corresponds to the addition of 11000 and 111001, which gives 1010001. The schema shows two carries generated when adding the columns of bits with weights 2 and 4. These carries are highlighted in red in the schema of Figure 5.56.

A half-adder is an elementary component that performs column-wise addition, similar to "manual" addition. This component takes two bits x and y as inputs and produces a sum S and a possible carry R as outputs. This component must follow the addition table shown in Figure 5.57.

x	y	S	R
0	0	0	0
0	1	1	0
1	0	1	0
1	1	0	1

Figure 5.57. *Truth tables for S and R*

Two Boolean functions S and R must be found that provide the correct results. These functions are not difficult to find:

$$S = \bar{x}.y + x.\bar{y} \text{ and } R = x.y$$

These functions were illustrated in a previous example, and the logic circuit corresponding to this half-adder is shown in Figure 5.58. Specifically, $S = x \oplus y$, where $\oplus$ denotes exclusive OR. Therefore, an XOR gate is used to obtain S, and an AND gate is used to obtain R.

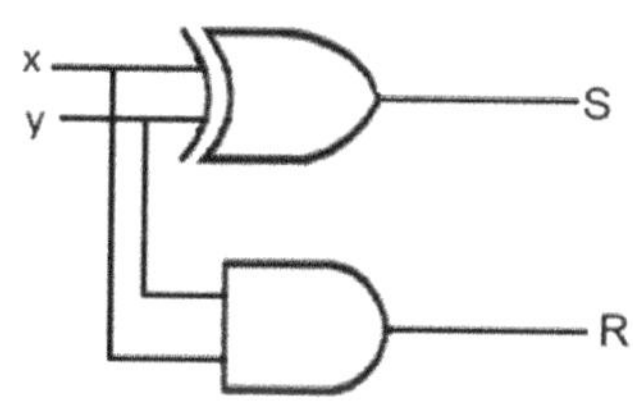

Figure 5.58. *Logic circuit for S and R*

A full adder is a combinatorial logic circuit used for column-wise additions, considering the carry from the previous step. Unlike the half-adder, the full adder takes three inputs: two bits to be added (x and y) and a carry input (c). The outputs are a sum S and a carry-out R for the next step. To construct a full adder, two half-adders can be combined with an OR logic gate. The Boolean functions for the result S and the carry R are as follows:

– $S = x \oplus y \oplus c$

– $R = (x \cdot y) + \big(c \cdot (x \oplus y)\big)$

where the symbol "$\oplus$" designates the XOR logic gate, the symbol "." Designates the AND logic gate and the symbol "+" designates the OR logic gate. The logic circuit corresponding to the full adder is given in Figure 5.59, where r_{n-1} plays the role of c in the previous equations.

Several full adders can be used in parallel to add two binary numbers represented by several bits, for example, a 4-bit number X and a 4-bit number Y. The bits of the numbers X and Y are denoted as X0, X1, X2, X3 for X and Y0, Y1, Y2, Y3 for Y, where index 0 indicates the lowest weight and index 3 the highest weight. Similarly, the bits of the sum S are also denoted as S0, S1, S2, S3. The parallel connection is shown in Figure 5.60.

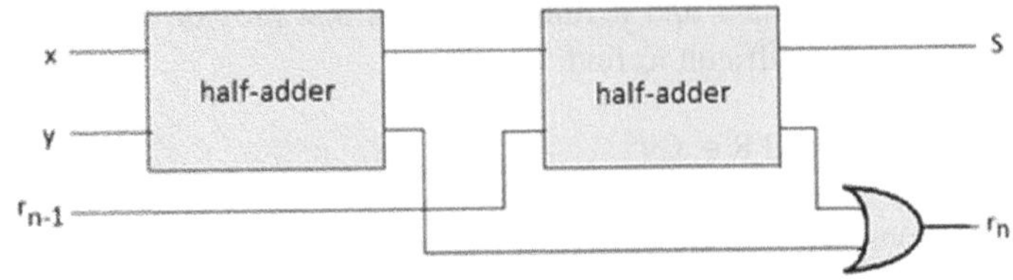

Figure 5.59. *Full adder*

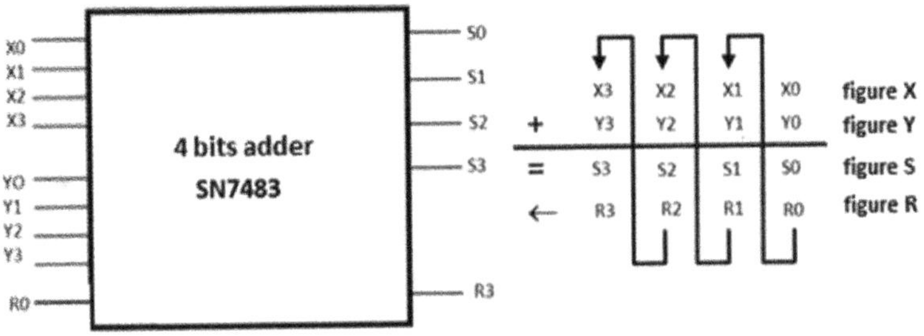

Figure 5.60. *Electronic adder schema*

Such a logic circuit is commercially implemented by an integrated circuit known by the code SN 7483 (see Figure 5.61).

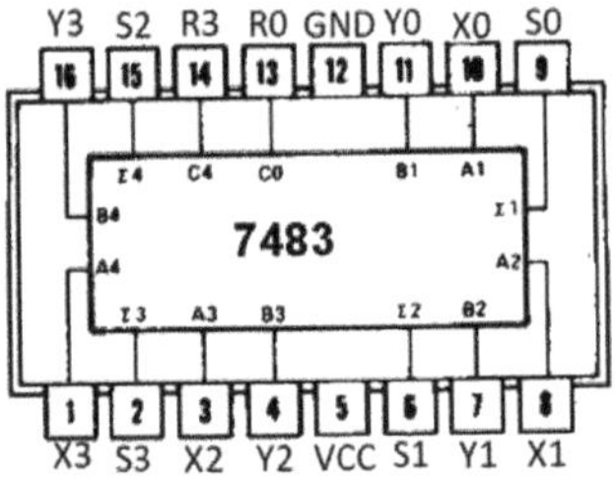

Figure 5.61. *The SN7483 box*

6

Arithmetic

CONCEPTS COVERED IN THIS CHAPTER.–

Arithmetic is often presented as the science of numbers and the study of their properties. The properties, especially those of integers, form the basis of encryption methods used to make information indecipherable by third parties. Methods of encrypting information will be explored in a future chapter, so it is important to first recall the basic notions of arithmetic, focusing specifically on integers.

The discussion will start with the Euclidean division, which leads to prime numbers, a topic that has engaged many mathematicians for centuries.

The discussion will also cover congruences, defined by a special operation called modulo, with an example of their use involving elliptic curves.

The exploration will conclude with Bézout's identity, theorem and algorithm.

References: [VEL 19, BUC 98, MUL 18, PRA 23].

6.1. Reminder about integers

Recalling some definitions and notations relating to *integers:*

– $\mathbb{N}$ represents the set of *natural numbers*, i.e. positive integers including zero: $\mathbb{N} = \{0, 1, 2, 3, \ldots, n, \ldots\}$.

– $\mathbb{N}^*$ represents the set of *non-zero natural numbers*, i.e. all natural numbers except zero: $\mathbb{N}^* = \{1, 2, 3, \ldots, n, \ldots\}$.

Mathematics for Digital Science 1,
by Gérard-Michel COCHARD and Mhand HIFI.

– $\mathbb{Z}$ represents the set of *relative integers*, i.e. positive, negative and zero integers: $\mathbb{Z} = \{\dots, -n, \dots., -3, -2, -1, 0, 1,2,3, \dots, n, \dots\}$ $(n > 0)$.

– $\mathbb{Z}^*$ represents the set of *non-zero relative integers*, i.e. all relative integers except zero: $\mathbb{Z}^* = \{\dots, -n, \dots, -3, -2, -1,1,2,3, \dots, n, \dots\}$ $(n > 0)$.

6.2. Euclidean division

DEFINITION.– *Euclidean division* in the set $\mathbb{Z}$: let a and b be two integers such that $b \neq 0$. There exist two unique integers q and r such that:

$$a = b \times q + r \qquad \text{with } 0 \; r < |b|$$

where q is called the *quotient* of the division of a by b, and r is called the *remainder* of this division.

DEFINITION.– *Euclidean division* in the set $\mathbb{N}$: let a and b be two natural integers such that b is different from 0. There exist two unique integers q and r such that:

$$a = b \times q + r \qquad \text{with } 0 \leq r < b$$

where q is called the *quotient* of the division of a by b, and r is called the *remainder* of this division.

DEFINITION.– If $r = 0$, then $a = b \times q$, indicating that b *divides* a. In this case, a is a *multiple* of b (i.e. there exists a natural number q such that $a = b \times q$), and b is a *divisor* of a (i.e. there exists a natural number k such that $a = k \times q$).

The quotient q is also a multiple of b and k is also a multiple of a. Let us return to the general case of the Euclidean division and demonstrate that q and r indeed exist and are unique.

1) Existence

Let us start by recalling two results known as Archimedes' properties:

PROPERTY 1.– (Archimedes) Let $a \in \mathbb{N}, b \in \mathbb{N}^*$. Then, $n \in \mathbb{N}$ exists such that $n \times b \geq a$.

Indeed, if $a = 0$, take $n = 1$; if $a \neq 0$, take $n = a$.

PROPERTY 2.– (Archimedes) Let $a \in \mathbb{N}, b \in \mathbb{N}^*$. Then, there *exists* $q \in \mathbb{N}$ such that $b \times q \leq a < b \times (q+1)$.

Either $E = \{n \in \mathbb{N} \mid n \times b > a\}$. According to Archimedes' first property, since $a + 1 \in \mathbb{N}$, there exists m such that $m \times b \geq a + 1$, which shows that E is not empty. Moreover, a is a lower bound of E, which means that there exists a smaller p of E such that $p - 1$ does not belong to E. Therefore, $(p-1) \times b \leq a < p \times b$, which can be written by positing $q = p - 1$: $b \times q \leq a < b \times q + b$, or equivalently $0 \leq a - b \times q < b$ which proves property 2.

If $r = a - b \times q$, then $a = b \times q + r$ and $0 \leq r < b$, which demonstrates the existence in the case of N.

Now, let us move on to $\mathbb{Z}$. Four cases can be considered:

– When $a > 0$ and $b > 0$: the division is straightforward, as it is the same as within $\mathbb{N}$.

– When $a > 0$ and $b < 0$: in $\mathbb{N}$, $a = (-b) \times q + r$ with $0 \leq r < -b$. The solution pair is $(-q, r)$.

– When $a < 0$ and $b > 0$: in $\mathbb{N}$, $-a = b \times q + r$ with $0 \leq r < b$. Therefore, $a = (-b) \times q - r = b \times (-q - 1) + (b - r)$. For $r = 0$, the solution pair is $(-q, 0)$. For $r \neq 0$, the solution pair is $(-q - 1, b - r)$.

– When $a < 0$ and $b < 0$: in $\mathbb{N}$, $-a = (-b) \times q + r$ with $0 \leq r < -b$. Therefore, $a = b \times q - r$ and $a = b \times (q+1) - b - r$. If $r = 0$, the solution pair is $(-q, 0)$. If $r \neq 0$, the solution pair is $(q + 1, -r - b)$.

Therefore, there is always a solution (q, r) in all cases for $\mathbb{N}$ and for $\mathbb{Z}$.

2) Uniqueness

Let us use proof by contradiction by assuming the existence of two different pairs (q, r) and (q', r') such that $a = b \times q + r = b \times q' + r'$ with $0 \leq r, r' < b$.

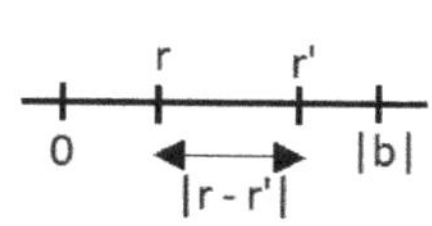

Figure 6.1. $|r - r'| < b$

Then, it follows that $b \times q - b \times q' = r' - r$, or $b \times (q - q') = r' - r$ which implies that b divides $r' - r$. Since $0 \leq r, r' < b$, $|r' - r| < |b|$ holds. Thus, $r' - r$ must equal 0, leading to $q = q'$ and $r = r'$. Therefore, the pair (q, r) is unique.

Therefore, this negates the initial hypothesis, conclusively proving the uniqueness of the pair (q, r).

6.3. Divisibility and prime numbers

In the case where the remainder of the Euclidean division is equal to 0, the relationship simplifies to $a = b \times q$. This expression establishes a relationship in $\mathbb{N}^*$ (the set of positive natural numbers) between a and b, which can be expressed by *b divides a*, denoted as $b \mid a$.

The properties of the relation "|" can be summarized as follows:

– Reflexivity: $a \mid a$ since $a = 1 \times a$.

– Antisymmetry: $b \mid a \nRightarrow a \mid b$. If $b \mid a$, then $a = b \times q$. However, $a \mid b$ would imply $b = a \times q'$, which leads to $a = a \times q \times q'$. This equality holds only if $q \times q' = 1$, implying $q = q' = 1$ and $a = b$, contradicting $a \neq b$ (the initial assumption).

– Transitivity: if $c \mid b$ and $b \mid a$, then $c \mid a$. Given $c \mid b$ and $b \mid a$, $b = c \times q$ and $a = b \times q'$. Thus, $a = c \times q \times q'$, demonstrating $c \mid a$.

With these three properties, the relationship "|" forms an *order relationship* in $\mathbb{N}^*$. Additionally:

– If $c \mid a$ and $c \mid b$, then $c \mid (l \times a + m \times b)$ for $l, m \in \mathbb{Z}$. In particular, $c \mid (a + b)$ and $c \mid (a - b)$.

DEFINITION.– A number $x \in \mathbb{N}^*$, greater than 1, is *prime* if it only has 1 and itself as divisors.

Prime numbers indeed play an important role in the encryption of digital information, a topic that will be explored further.

Let us cite Euclid's definition of prime numbers: a prime number is a number which has exactly *2* divisors. All integers admit 1 as a divisor; any non-zero integer has itself as a divisor. Therefore:

– 0 is not prime because it has infinitely many divisors;

– 1 is not prime because it has only one divisor: itself;

– any integer other than 0 and 1 has at least 2 divisors (1 and itself). It may have more divisors, but if it does not, it is prime.

The first 10 prime numbers are 2, 3, 5; 7; 11, 13, 17, 19, 23.

Prime numbers have many properties, some of which we examine.

THEOREM 6.1.– If an integer is strictly greater than 1, then its smallest divisor strictly greater than 1 is a prime number.

The proof of the above theorem is done by reasoning through contradiction. Let a be an integer greater than 1. It has at least two divisors: itself and 1. It may have other divisors as well, but let b denote the smallest divisor greater than 1.

Assume, for contradiction, that b is not a prime number, meaning it has a divisor c other than 1 and itself, and $c < b$. Therefore, c divides b. Since b divides a, by transitivity, c also divides a. However, this contradicts the assumption that b is the smallest divisor of a.

Therefore, b cannot be non-prime and must be a prime number. Hence, the smallest divisor of any integer greater than 1 is a prime number.

THEOREM 6.2.– Any number in $\mathbb{N}^*$, greater than or equal to 2, is either a prime number or a product of prime numbers.

Let us prove this theorem by mathematical induction:

– Base case: the proposition is true for number 2, which is a prime number.

– Inductive step: assume that the proposition is true for all numbers from 2 to n. That is, each integer k (where $2 \leq k \leq n$) is either a prime number or a product of prime numbers.

– To prove: show that the proposition holds for $n + 1$.

According to theorem 6.1, $n + 1$ has a prime divisor p, which is the smallest divisor of $n + 1$. Consider $\left(\frac{n+1}{p}\right)$. Let us form the integer $\frac{(n+1)}{p}$, which is less than n + 1 since p is greater than 1. Thus, $\frac{(n+1)}{p}$ is between 2 and n.

By inductive hypothesis, $\left(\frac{n+1}{p}\right)$ is either a prime number or a product of prime numbers. Therefore, $n+1$ itself is either prime (if $\left(\frac{n+1}{p}\right)=1$) or a product of p and $\left(\frac{n+1}{p}\right)$, where both p and $\left(\frac{n+1}{p}\right)$ are prime numbers (by the induction assumption).

Thus, the proposition is true for $n+1$.

Therefore, mathematical induction, the proposition is true for all integers $\mathbb{N}^*$ greater than or equal to 2.

Theorem 6.2 states that an integer can be decomposed into a product of prime numbers of the form $n = p^a \times q^b \times ... \times r^c$, where $p, q, ..., r$ are prime numbers. For example, the decomposition of 245 000 is $245\,000 = 2^3 \times 5^4 \times 7^2$.

We may wonder how many prime numbers are there in the set $\mathbb{N}^*$, which has infinite cardinality.

THEOREM 6.3.– There are countably infinitely many prime numbers.

Let us consider a sequence S_k of prime numbers $p_1, p_2, p_3, ..., p_k$ and form the number $m = p_1 \times p_2 \times p_3 \times ... \times p_{k+1}$. According to theorem 6.1, m has a smallest prime divisor q strictly greater than 1. This prime number q is different from the numbers in the sequence S_k.

If q were not different from the primes $p_1, p_2, ..., p_k$, then q would divide both m and $p_1 \times p_2 \times p_3 \times ... \times p_k$. Consequently, q would also divide their difference, which is 1. However, since $q > 1$, this scenario contradicts the property that only 1 and itself can divide 1. Therefore, q must indeed be a prime number distinct from $p_1, p_2, ..., p_k$.

Therefore, an additional prime number is obtained that can be added to the sequence S_k. This construction can be repeated indefinitely, showing that the sequence S_k grows continuously, indicating that prime numbers form a countably infinite set. Eratosthenes proposed a method for finding all prime numbers less than a given integer $N > 1$: the *sieve of Eratosthenes*.

The sieve of Eratosthenes method consists of the following steps:

1) Create a list of all integers from 2 to N (excluding 1, which is not prime).

2) Start with the first number in the list, which is 2, and mark it as prime.

3) Eliminate from the list all multiples of this prime number (2) by marking them as non-prime.

4) Move to the next unmarked number in the list (3), which will be the next prime number, and repeat step 3.

5) Continue this process until all numbers in the list are processed.

At the end of the method, the only unmarked numbers in the list will be prime numbers less than *N*. The sieve of the Eratosthenes algorithm efficiently identifies prime numbers up to a specified limit *N*, although its computational complexity increases with larger values of *N*. For very large values of *N*, more advanced methods may be employed to find prime numbers.

The main steps of the sieve of the Eratosthenes algorithm are as follows:

```
L = an array of Booleans initialized to the value 1 (True) and of
size N.
Set L[1] = 0 (False)
For i = 2 to N
    If L[i] = 1
            For j = i² to N in steps of i
            L[j] = 0
            EndFor
    EndIf
EndFor
Return L
```

The algorithm notes the square of i in the loop for j. Starting from i^2 $(i.e., i \times i)$ ensures that multiples of i less than i^2 have already been eliminated (determined to be non-prime). The list L maintains information about whether numbers from 1 to N are prime or non-prime.

Theorems 6.1 and 6.2 outline the procedure to decompose a number n into its prime factors. Begin by identifying p_1, the smallest prime divisor of the number n. Divide n by p_1 to obtain m, the quotient. Repeat the process by finding p_2, the smallest prime divisor of m. Continue this iterative process m becomes a prime number (i.e. non-decomposable), which indicates that n has been completely decomposed into its prime factors. These steps can also be summarized as follows:

1) Identify the smallest prime divisor p_1 of n.

2) Compute the quotient $m = n/p_1$.

3) Find the smallest prime divisor p_2 of m.

4) Calculate the quotient $m = m/p_2$.

5) Repeat steps 3 and 4 until m is a prime number.

The iterative process makes it possible to decompose n into a product of prime factors. Each prime divisor p found is a distinct prime factor of n.

Note that theorems 6.1 and 6.2 guarantee the existence of this decomposition into prime factors for any integer n greater than 1.

EXAMPLE 6.1.– 245000 can be decomposed into its prime factors as follows:

245000 is even (divisible by 2):

$$245000 = 2 \times 122500$$

122500 is even (divisible by 2):

$$122500 = 2 \times 61250$$

61250 is even (divisible by 2):

$$61250 = 2 \times 30625$$

30625 is divisible by 5:

$$30625 = 5 \times 6125$$

6125 is divisible by 5:

$$6125 = 5 \times 1225$$

1225 is divisible by 5:

$$1225 = 5 \times 245$$

245 is divisible by 5:

$$245 = 5 \times 49$$

49 is divisible by 7:

$$49 = 7 \times 7$$

Since 7 is a prime number:

$$245000 = 2 \times 2 \times 2 \times 5 \times 5 \times 5 \times 5 \times 7 \times 7$$

Therefore, the prime factorization of 245000 is $2^3 \times 5^4 \times 7^2$

To construct a number from prime numbers, simply multiply them together. For instance, using prime numbers p and q, the number m is formed as $m = p \times q$, which typically is not a prime number itself.

However, the inverse problem of determining the prime factors p and q starting from a given number m is much more difficult, especially when m is a large number. Finding the prime factors of a number is a complex problem and often requires significant computational resources in terms of capacity and time. Even with computer assistance, effectively factoring large numbers can be arduous. Computers have limitations in computational capacity and may require specific methods, such as advanced factorization algorithms, to factor very large numbers. Additionally, proving that a number is prime is also a complex problem, particularly for large numbers.

For example, Mersenne numbers, which are of the form $2^n - 1$, were once "likely" to be prime. However, not all Mersenne numbers are prime. For instance, $(2^{11} - 1) = 23 \times 89$ is not prime, and $(2^{67} - 1) = 193707721 \times 761838257287$ is not prime (Cole - 1903). Some Mersenne numbers are indeed prime, such as $(2^{61} - 1)$ (Pervouchine - 1883). The challenge lies in proving or disproving the primality of Mersenne numbers. With older computers, which had limitations in terms of number representation (32-bit coding quickly reaches limits: $(2^{257} - 1)$ has 257 "1s" in binary), this task was particularly challenging.

The study and exploration of prime numbers have been, and continue to be, subject to significant mathematical research. Some of the greatest mathematicians have made remarkable contributions to this field, including Eratosthenes, Euclid, Euler, Fermat, Gauss, Dirichlet, Riemann and others.

Thus, the prime numbers constitute an infinity within the infinity of integers. There are 168 prime numbers among integers up to 1,000 (16.8%), 9,592 among integers up to 100,000 (9.6%), 33,860 among integers up to 400,000 (8.5%), 74,498 among integers up to 1,000,000 (7.8%), 5761455 among integers up to 100,000,000 (5.8%). They become less frequent among large integers. The largest known prime number, as of 2023, is $(2^{82589933} - 1)$.

6.4. GCD and LCM

The abbreviations used in the title of this section mean greater common divisor (GCD) and least common multiple (LCM), respectively.

Given two numbers a and b belonging to $\mathbb{N}^*$ (the set of strictly positive integers), let E be the set of their common divisors. The smallest element of E is obviously 1, and the largest element is less than or equal to $\min(a, b)$. This indicates that E is a finite set. There is a largest element in E, called the *GCD*.

EXAMPLE 6.2.– The numbers 28 and 36 have 1, 2 and 4 as common divisors; the GCD is therefore 4: $\text{GCD}(28, 36) = 4$.

DEFINITION.– The numbers a and b are *coprime* if $\text{GCD}(a, b) = 1$

This means that two prime numbers have no common divisor other than 1.

EXAMPLE 6.3.– The numbers 28 and 27 are coprime. Indeed, $28 = 2^2 \times 7$ and $27 = 3^3$ have no common factor (other than 1 of course).

Euclid's algorithm finds the GCD of two numbers iteratively. This algorithm is based on the following property:

THEOREM 6.4.– Let a and b be in $\mathbb{N}^*$ and r is the remainder of the Euclidean division of a by b , then $\text{GCD}(a, b) = \text{GCD}(b, r)$.

The Euclidean division of a by b, denoted as $a = b \times q + r$, is a fundamental property. If d is a divisor of both b and r, then it also divides a. Therefore, d is a common divisor of a and of b.

Conversely, if d is a common divisor of a and b, then it divides $r = a - b \times q$ as well. Hence, d is a common divisor of b and r.

In other words, the set of common divisors of a and b is identical to the set of common divisors of b and r. Specifically, this equivalence implies that $\text{GCD}(a, b) = \text{GCD}(b, r)$.

Assuming $a > b$, the Euclidean algorithm proceeds by repeating applying the division: $a = p_1 \times b + r_1$, $b = p_2 \times r_1 + r_2$, $r_1 = p_3 \times r_2 + r_3$, etc., until the remainder r_n is zero. The last non-zero remainder obtained in this process is indeed the GCD of the numbers a and b.

EXAMPLE 6.4.– Find the GCD of 1680 and 1485

$a = p_1 \times b + r_1$:
$1680 = 1 \times 1485 + 195$, with $a = 1680, p_1 = 1, b = 1485$ and $\boldsymbol{r_1} = \mathbf{195}$
$b = p_2 \times r_1 + r_2$:
$1485 = 195 \times 7 + 120$, with $b = 1485, p_2 = 195, r_1 = 7$ and $\boldsymbol{r_2} = \mathbf{120}$

$$195 = 120 \times 1 + \mathbf{75}$$
$$120 = 75 \times 1 + \mathbf{45}$$
$$75 = 45 \times 1 + \mathbf{30}$$
$$45 = 30 \times 1 + \mathbf{15}$$
$$30 = 15 \times 2 + 0$$

Since the remainder has become 0, the last non-zero remainder obtained in the Euclidean algorithm is 15, i.e. $\text{GCD}(1680, 1485) = 15$.

Prime factor decomposition allows common factors to be identified between two numbers. Multiplying these common factors gives the GCD.

EXAMPLE 6.5.– $1680 = 2^4 \times 3 \times 5 \times 7$ and $1485 = 3^3 \times 5 \times 11$. The product of the common factors is $3 \times 5 = 15$, which is the GCD of 1680 and 1485.

Let us look at some of the properties of GCD.

THEOREM 6.5.– The common divisors of the numbers a and b belonging to $\mathbb{N}^*$ are all the divisors of the GCD of a and b. In other words, the set of divisors common to a and b is equal to the set of divisors of the $\text{GCD}(a, b)$.

Let $p = \text{GCD}(a, b)$. Let d be a divisor of p. Therefore, d divides both a and b. Thus, d is a common divisor of a and b. In conclusion, every divisor of $\text{GCD}(a, b)$ is also a common divisor of a and b.

Conversely, let d be a common divisor of a and b. Suppose $a > b$. There are two possible cases:

1) If b divides a, then $\text{GCD}(a, b) = b$. Therefore, b divides $\text{GCD}(a, b)$.

2) If b does not divide a, Euclid's algorithm (theorem 6.1) is applied to obtain $\text{GCD}(a, b) = \text{GCD}(b, r)$, where r is the remainder of the division of a by b.

Continuing Euclid's algorithm, $\mathrm{GCD}(a,b) = \mathrm{GCD}(b,r_1) = GCD(r_1,r_2) = \cdots = \mathrm{GCD}(r_{n-1},r_n)$, where r_n is the last non-zero remainder.

In both cases, d divides each remainder r_i of the iterations of Euclid's algorithm, including the last remainder r_n, which is equal to $\mathrm{GCD}(a,b)$. Thus, any common divisor of a and b also divides $\mathrm{GCD}(a,b)$.

THEOREM 6.6.– a, b, $c \in \mathbb{N}^*$, $\mathrm{GCD}(a \times c, b \times c) = c \times \mathrm{GCD}(a,b)$.

Euclid's algorithm is used to prove this theorem. It proceeds as follows:

1) Define $c = \mathrm{GCD}(a,b)$.

2) Perform Euclidean division by $c \times a$, and $c \times b$, yielding $c \times a = c \times b \times q_1 + c \times r_1$, where q_1 is the quotient and $c \times r_1$ is the remainder.

3) By Euclid's algorithm property, $\mathrm{GCD}(c \times a, c \times b) = \mathrm{GCD}(c \times b, c \times r_1)$.

4) Repeat the division $c \times b$ by $c \times r_1$, resulting in $c \times b = c \times r_1 \times q_2 + c \times r_2$.

5) Continue this process until $c \times r_n = 0$.

6) Therefore, $\mathrm{GCD}(c \times r_{n-1}, 0) = c \times r_{n-1}$, which equals c.

Thus, it was demonstrated that $\mathrm{GCD}(a \times c, b \times c) = c \times \mathrm{GCD}(a,b)$ from Euclid's algorithm. This shows that the GCD of the numbers $c \times a$ and $c \times b$ is equal to c multiplied by the GCD of a and b.

THEOREM 6.7.– a, b, c, $d \in \mathbb{N}^*$, the GCD (a,b) divides the GCD $(a \times c, b \times d)$.

A divisor of a divides $a \times c$ and a divisor of b divides $b \times d$. Therefore, any common divisors of a and b also divide $a \times c$ and $b \times d$. In particular, $\mathrm{GCD}(a,b)$ is a divisor of $a \times c$ and $b \times d$.

The LCM of two natural numbers are the *smallest number* that is a multiple of both of them. So,

LCM(18, 32) = 288

LCM(2, 8) = 8

LCM(72, 132) = 792

To find the LCM of two numbers from their prime decompositions, select each prime factor that appears in either number's decomposition. For each prime factor, choose the highest exponent that appears in either number's decomposition.

EXAMPLE 6.6.– LCM(24, 36)?

$24 = 2^3 \times 3$

$36 = 2^2 \times 3^2$

$\text{LCM}(24, 36) = 2^3 \times 3^2 = 72$

There is a very simple relationship between the GCD and the LCM of two natural integers a and b:

THEOREM 6.8.– $\text{GCD}(a, b) \times \text{LCM}(a, b) = a \times b$.

This relationship is very useful because it allows the calculation of the LCM of a number from its GCD with another number, or vice versa, without the need to directly perform the decomposition into prime factors.

To prove this property, the analysis focuses on the exponents of prime factors in number decompositions. Consider p as any prime number. Let g represent the exponent of p in the prime factorization of a, and h the exponent of p in the prime factorization of b. The exponent of p in the prime factorization of $(a \times b)$ is therefore $(g + h)$.

Let $u = \text{GCD}(a, b)$. The exponent of p in the prime factorization of u is denoted as e, where $e = \min(g, h)$. Define $v = \text{LCM}(a, b)$. The exponent of p in the prime factorization of v is denoted as f, where $f = \max(g, h)$. Consequently, the exponent of p in the prime factorization of $(u \times v)$ is $(e + f)$.

If $g \geq h$, then $(e + f) = (h + g)$; if $g \leq h$, then $(e + f) = (g + h)$. In either scenario, $(e + f = g + h)$. This implies that the exponent of p remains consistent between the prime factorization of $(a \times b)$ and $(u \times v)$. Due to the uniqueness of prime decomposition, it follows that $(a \times b) = (u \times v)$.

EXAMPLE 6.7. Let $a = 12$ and $b = 18$. The $\text{GCD}(12, 18) = 6$. Using the relation, the LCM can be calculated as follows:

$$\text{LCM}(12, 18) = \frac{(a \times b)}{\text{GCD}(a,b)} = \frac{(12 \times 18)}{\text{GCD}(12,18)} = \frac{(12 \times 18)}{6} = 36.$$

So, in this example, the LCM of 12 and 18 is equal to 36.

6.5. Congruencies

Modulo division is complex but involves the use of modulus. Modulo division is defined as follows:

DEFINITION.– Let a and b be two integers and n be an integer greater than or equal to 2. The operation $a \equiv b\ [n]$ is called a *congruence* of *module* n (read as "a is congruent to b modulo n"). This operation means that the Euclidean division of a by n gives a quotient q and the remainder b. Formally, this is expressed as: $a = q \times n + b$, and

$$a \equiv b\ [n] \Leftrightarrow \exists\ q \text{ such that } a = q \times n + b.$$

Additionally, n divides $a - b$ since $a - b = q \times n$.

EXAMPLE 6.8.– To illustrate congruence modulo n, consider the example involving the number 28. Since $28 = 3 \times 9 + 1$, then $28 \equiv 1\ [3]$ or even $28 \equiv 1\ [9]$. Verify that 3 divides $28 - 1 = 27$ and that 9 divides $28 - 1 = 27$.

THEOREM 6.9.– For a fixed n, the congruence $a \equiv b\ [n]$ is an equivalence relation in $\mathbb{Z}$.

Recall that an equivalence relation $\mathcal{R}$ is a map having the properties of reflexivity, symmetry and transitivity:

Reflexivity: $a\mathcal{R}a$

Symmetry: $a\mathcal{R}b \Rightarrow b\mathcal{R}a$

Transitivity: $a\mathcal{R}b$ and $b\mathcal{R}c \Rightarrow a\mathcal{R}c$

For congruence, an equivalence relation $a \equiv b\ [n]$ is a relationship that satisfies the properties of reflexivity, symmetry and transitivity:

– *Reflexivity*: $a \equiv a\ [n]$ means that a can be expressed in the form $a = q \times n + a$, where $q = 0$.

– *Symmetry*: $a \equiv b\ [n]$ means that a can be expressed in the form $a = q \times n + b$. The equation can also be written as $b = (-q) \times n + a$, which shows that $b \equiv a\ [n]$.

– *Transitivity*: if $a \equiv b\ [n]$ and $b \equiv c\ [n]$, it means that a can be expressed as $a = p \times n + b$ and b can be expressed as $b = q \times n + c$. Substituting the second equation into the first one, $a = (p + q) \times n + c$ is obtained, which shows that $a \equiv c\ [n]$.

Thus, the congruence relation satisfies the properties of reflexivity, symmetry and transitivity, making it an equivalence relation.

EXAMPLE 6.9.– If $n = 3$, the following three disjoint sets, called equivalence classes, can be found:
$E_0 = \{..., -3, 0, 3, 6, 9, ...\}$
$E_1 = \{..., -2, 1, 4, 7, 10, ...\}$
$E_2 = \{..., -1, 2, 5, 8, 11, ...\}$
$E_3 = \{..., 0, 3, 6, 9, 12, ...\} = E_0$

Addition and multiplication "modulo n" can be defined. Let, for example, $a \equiv b\ [n]$ and $a' \equiv b'\ [n]$. So:

$$a + a' \equiv b + b'\ [n] \text{ and } a \times a' \equiv b \times b'\ [n]$$

Indeed, $a = p \times n + b$ and $a' = q \times n + b'$; therefore, $a + a' = (p + q) \times n + b + b'$ or $a + a' \equiv b + b'\ [n]$.

Likewise, $a \times a' = (p \times n + b)(q \times n + b') = (p \times q + b \times q + b' \times p)n + b \times b'$ or $a \times a' \equiv b \times b'\ [n]$.

THEOREM 6.10.– (*Fermat's Little Theorem*): let a be an integer and n be a prime number.
1) $a^n \equiv a[n]$.
2) If n and a are coprime ($\text{GCD}(a, n) = 1$), then $a^{n-1} \equiv 1[n]$.

The above theorem 6.10 was stated by Fermat who, perhaps, did not prove it. The first known demonstrations are due to Euler and Leibnitz. It should not be confused with Fermat's "Great" Theorem (only proven in 1995 by Andrew Wiles), which shows the impossibility of writing $a^n + b^n = c^n$ with a, b, c as positive integers and $n > 2$. The only valid relationship for $n = 2$ is given by the Pythagorean theorem: $a^2 + b^2 = c^2$.

Let us move on to the proof of the "Little" Theorem using induction. Note that for $a = 0$, $0^n = q \times n + 0$ (with $q = 0$), which is trivially congruent to 0[n]. Therefore, $0^n \equiv 0[n]$.

Assume the theorem is true for some integer a, i.e. $a^n \equiv a[n]$. Now, show that $(a+1)^n \equiv (a+1)\,[n]$.

Expand $(a+1)^n$ using Newton's binomial formula:

$$(a+1)^n = (1+a)^n = \sum_{k=0}^{n} C_n^k a^k = a^n + \sum_{k=1}^{n-1} C_n^k a^k + 1$$

by separating the first and last terms. The binomial coefficient has the following expression:

$$C_n^k = \frac{n!}{k!(n-k)!} \quad \text{or equivalently,} \quad k!\,C_n^k = n(n-1)(n-2)\,....\,(n-k+1).$$

The above expression shows that n divides $k!\,C_n^k$.. In the sum of 1 to $n-1$, $0 < k < n$. It follows that n does not divide $! = 1.2.3 \dots\dots k$, so n divides C_n^k . The expansion of the binomial, as formulated above, can be written as:

$$(a+1)^n - a^n - 1 = \sum_{k=1}^{n-1} C_n^k a^k$$

Since n divides each term $a^k C_n^k$ for $1 \le k \le n-1$, it follows that n divides $(a+1)^n - a^n - 1$, which can be written as:

$$(a+1)^n - a^n - 1 \equiv 0\,[n]$$

Given the induction hypothesis $a^n \equiv a[n]$, it follows that:

$$(a+1)^n - a - 1 \equiv 0\,[n] \text{ or } (a+1)^n \equiv (a+1)[n].$$

This completes the induction step and hence the proof of Fermat's Little Theorem.

The second part of the theorem is deduced from the first. $a^n \equiv a[n]$ can be written as:

$$a^n - a \equiv 0\,[n]$$

or even:

$$a \times (a^{n-1} - 1) \equiv 0\,[n].$$

In this second part of the theorem, the additional hypothesis, a and n are relatively prime, means that n does not divide a. Therefore, n divides $a^{n-1} - 1$ which can be written as:

$$a^{n-1} - 1 \equiv 0\,[n] \quad \text{or} \quad a^{n} - 1 \equiv 1\,[n].$$

EXAMPLE 6.10.– Fermat's test: in the second part of Fermat's Little Theorem, take $a = 2$. Every prime number n is obviously coprime with 2. Therefore, $2^{n-1} \equiv 1\,[n]$ according to Fermat, which means that $2^{n-1} - 1 = q \times n$ and that any prime number n is a divisor of $2^{n-1} - 1$. Conversely, if a number n is not a divisor of $2^{n-1} - 1$, it is not prime. This constitutes the "Fermat test" which allows the determination of whether a number is prime or not. So,

Is 3 prime? It is in fact a divisor of $2^2 - 1 = 3$, therefore prime.
Is 6 prime? $2^5 - 1 = 31$; 6 does not divide 31, so 6 is not prime.
Is 54321 prime? Calculation of $2^{54320} - 1$ is lengthy. Another method similar to decimal-binary conversion is required. Decompose 54320 according to the powers of 2 (see Figure 6.2):

$$54320 = 2^{15} + 2^{14} + 2^{12} + 2^{10} + 2^{5} + 2^{4}$$

So,

$$2^{54320} = 2^{2^{15}} \times 2^{2^{14}} \times 2^{2^{12}} \times 2^{2^{10}} \times 2^{2^{5}} \times 2^{2^{4}}$$

Next, perform the integer division of 2^{54320} by 54321, or the division of each factor 2^{2^i} by 54321. According to the modulo multiplication rule, compute the product of the remainders:

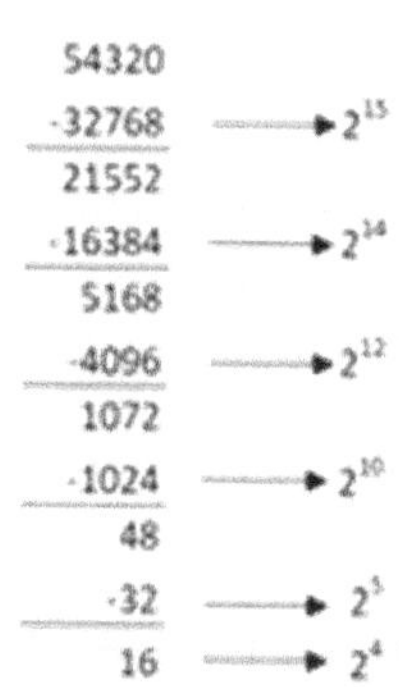

Figure 6.2. *Decomposition into powers of 2*

To calculate the remainders r_{i+1} of $2^{2^{i+1}}$ modulo 54321, r_{i+1} is derived from r_i the remainder of 2^{2^i}, by squaring r_i (i.e. r_i^2) and taking modulo 54321. Such a process generates the following table:

$2^0 = 1$	the remainder of 2^1 by 54321 is	2
$2^1 = 2$	the remainder of 2^2 by 54321 is	4
$2^2 = 4$	the remainder of 2^4 by 54321 is	16
$2^3 = 8$	the remainder of 2^8 by 54321 is	256
$2^4 = 16$	the remainder of 2^{16} by 54321 is	11215
$2^5 = 32$	the remainder of 2^{32} by 54321 is	23110
$2^6 = 64$	the remainder of 2^{64} by 54321 is	42349
$2^7 = 128$	the remainder of 2^{128} by 54321 is	29986
$2^8 = 256$	the remainder of 2^{256} by 54321 is	39004
$2^9 = 512$	the remainder of 2^{512} by 54321 is	52411
$2^{10} = 1024$	the remainder of 2^{1024} by 54321 is	8593
$2^{11} = 2048$	the remainder of 2^{1048} by 54321 is	17410
$2^{12} = 4096$	the remainder of 2^{4096} by 54321 is	51241
$2^{13} = 8192$	the remainder of 2^{8192} by 54321 is	34546
$2^{14} = 16384$	the remainder of 2^{16384} by 54321 is	48067
$2^{15} = 32768$	the remainder of 2^{32768} by 54321 is	1396

By considering the powers in the decomposition of 54320, it is found that $2^{54320} \equiv (16 \times 23110 \times 8593 \times 51241 \times 48067 \times 1396)\ [54321]$
$\equiv (16 \times 23110 \times 8593 \times 51241 \times 15097)\ [54321]$
$\equiv (16 \times 23110 \times 8593 \times 16)\ [54321]$
$\equiv (16 \times 23110 \times 28446)\ [54321]$
$\equiv (16 \times 48639)\ [54321]$
$\equiv 17730\ [54321]$

Therefore, it is concluded that 54321 is not prime (it is divisible by 3).

Finding a prime factor of a large number can be challenging. For instance, in the case of 54321 from the previous example, it is evident that 3 is a prime factor. However, for a number such as 58483, by trying successive prime numbers, it equals 233 x 251 after testing successive prime numbers up to 233.

Consider the difficulty of factoring 114 381 825 757 888 867 669 235 779 976 146 612 010 218 296 721 242 362 562 561 842 935 706 935 245 733 897 830 597 123 563 958 705 058 989 075 147 599 290 026 979 543 541. Good luck!

This challenge highlights the foundation of certain encryption systems. Given primes p and q, their product $Z = p \times q$ is straightforward to calculate. However, from Z, *determining* p and q is computationally difficult. This characteristic forms the basis of the RSA (Rivest–Shamir–Adleman) encryption system, which will be discussed in Chapter 8.

6.6. Elliptic curves

Fermat's little theorem is well-known, but Fermat's "big" theorem was only demonstrated by Andrew Wiles in 1994. He used the properties of a family of curves called elliptic curves. The details of this demonstration are not covered here, but the focus is on the elliptic curves themselves.

In the family of elliptic curves, consider particularly the subfamily of equations of the form $y^2 = x^3 + ax + b$, where a and b are integers such that $4a^3 + 27b^2 \neq 0$. The discriminant $\Delta = 4a^3 + 27b^2$ is used to characterize these curves. It allows the determination of properties of the roots of the cubic equation $x^3 + ax + b$. If $\Delta \neq 0$, the equation has no multiple roots. If $\Delta < 0$, the equation has three distinct real roots. Conversely, if $\Delta > 0$, the equation has only one unique real root.

The curves in this family display symmetry with respect to the x-axis (the horizontal axis O_x). An example is the curve represented by the equation $y^2 = x^3 - 5x + 3$, where $a = -5$ and $b = 3$. This equation has a discriminant $\Delta = -257$.

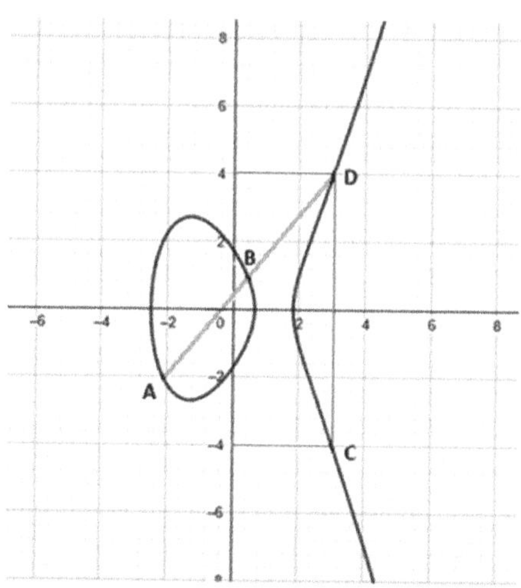

Figure 6.3. *Example of an elliptic curve*

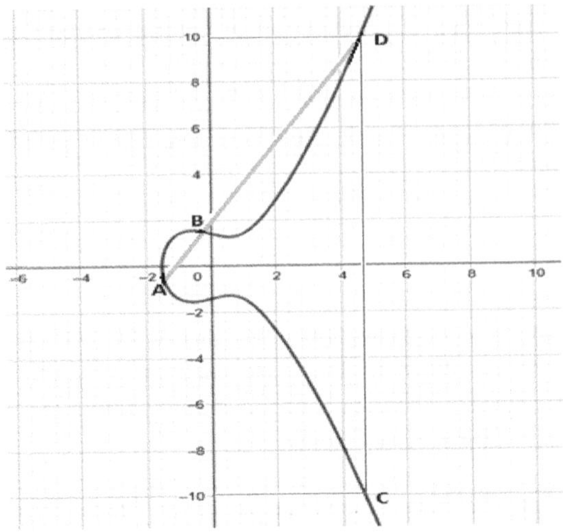

Figure 6.4. *Another example of an elliptic curve*

Figure 6.3 graphically shows the curve, highlighting its characteristic shape. This curve consists of two components. With this family, curves with a single component can also be found, such as the equation $y^2 = x^3 - x + 2$ (with discriminant $\Delta = 104$), shown in Figure 6.4. These (elliptic) curves possess remarkable properties. One of these properties concerns the so-called addition operation between the points of the curve. Consider two points A and B on the elliptic curve. Drawing a straight line passing through these two points generally intersects the curve at a third point D. Let C be the symmetric point of D with respect to the abscissa axis (horizontal axis O_x). An addition operation between points A and B is defined as A + B = C. This addition operation on elliptic curves has properties similar to those of number addition:

– It is commutative: A + B = B + A, regardless of the order of the points.

– It is associative: (A + B) + C = A + (B + C), regardless of how the points are grouped.

When the abscissa of points A and B are equal ($x_A = x_B$), the line AB becomes parallel to the ordinate axis (O_y axis) due to the curve's symmetry with respect to O_x. In this case, the ordinate of point B is the negative of the ordinate of point A ($y_B = -y_A$), and point D (and therefore point C) can be considered to be at infinity, indicating that the line AB intersects the curve at infinity. This specific point C is denoted as Ω. Therefore, in this case, A + B = Ω, where B is considered the reflexion of A across the O_x axis. Considering the line AΩ (a vertical line), it intersects the curve at B, and point C coincides with A, resulting in A + Ω = A. These observations lead us to conclude the following:

– Addition on elliptic curves has a neutral element denoted as Ω (the "infinity" point), such that A + Ω = Ω + A = A.

– Each point A has an "inverse" or "opposite" denoted as A', such that A + A' = Ω, where A' is the opposite of A. Alternatively, A' = -A.

Note that if points A and B coincide, line AB becomes tangent to the curve, and A + A = 2A. This property is particular when points are repeated on the curve.

The relations allowing the coordinates of a "sum" point to be found can be analytically derived by solving the following system of equations, which represents the intersection of a line AB with the elliptic curve:

$$\begin{cases} y^2 = x^3 + ax + b \\ y = cx + d \end{cases}$$

where the second equation represents the line AB. Given the coordinates of points A(x_A, y_A) and B(x_B, y_B), the coefficients c and d are deduced as follows:

$$c = \frac{y_B - y_A}{x_B - x_A} \quad if\ x_A \neq x_B \qquad c = \frac{3x_A^2 + a}{2y_A} \quad if\ x_A = x_B$$

In the second case, where the line AB is tangent to the curve and points A and B coincide:

$$d = \frac{y_A x_B - y_B x_A}{x_B - x_A} \quad if\ x_A \neq x_b \qquad d = y_A - \frac{3x_A^2 + a}{2y_A} x_A \quad if\ x_A = x_B$$

In the general case, solving the system leads to a third-degree equation in x. As two points of intersection (providing two roots of the equation), and because a cubic equation has three roots, finding the third root gives the x-coordinate of the desired point A+B:

$$x = \left(\frac{y_B - y_A}{x_B - x_A}\right)^2 - x_A - x_B \quad y = \frac{y_B - y_A}{x_B - x_A} x + \frac{y_A x_B - y_B x_A}{x_B - x_A} \qquad if\ x_A \neq x_B$$

$$x = \left(\frac{3x_A^2 + a}{2y_A}\right)^2 - 2x_A \qquad y = \frac{3x_A^2 + a}{2y_A}(x - x_A) + y_A \qquad if\ \ x_A = x_B$$

Thus, the coordinates of point A + B are $(x, -y)$.

Now, considering a bit of abstraction, replace the equation of the elliptic curve C with an equation modulo p representing an (abstract) elliptic curve Γ, where x and y are integers:

$$y^2 \equiv x^3 + ax + b\ [p]$$

with $4a^3 + 27b^2 \neq 0\ [p]$, which ensures that the point (x, y) belongs to Γ if x and y satisfy the above equation. Here, all points of Γ no longer have any relation to the previous "continuous" elliptic curves.

Let us outline the following rules of addition:

a) $(x_\Omega + y_\Omega) + (x_\Omega, y_\Omega) = (x_\Omega, y_\Omega)$, where Ω is the "infinity" point.

b) $(x, y) + (x_\Omega, y_\Omega) = (x, y)$.

c) $(x, y) + (x, -y) = (x_\Omega, y_\Omega)$ $(x, -y)$ is thus the "opposite" point of (x, y).

d) For distinct points, $(x_1, y_1) + (x_2, y_2) = (x_3, y_3)$ with

$$x_3 \equiv k^2 - x_1 - x_2 [p]$$
$$y_3 \equiv k(x_1 - x_3) - y_1 [p]$$

where k is defined by $k \equiv \frac{y_2 - y_1}{x_2 - x_1}\ [p]$.

e) For points at the same location, $(x_1, y_1) + (x_1, y_1) = 2(x_1, y_1) = (x_3, y_3)$ with

$$x_3 \equiv k^2 - 2x_1 [p]$$
$$y_3 \equiv k(x_1 - x_3) - y_1 [p]$$

where k is defined by $k \equiv \frac{3x_1^2 + a}{2y_1}\ [p]$.

The relationships observed here resemble those in the previous calculations involving points of the elliptic curve (denoted as Γ). The main distinction is that all operations are performed modulo a specific number. If two points $P_1(x_1, y_1)$ and $P_2(x_2, y_2)$ are points on the elliptic curve Γ, then their sum $P_1 + P_2$ also belongs on the curve Γ, based on the previous defined relationships.

Moreover, this addition operation forms the basis for multiplying by an integer n. Specifically, nP is obtained by adding P to itself n times: $nP = P + P +$

$\ldots. + P$ (n times), where P is a point on the elliptic curve. The result nP is likewise a point on the curve Γ.

EXAMPLE 6.11.– Consider the curve Γ, given by $y^2 \equiv x^3 + x + 2\ [11]$.

If $x = 2$, then $y^2 \equiv x^3 + x + 2\ [11] \equiv 12\ [11] \equiv 1[11]$; therefore, $y = 1$. The point $(2, 1)$ belongs to Γ.

Now, calculate $(2,1) + (2,1) = 2(2,1)$ using the addition formulas:

$$k \equiv \frac{13}{2}[11] \text{ or } 2k \equiv 13\ [11] \equiv 2\ [11], \text{therefore } k \equiv 1\ [11]$$
$$x \equiv 1 - 4[11] \equiv -3[11] \equiv 11 - 3[11] \equiv 8[11]$$
$$y \equiv 2 - 8 - 1[11] \equiv -7[11] \equiv 11 - 7[11] \equiv 4[11]$$

Let us check that the point $(8, 4)$ indeed belongs to Γ:

$x^3 + x + 2 \equiv 512 + 8 + 2\ [11] \equiv 522\ [11] \equiv 5\ [11]$ and $y^2 \equiv 16\ [11] \equiv 5[11]$. Therefore, the point $(8, 4)$ indeed belongs to Γ.

The two points $(2, 1)$ and $(8, 4)$ belong to Γ. Let us calculate $(2,1) + (8,4) = 3(2,1)$:

$$k \equiv \frac{3}{6}\ [11] \quad \text{or} \quad 6k \equiv 3\ [11]$$

If $k \equiv 0[11]$ then $6k \equiv 0[11]$.
If $k \equiv 1[11]$ then $6k \equiv 6[11]$.
If $k \equiv 2[11]$ then $6k \equiv 12[11] \equiv 1[11]$.
If $k \equiv 3[11]$ then $6k \equiv 18[11] \equiv 7[11]$.
If $k \equiv 4[11]$ then $6k \equiv 24[11] \equiv 2[11]$.
If $k \equiv 5[11]$ then $6k \equiv 30[11] \equiv 8[11]$.
If $k \equiv 6[11]$ then $6k \equiv 36[11] \equiv 3[11]$, therefore $k \equiv 6[11]$.

The values of the couple (x, y):

$$x \equiv 36 - 2 - 8[11] \equiv 26[11] \equiv 4[11]$$
$$y \equiv 6(2 - 4) - 1[11] \equiv -13[11] \equiv 22 - 13[11] \equiv 9[11]$$

Let us check that the point $(9, 7)$ indeed belongs to Γ:

$x^3 + x + 2 \equiv 64 + 4 + 2[11] \equiv 70[11] \equiv 4[11]$ and $y^2 \equiv 81[11] \equiv 4[11]$

The point $(4,9) = 3(2,1)$ belongs to Γ.

Now, determine how many points (x,y) are on the curve Γ. It is shown that there exists a finite number between $p - 2\sqrt{p} + 1$ and $p + 2\sqrt{p} + 1$. For this example where $p = 12$:

$$12 - 2\sqrt{12} = 5.07 \text{ and } 12 + 2\sqrt{12} = 18.9.$$

In fact, there are 16 points (including the infinite point Ω):

(2.1), (8.4), (2.10), (10.0), (8.7), (4.2), (4.9), (1.2), (5,0), (6.2), (7.0), (9.6), (9.5), (1.9), (6.9) and Ω.

The graphical representation of the elliptic curve Γ, including the point Ω, is shown in Figure 6.5. Note that the graphical representation of elliptic curves with modulo operations differs significantly from the curves used in calculations without modulo.

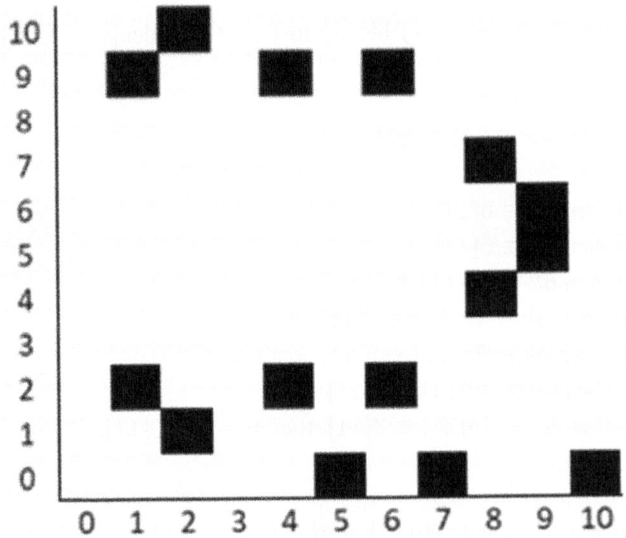

Figure 6.5. *Graphical representation of the Γ curve*

Figure 6.6 presents the addition of points on the elliptic curve Γ.

	(2,1)	(8,4)	(2,10)	(10,0)	(8,7)	(4,2)	(4,9)	(1,2)	(5,0)	(6,2)	(7,0)	(9,6)	(9,5)	(1,9)	(6,9)	Ω
(2,1)	(8,4)	(2,10)	Ω	(4,2)	(2,10)	(8,7)	(10,0)	(9,6)	(9,5)	(1,2)	(6,2)	(5,0)	(1,9)	(6,9)	(7,0)	(2,1)
(8,4)	(2,10)	(10,0)	(2,1)	(8,7)	Ω	(2,10)	(4,2)	(5,0)	(1,9)	(9,6)	(1,2)	(9,5)	(6,9)	(7,0)	(6,2)	(8,4)
(2,10)	Ω	(2,1)	(8,7)	(4,9)	(4,2)	(10,0)	(8,4)	(6,2)	(9,6)	(7,0)	(6,9)	(1,2)	(5,0)	(9,5)	(1,9)	(2,10)
(10,0)	(4,2)	(8,7)	(4,9)	Ω	(8,4)	(2,1)	(2,10)	(1,9)	(7,0)	(9,5)	(5,0)	(6,9)	(6,2)	(1,2)	(9,6)	(10,0)
(8,7)	(2,10)	Ω	(4,2)	(8,4)	(10,0)	(4,9)	(2,1)	(7,0)	(1,2)	(6,9)	(1,9)	(6,2)	(9,6)	(5,0)	(9,5)	(8,7)
(4,2)	(8,7)	(2,10)	(10,0)	(2,1)	(4,9)	(8,4)	Ω	(6,9)	(6,2)	(1,9)	(9,5)	(7,0)	(1,2)	(9,6)	(5,0)	(4,2)
(4,9)	(10,0)	(4,2)	(8,4)	(2,10)	(2,1)	Ω	(8,7)	(9,5)	(6,9)	(5,0)	(7,0)	(1,9)	(7,0)	(6,2)	(1,2)	(4,9)
(1,2)	(9,6)	(5,0)	(6,2)	(1,9)	(7,0)	(6,9)	(9,5)	(10,0)	(8,7)	(4,9)	(8,4)	(4,2)	(2,10)	Ω	(2,1)	(1,2)
(5,0)	(9,5)	(1,9)	(9,6)	(7,0)	(1,2)	(6,2)	(6,9)	(8,7)	Ω	(6,9)	(10,0)	(2,10)	(2,1)	(8,4)	(4,9)	(5,0)
(6,2)	(1,2)	(9,6)	(7,0)	(9,5)	(6,9)	(1,9)	(5,0)	(4,9)	(6,9)	(8,4)	(2,1)	(10,0)	(8,7)	(2,10)	Ω	(6,2)
(7,0)	(6,2)	(1,2)	(6,9)	(5,0)	(1,9)	(9,5)	(7,0)	(8,4)	(10,0)	(2,1)	Ω	(4,9)	(4,2)	(8,7)	(2,10)	(7,0)
(9,6)	(5,0)	(9,5)	(1,2)	(6,9)	(6,2)	(7,0)	(1,9)	(4,2)	(2,10)	(10,0)	(4,9)	(8,7)	Ω	(2,1)	(8,4)	(9,6)
(9,5)	(1,9)	(6,9)	(5,0)	(6,2)	(9,6)	(1,2)	(7,0)	(2,10)	(2,1)	(8,7)	(4,2)	Ω	(8,4)	(4,9)	(10,0)	(9,5)
(1,9)	(6,9)	(7,0)	(9,5)	(1,2)	(5,0)	(9,6)	(6,2)	Ω	(8,4)	(2,10)	(8,7)	(2,1)	(4,9)	(10,0)	(4,2)	(1,9)
(6,9)	(7,0)	(6,2)	(1,9)	(9,6)	(9,5)	(5,0)	(1,2)	(2,1)	(4,9)	Ω	(2,10)	(8,4)	(10,0)	(4,2)	(8,7)	(6,9)
Ω	(2,1)	(8,4)	(2,10)	(10,0)	(8,7)	(4,2)	(4,9)	(1,2)	(5,0)	(6,2)	(7,0)	(9,6)	(9,5)	(1,9)	(6,9)	Ω

Figure 6.6. *Addition table*

What is the purpose of all this? Elliptic curves are particularly useful in cryptography systems, a topic that will be covered in a future chapter.

6.7. Identity, theorem and Bézout algorithm

The following proposition is known as the Bézout identity or the Bézout–Bachet theorem:

THEOREM 6.11.– (Bézout identity) Let a and b be two non-zero integers. There exists a pair (u, v) of integers such that $\text{GCD}(a, b) = a \times u + b \times v$.

Denote E *as* the set of non-zero natural integers of the form $k \times a + h \times b$, where $k, h \in \mathbb{Z}$ and $E \subseteq \mathbb{N}$. Since $a \in E$, E is not empty and contains a smallest element p such that $p = a \times u + b \times v$.

The $\text{GCD}(a, b)$ divides both a and b; hence, it also divides p, implying that $\text{GCD}(a, b) \leq p$.

Performing the Euclidean division of a by p gives $a = p \times q + r$ with $0 \leq r < p$. So,

$$r = a - p \times q = a - a \times u \times q - b \times v \times q = a \times (1 - u \times q) - b \times v \times q.$$

If $r \neq 0$, then $r \in E$. However, this contradicts the minimality of p in E because $r < p$. Therefore, $r = 0$.

Thus, p divides a, and similarly, p divides both a and b, implying $p \leq \text{GCD}(a, b)$.

Since $\text{GCD}(a, b) \leq p$ and $p \leq \text{GCD}(a, b)$, then $p = \text{GCD}(a, b) = a \times u + b \times v$.

THEOREM 6.12.– (*Bézout's theorem*) $a, b \in \mathbb{Z}$ are relatively prime if and only if there exist $u, v \in \mathbb{Z}$ such that $a \times u + b \times v = 1$.

If a and b are coprime, by definition, $\text{GCD}(a, b) = 1$. According to Bézout's identity, there exist integers u and v such that $a \times u + b \times v = 1$. Therefore, $a \times u + b \times v = 1$.

Conversely, if there exist integers u and v such that $a \times u + b \times v = 1$, then the $\text{GCD}(a, b)$ divides both a and b. Therefore, it divides any linear combination of a and b, including $a \times u + b \times v$. Since $a \times u + b \times v = 1$, this implies that the $\text{GCD}(a, b)$ divides 1.

Therefore, $\text{GCD}(a, b) = 1$ if and only if there exist integers u and v such that $a \times u + b \times v = 1$. This establishes that a and b are relatively prime.

EXAMPLE 6.12.– The numbers 28 and 27 are relatively prime. Integers $u = 1$ and $v = -1$ exist such that $28 \times u + 27 \times v = 28 - 27 = 1$.

Example 6.12 illustrates the use of *Bézout's algorithm* to find the numbers u and v, given the numbers a and b. This algorithm is used to compute the GCD of two numbers and determine the Bézout coefficients. It proceeds through a sequence of iterative steps

```
a, b integers
r = 0
u = 0
While r ≠ 1
     u = u + 1
     m = au
     If b > 0
             r = m-E(m/b)b
     Otherwise
             r=m-E(m/b - 1)b
     EndIf
EndWhile
v = (1-m)/b
```

EXAMPLE 6.13.– Consider the numbers $a = 23$ and $b = 34$.

Since 23 is a prime number and 34 is not a multiple of 23, the numbers a and b are co-prime. Let us apply Bézout's algorithm:

$r = 0$
$u = 1$
$v = 0$

Performing successive steps:

$m = 23,\ r = 23 - E(23/34).34 = 23$
$u = 2$
$m = 46,\ r = 46 - E(46/34).34 = 12$
$u = 3$
$m = 69, r = 69 - E(69/34).34 = 12$

Therefore, $u = 3$ and $v = (1 - 69)/34 = -2$

In fact, $3 \times 23 - 2 \times 34 = 1$.

THEOREM 6.13.– (*Bézout's Corollary*) The equation $a \times x + b \times y = c$ admits integer solutions if and only if c is a multiple of $\mathrm{GCD}(a, b)$.

Suppose the equation $a \times x + b \times y = c$ has a solution (x_0, y_0). Since $\mathrm{GCD}(a, b)$ divides both a and b, it also divides their linear combination $a \times x_0 + b \times y_0$. Therefore, $\mathrm{GCD}(a, b)$ divides c, making c a multiple of $\mathrm{GCD}(a, b)$.

Conversely, suppose that c is a multiple of the $\mathrm{GCD}(a, b)$, say $c = k \times \mathrm{GCD}(a, b)$ where k is an integer. Then, by Bézout's identity, there exist integers u and v such that $a \times u + b \times v = \mathrm{GCD}(a, b)$. Multiplying this identity by k, the following result holds: $k(a \times u + b \times v) = k \times \mathrm{GCD}(a, b) = c$. Thus, $(x_0, y_0) = (k \times u, k \times v)$ is a solution satisfying $a \times x_0 + b \times y_0 = c$.

Therefore, the equation $a \times x + b \times y = c$ has a solution if and only if c is a multiple of the $\mathrm{GCD}(a, b)$. In such cases, (x_0, y_0), obtained using the coefficients u and v of Bézout's identity, provides a particular solution.

EXAMPLE 6.14.– The equation $3x + 5y = 2$ has integer solutions because $\mathrm{GCD}(3,5) = 1$ and 2 is a multiple of 1.

The equation $28x + 36y = 3$ has no integer solution because $\mathrm{GCD}(28, 36) = 4$ and 3 is not a multiple of 4.

THEOREM 6.14.– (*Gauss' Theorem*): for integers a, b, c, where $a, b, c \in \mathbb{Z}^*$, If a divides $b \times c$ and if a and b are coprime ($\text{GCD}(a, b) = 1$), then a divides c.

If a divides $b \times c$, then there exists an integer k such that $k \times a = b \times c$.

Now, suppose that a and b are coprime to each other, meaning $\text{GCD}(a, b) = 1$. According to Bézout's theorem, there exist integers u and v such that $a \times u + b \times v = 1$. By multiplying this equality by c, the following equation holds: $c \times a \times u + c \times b \times v = c$. Using the first relation ($k \times a = b \times c$) from a previous relation, and by replacing $c \times b \times v$ with $k \times a \times v$, it gives $c \times a \times u + k \times a \times v = c$. Therefore, $(c \times u + k \times v) \times a = c$. Consequently, a divides c because $(c \times u + k \times v)$ is an integer.

THEOREM 6.15.– Relationship between GCD and LCM: $d = \text{GCD}(a, b)$ and $m = \text{LCM}(a, b)$

1) There exist integers a' and b' such that $a = d \times a'$, $b = d \times b'$, and $\text{GCD}(a', b') = 1$. Then, $m = d \times a' \times b$.

2) $m \times d = a \times b$.

Suppose $d = GCD(a, b)$. Since d divides both a and b, there exist integers a' and b' such that $a = d \times a'$ and $b = d \times b'$. Let us form the product $M = d \times a' \times b'$, which is a multiple of both a and b.

Now, consider c to be any common multiple of a and b. Since c is a multiple of a, there exists an integer p such that $c = p \times a = p \times (d \times a')$. Similarly, since c is a multiple of b, there exists an integer q such that $c = q \times b = q \times (d \times b')$. Thus, $c = p \times (d \times a') = q \times (d \times b')$.

This implies that $p \times a' = q \times b'$. Since a' and b' are coprime ($\text{GCD}(a', b') = 1$), according to Gauss' theorem, this means that a' divides q. Therefore, there exists an integer r such that $a' \times r = q$. Thus, $c = q \times b' = a' \times r \times b' = (a' \times b') \times r$, which shows that c is a multiple of $a' \times b'$. Therefore, $M = a' \times b' \times d$ is the $LCM(a, b)$. Since $M = a' \times b' \times d$, then $d \times m = d \times a' \times b' \times d = (a' \times d) \times (b' \times d) = a \times b$.

7

Error Protection

CONCEPTS COVERED IN THIS CHAPTER.–

Nothing is more common than the accidental transformation of a 0 into a 1, or vice versa, when transmitting information. The causes are diverse. Regardless, it is essential to have ways to detect errors and, if possible, correct them automatically. This is the subject of this chapter.

After discussing the general principles of error detection and correction in binary sequences, linear codes are defined, where the bits of the coded information are expressed linearly as a function of the useful information bits and the control bits.

Current transmission methods use a variety of linear codes, including polynomial codes. Their principle will be explained in a section dedicated to them.

The chapter concludes with an overview of convolutional codes, which involve temporal sequences of bits.

References: [BAD 02, LAH 95; MAC 87].

7.1. General context

It is well known that imperfections in transmission media result in noise, which can cause a number of errors due to misinterpretation of signals. The error rate is defined as the ratio between the number of erroneous bits and the total number of bits transmitted, such that:

Mathematics for Digital Science 1,
by Gérard-Michel COCHARD and Mhand HIFI.

$$\tau = \frac{\text{number of erroneous bits}}{\text{number of transmitted bits}}$$

Depending on the media used, the error rate generally varies from 10^{-5} to 10^{-7}. For many applications, this rate is still considered too high. Therefore, it is necessary to provide a system for detecting and possibly correcting errors.

Statistical studies have shown that 88% of errors are due to a single wrong bit, while 10% of errors are due to two consecutive wrong bits. This indicates that the main problem is the correction of "isolated" errors (an erroneous bit occurring occasionally). However, to be more precise, an error packet is defined according to the following convention: two successive error bits belong to distinct error packets if the number of correct bits separating them is greater than or equal to a fixed number N (generally $N = 10$, as shown in Figure 7.1).

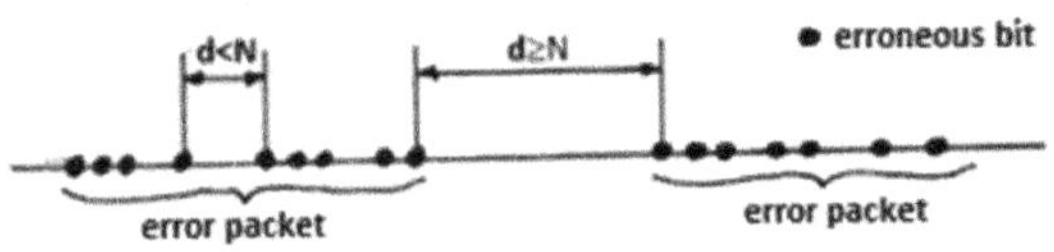

Figure 7.1. *Error packets*

Figure 7.2 shows the statistical distribution of error packets as a function of their length.

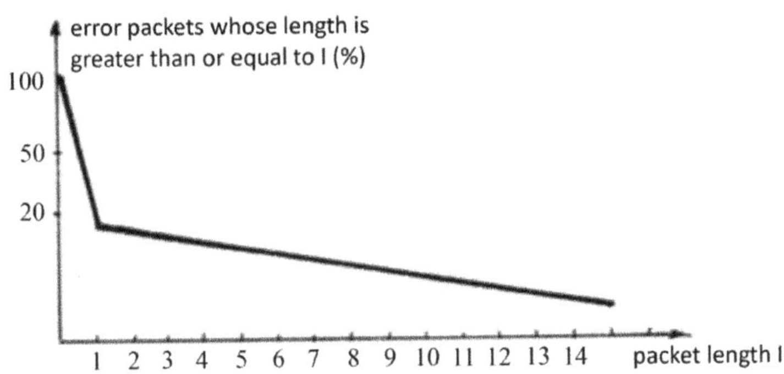

Figure 7.2. *Distribution of error packets*

It should be noted that less than 20% of packets have a length greater than or equal to 3.

To protect against errors, a simple process is to first use the *echoplex mode* (as shown in Figure 7.3). Such a mode consists of sending the received message back to the transmitter. The comparison, at the transmitter level, between the transmitted message that has been stored in memory and the echo of the received message allows the detection of possible errors. Note that echo transport can be prone to errors, even if the message received is correct.

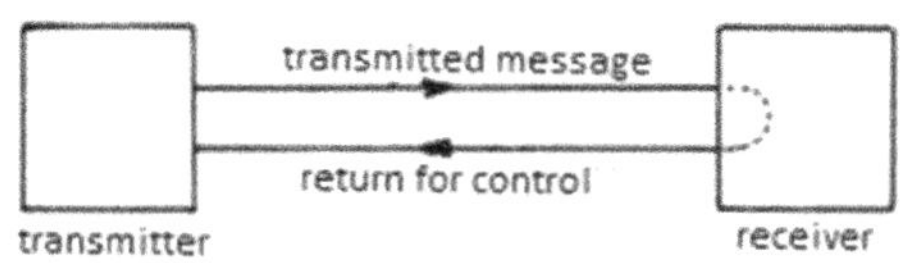

Figure 7.3. *Echoplex mode*

If there is a difference between the message sent and the echo of the message received, the initial message will be retransmitted. Such a technique is often used for small, low-speed transmissions. However, for faster or larger transmissions, *echoplex mode* can become expensive and reduce the effective throughput. An improvement involves sending two successive blocks of information: the first block contains the information to be transmitted, and the second one is a copy of the first block. The verification is then performed at the receiver level. If the two blocks are identical, it is assumed that there is no error. The receiver then sends a brief acknowledgment signal to the transmitter. In the event of a discrepancy, the receiver sends a non-acknowledgment signal to the transmitter, prompting it to retransmit the information. However, such a method still involves redundancy in transmitted information.

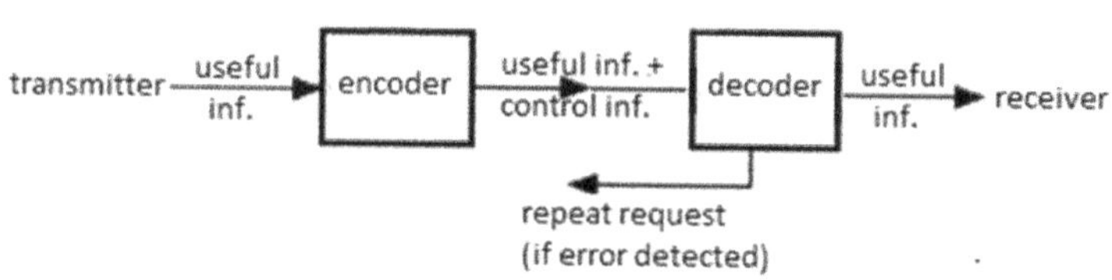

Figure 7.4. *Encoding and decoding*

A more effective solution is to use error detection codes. The principle of such a code involves adding supplementary control information to the useful information (see Figure 7.4). The receiver then decodes and examines the control information to determine whether the information is correct or erroneous. Upon detecting an error, a request to retransmit the information is sent.

Two main categories of codes can be distinguished:

– *Block codes*: in block codes, the control information and the useful information form an autonomous block from the point of view of detection and possibly correction of errors.

– *Convolutional or recurrent codes*: these codes also operate with blocks, but error detection and correction within a block are influenced not only by the current block but also by preceding blocks.

Now, focusing on block codes, they are classified as *systematic codes* when the useful information and the control information are grouped into separate fields within the block, as shown in Figure 7.5.

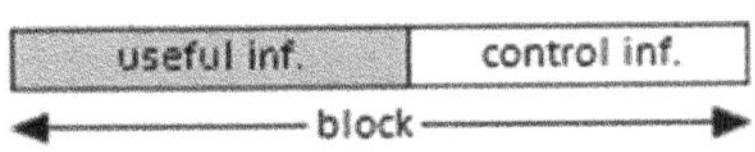

Figure 7.5. *Systematic code*

Consider a set of m-bit words representing useful information, allowing for a total of 2^m distinct words. Introducing control information in the form of r bits derived uniquely from the m useful bits, the resultant "encoded" information contains $n = m + r$ bits. Given the unique definition of the control bits, there are 2^m valid n bits words. However, with n bits, there are 2^n different combinations of words. The difference, $2^n - 2^m$, indicates the number of erroneous words (see Figure 7.6).

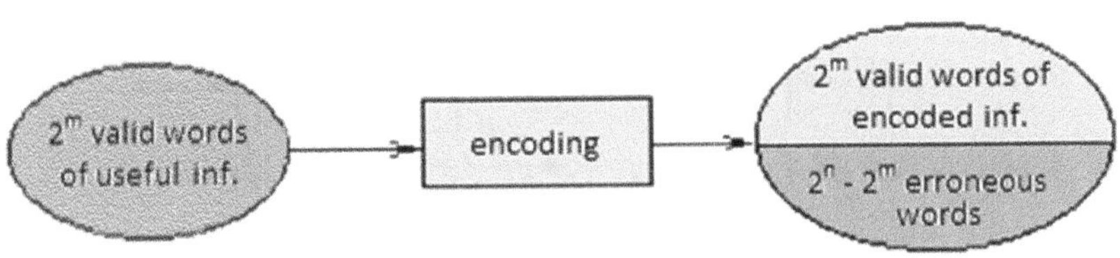

Figure 7.6. *Valid words and incorrect words*

To avoid errors, it is advantageous for the 2^m codewords to be maximally distinct from each other. Such an approach prevents healthy codewords from being transformed into other healthy codewords due to errors, thereby ensuring effective error detection.

Generally speaking, the Hamming distance $d(m_1, m_2)$ between two codewords m_1 and m_2 is defined as the number of different bits of the same rank by which they differ. For example, if $m_1 = 10110010$ and $m_2 = 10000110$, then $d(m_1, m_2) = 3$. The greater the Hamming distance between two codewords, the easier it is to distinguish them. Specifically, if $d(m_1, m_2) = a$, it takes a *properly* placed errors to transform m_1 into m_2.

For each code, it is important to define the minimum Hamming distance d_{min} that allows the codewords to be sufficiently separated. Suppose that p errors occur in a codeword. These p errors will be detected if $d_{min} > p$, i.e. $d_{min} \geq p + 1$. This establishes the primary rule to follow in the quest for error detection and correction codes.

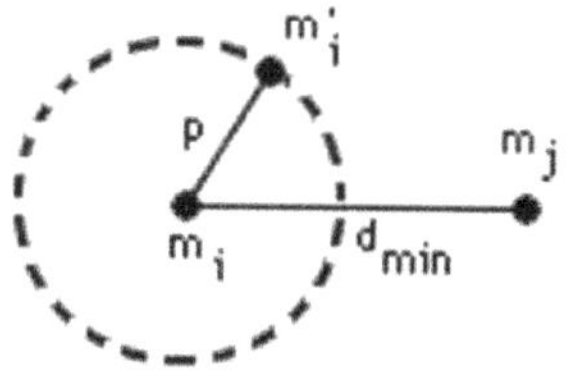

Figure 7.7. *Rule R1*

Rule R1: to detect p errors, it is necessary that $d_{min} \geq p + 1$ (see Figure 7.7).

Thus, to detect simple errors (only one erroneous bit), d_{min} must be $d_{min} \geq 2$.

Let us examine error correction up to an order q. Each codeword m_i has a sphere of influence where any erroneous codeword m_i' (with $d(m_i, m_i') \leq q$) is considered to be m_i. These spheres of influence must not touch or overlap, necessitating that the minimum distance between any two codewords be at least $2q + 1$.

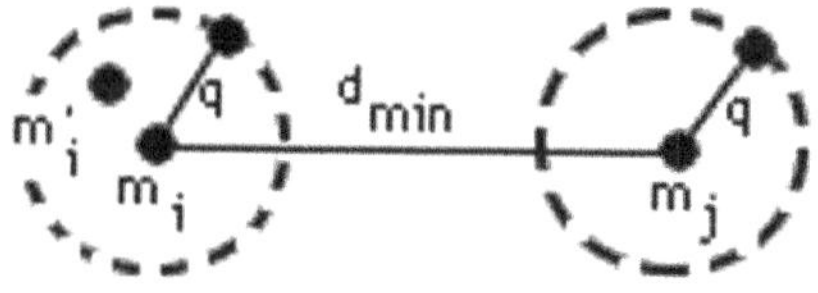

Figure 7.8. *Rule R2*

Rule R2: to correct errors up to order q, it is necessary that $d_{min} \geq 2q + 1$ (see Figure 7.8).

Thus, correcting single errors (only one incorrected bit) requires a code with $d_{min} \geq 3$.

Consider now the case where errors up to order p need detection, and errors up to order q (where $p > q$) require correction. An erroneous codeword m'_i, resulting from the introduction of q errors in a codeword m_i, must not be mistaken for a codeword m_j such that $d(m'_i, m_j) \leq p$. Therefore, it is deduced that d_{min} must be greater than or equal to $p + q + 1$ to prevent confusion between codewords.

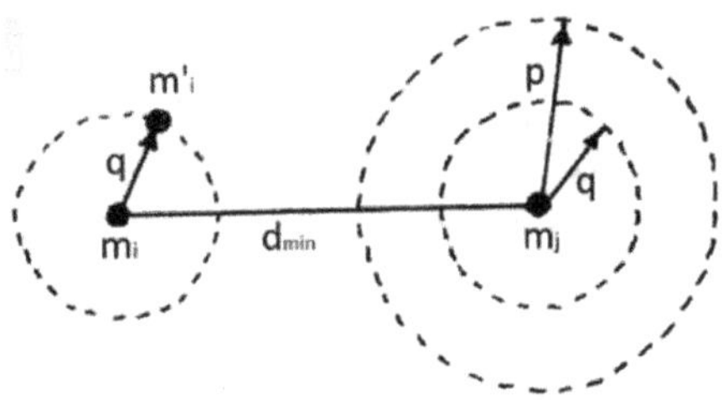

Figure 7.9. *Rule R3*

Rule R3: to correct errors up to order q and detect errors up to order p (where $p > q$), it is necessary that $d_{min} \geq p + q + 1$ (see Figure 7.9).

7.2. Linear codes

A linear code is a systematic (n, m) block code in which the $r = n - m$ control bits depend linearly on the m bits of useful information:

Useful information: $\widetilde{X} = [x_1 \; x_2 \; \ldots\ldots \; x_m]$

Coded information: $\widetilde{Y} = [y_1 \; y_2, \ldots\ldots \; y_m, y_{m+1} \; \ldots\ldots \; y_{m+r}]$

with

$$y_1 = x_1, y_2 = x_2, \ldots y_m = x_m, y_{m+1} = a_1, y_{m+2} = a_2, \ldots, y_{m+r} = a_r.$$

The representation $\widetilde{Y} = [x_1, x_2 \; \ldots\ldots \; x_m \; a_1 \; a_2 \; \ldots\ldots, a_r]$, where the a_i are the control bits. The following relationship therefore defines the code used:

$$\tilde{Y} = \tilde{X}G$$

where G is the *code generating matrix*. The structure of G is specified in Figure 7.10.

$$G = [1, g] = \overbrace{\begin{bmatrix} \underbrace{\begin{matrix} 1 & 0 & \cdots & 0 \\ 0 & 1 & \cdots & 0 \\ \vdots & \vdots & \ddots & \vdots \\ 0 & 0 & \cdots & 1 \end{matrix}}_{m} & \underbrace{\begin{matrix} g_{11} & g_{12} & \cdots & g_{1r} \\ g_{21} & g_{22} & \cdots & g_{2r} \\ \vdots & \vdots & \ddots & \vdots \\ g_{m1} & g_{m2} & \cdots & g_{mr} \end{matrix}}_{r} \end{bmatrix}}^{n} \Big\updownarrow m$$

Figure 7.10. *Structure of the* **G** *matrix*

The control bits are given by:

$$a_i = \sum_{j=1}^{m} x_j g_{ji}$$

$$H = \underbrace{\begin{bmatrix} \overbrace{\begin{matrix} g_{11} & g_{21} & \cdots & g_{m1} \\ g_{12} & g_{22} & \cdots & g_{m2} \\ \vdots & \vdots & \ddots & \vdots \\ g_{1r} & g_{2r} & \cdots & g_{mr} \end{matrix}}^{m} & \overbrace{\begin{matrix} 1 & 0 & \cdots & 0 \\ 0 & 1 & \cdots & 0 \\ \vdots & \vdots & \ddots & \vdots \\ 0 & 0 & \cdots & 1 \end{matrix}}^{r} \end{bmatrix}}_{n} \Big\updownarrow r$$

Figure 7.11. *Structure of the* **H** *matrix*

Consider the matrix H defined in Figure 7.11. H is constructed using the transpose of the matrix g and the identity matrix. Calculate the matrix product $H.Y$:

$$H.Y = \begin{bmatrix} g_{11} & g_{21} & \cdots & g_{m1} & 1 & 0 & \cdots & 0 \\ g_{12} & g_{22} & \cdots & g_{m2} & 0 & 1 & \cdots & 0 \\ \vdots & \vdots & \ddots & \vdots & \vdots & \vdots & \ddots & \vdots \\ g_{1r} & g_{2r} & \cdots & g_{mr} & 0 & 0 & \cdots & 1 \end{bmatrix} \cdot \begin{bmatrix} x_1 \\ x_2 \\ \vdots \\ x_m \\ a_1 \\ a_2 \\ \vdots \\ a_r \end{bmatrix} = \begin{bmatrix} \sum_{j=1}^{m} g_{j1}x_j + a_1 \\ \sum_{j=1}^{m} g_{j2}x_j + a_2 \\ \vdots \\ \sum_{j=1}^{m} g_{jr}x_j + a_r \end{bmatrix} = \mathbf{0}$$

Since addition or subtraction modulo 2 are equivalent, the matrix $\boldsymbol{H}$ serves as the *parity check matrix*.

EXAMPLE 7.1.– Consider a code (6,3) whose matrices $\boldsymbol{G}$ and $\boldsymbol{H}$ are:

$$\boldsymbol{G} = \begin{bmatrix} 1 & 0 & 0 & 0 & 1 & 1 \\ 0 & 1 & 0 & 1 & 0 & 1 \\ 0 & 0 & 1 & 1 & 1 & 0 \end{bmatrix} \qquad \boldsymbol{H} = \begin{bmatrix} 0 & 1 & 1 & 1 & 0 & 0 \\ 1 & 0 & 1 & 0 & 1 & 0 \\ 1 & 1 & 0 & 0 & 0 & 1 \end{bmatrix}$$

Useful information: $\widetilde{\boldsymbol{X}} = [x_1 \quad x_2 \quad x_3]$ (eight possible words)

Coded information: $\widetilde{\boldsymbol{Y}} = [x_1 \quad x_2 \quad x_3 \quad a_1 \quad a_2 \quad a_3]$

The relationship $\widetilde{\boldsymbol{Y}} = \widetilde{\boldsymbol{X}}.\boldsymbol{G}$ leads to:

$$a_1 = x_2 + x_3$$
$$a_2 = x_1 + x_3$$
$$a_3 = x_1 + x_2$$

The codewords are therefore

000000 001110 010101 011011 100011 101101 110110 111000

It is evident that $d_{min} = 3$, thus allowing the correction of all single errors and the detection of all double errors.

EXAMPLE 7.2.– Let the code be (8,7) with $\boldsymbol{H} = [1 \quad 1 \quad 1 \quad 1 \quad 1 \quad 1 \quad 1 \quad 1]$

Useful information: $\widetilde{\boldsymbol{X}} = [x_1 \quad x_2 \quad x_3 \quad x_4 \quad x_5 \quad x_6 \quad x_7]$

Coded information: $\widetilde{\boldsymbol{Y}} = [x_1 \quad x_2 \quad x_3 \quad x_4 \quad x_5 \quad x_6 \quad x_7 \quad a_1]$

The relation $\boldsymbol{H}.\boldsymbol{Y} = \boldsymbol{0}$ leads, with the modulo 2 operation, to $a_1 = x_1 + x_2 + x_3 + x_4 + x_5 + x_6 + x_7$.

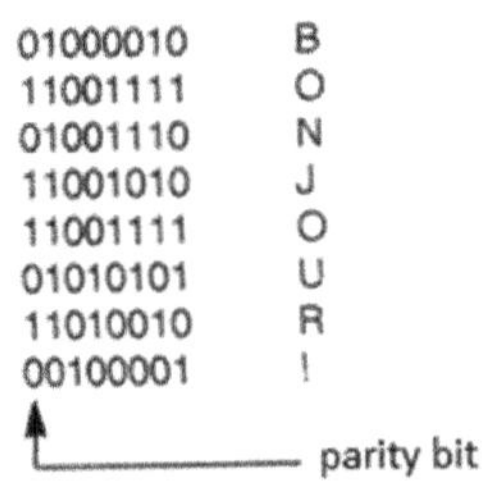

Figure 7.12. *Parity bit*

In this example, the bit a_1 is known as the parity bit, and the codewords have the characteristic of having an even number of 1. This allows 8-bit characters to be represented using 7 bits for the ASCII code, with the eighth bit being the parity bit. Note that the parity bit can be placed wherever desired (see Figure 7.12).

With this system, if two characters in the ASCII code differ by only one bit, they will also differ in the parity bit. Therefore, the minimum distance d_{min} is equal to 2, which only allows the detection of simple errors.

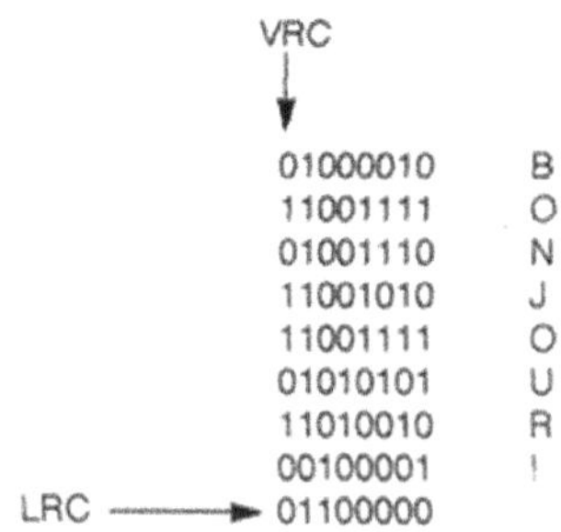

Figure 7.13. *VRC and LRC*

However, it is possible to improve protection against errors by adding a "longitudinal" parity check in addition to the "vertical" parity check described previously (LRC: Longitudinal Redundancy Check; VRC: Vertical Redundancy Check). Such a method consists of adding a control character after every group of b blocks (see Figure 7.13).

Figure 7.14 shows the serial transmission of these blocks.

Figure 7.14. *Serial transmission*

With the addition of the VRC + LRC parity check method, groups of blocks differing by just 1 useful information bit will also differ in VRC bit, LRC bit and VRC + LRC bit. Consequently, the minimum distance $d_{min} = 4$ allows for the detection of single and double errors, as well as the correction of single errors.

7.3. Polynomial codes

Polynomial codes are systematic linear codes commonly used in data transmission for error detection, widely used in modern applications.

Consider an m-bit message X written as follows:

$$\widetilde{X} = [x_0 \quad x_1 \quad \cdots \quad x_{m-1}],$$

where the bit numbering used here is slightly different from before but traditional in polynomial code applications. A polynomial $X(z)$ is associated with the message X:

$$X(z) = x_0 z^0 + x_1 z^1 + x_2 z^2 + \dots \dots \dots + x_{m-1} z^{m-1}$$

For polynomial codes, used to represent binary data, allow these polynomials to be added (modulo 2) and multiplied according to standard algebraic rules. In a polynomial code, all codewords are represented by polynomials $Y(z)$ that are multiples of a *generating polynomial* $H(z)$. This type of code is known as linear and systematic because each polynomial $Y(z)$ can be expressed in the form:

$$Y(z) = Q(z).H(z)$$

where $Q(z)$ is a polynomial whose degree is less than that of $H(z)$ and represents the message to be transmitted. The generator polynomial $H(z)$ is used for detecting and correcting transmission errors.

To transition from the useful information represented by the polynomial $X(z)$ to the coded information represented by the polynomial $Y(z)$, a polynomial code (n, m) is defined where $r = n - m$ control bits are added. The definitions are as follows:

$$X(z) = x_0 + x_1 z + x_2 z^2 + \dots \dots \dots + x_{m-1} z^{m-1}$$

$H(z) = h_0 + h_1 z + h_2 z^2 + \dots \dots + z^r$ is the generator polynomial of degree r.

The polynomial $z^r X(z)$ is of degree $m + r - 1 = n - 1$, containing n terms where the first r terms are zero. Performing polynomial division $z^r X(z)$ by $H(z)$:

$$z^r X(z) = Q(z).H(z) + R(z).$$

Here, $R(z)$ is the remainder polynomial of degree $r - 1$. Since modulo 2 addition is equivalent to subtraction, it follows that:

$$Y(z) = z^r X(z) + R(z) = Q(z).H(z)$$

Note that $Y(z)$ represents the polynomial associated with the codeword, which is of degree $n - 1$ and includes n terms.

EXAMPLE 7.3.– Polynomial code (7,4) with generator polynomial $H(z) = 1 + z + z^3$. The useful information corresponds to the polynomial $X(z) = x_0 + x_1 z + x_2 z^2 + x_3 z^3$. Performing the division $z^3 X(z)$ by $H(z)$ yields the following results (see Figure 7.15):

$$\begin{array}{l|l}
x_3z^6+x_2z^5+x_1z^4 \quad + x_0z^3 & z^3+z+1 \\
\hline
\quad 0+x_2z^5+(x_1+x_3)z^4+(x_0+x_3)z^3 & x_3z^3\P \\
\qquad 0 \; +(x_1+x_3)z^4+(x_0+x_2+x_3)z^3+x_2z^2 & +x_2z^2 \\
\qquad\quad 0 \quad +(x_0+x_2+x_3)z^3+(x_1+x_2+x_3)z^2+(x_1+x_3)z & +(x_1+x_3)z \\
\qquad\qquad 0 \quad +(x_1+x_2+x_3)z^2+(x_0+x_1+x_2)z+x_0+x_2+x_3 & +x_0+x_2+x_3
\end{array}$$

Figure 7.15. *Division of $z^3 X(z)$ by $H(z)$*

$Q(z) = x_3 z^3 + x_2 z^2 + (x_1 + x_3)z + x_0 + x_2 + x_3$
$R(z) = (x_1 + x_2 + x_3)z^2 + (x_0 + x_1 + x_2)z + x_0 + x_2 + x_3$
$Y(z) = x_3 z^6 + x_2 z^5 + x_1 z^4 + x_0 z^3 + (x_1 + x_2 + x_3)z^2 + (x_0 + x_1 + x_2)z + x_0 + x_2 + x_3$
Hence, the codeword is $[x_0\ x_1\ x_2\ x_3\ a_0\ a_1\ a_2]$ with

$$a_2 = x_1 + x_2 + x_3$$
$$a_1 = x_0 + x_1 + x_2$$
$$a_0 = x_0 + x_2 + x_3$$

The generating matrix $\boldsymbol{G}$ of the code is deduced as follows:

$$\boldsymbol{G} = \begin{bmatrix} 1 & 0 & 0 & 0 & 1 & 1 & 0 \\ 0 & 1 & 0 & 0 & 0 & 1 & 1 \\ 0 & 0 & 1 & 0 & 1 & 1 & 1 \\ 0 & 0 & 0 & 1 & 1 & 0 & 1 \end{bmatrix}$$

Polynomial codes are widely used in telecomputing, with several well-known examples:

– CCITT V.41 code: the generating polynomial is $H(z) = z^{16} + z^{12} + z^5 + 1$, used in the HDLC procedure.

– CRC code 16: the generating polynomial is $H(z) = z^{16} + z^{15} + z^2 + 1$, used in the BSC procedure with EBCDIC coding.

– CRC code 12: the generating polynomial is $H(z) = z^{12} + z^{11} + z^3 + z^2 + z + 1$, used in the BSC procedure with 6-bit character coding.

– ARPA code: the generating polynomial is $H(z) = z^{24} + z^{23} + z^{17} + z^{16} + z^{15} + z^{13} + z^{12} + z^{11} + z^{10} + z^9 + z^8 + z^5 + z^3 + 1$.

– Ethernet code: the generating polynomial is $H(z) = z^{32} + z^{26} + z^{23} + z^{22} + z^{16} + z^{12} + z^{11} + z^{10} + z^8 + z^7 + z^5 + z^4 + z^2 + z + 1$.

A significant subset of polynomial codes consists of cyclic codes. A polynomial code (n, m) is cyclic if its generator polynomial $H(z)$ divides $z^n + 1$:

$$z^n + 1 = H(z).\Omega(z),$$

where $\Omega(z)$ is a polynomial of degree $(n - m)$.

Cyclic codes possess a fundamental property that makes them particularly advantageous: any circular shift (i.e. any circular permutation of bits) of a codeword remains a valid codeword. This property greatly simplifies their practical implementation and enhances their utility across various telecommunications applications. Cyclic codes find widespread application in wireless communication protocols such as Wi-Fi and Bluetooth.

7.4. Convolutional codes

A *convolutional code*, also known as a *recurrent* (n, m) code, encodes the block based on the m bits of information in the block and the $m - 1$ previous bits or blocks.

Let us consider the example of the convolutional code (3,1) where 1 bit of useful information corresponds to 3 bits of encoded data. The encoder is shown in Figure 7.16, where the addition is performed using the "exclusive or" operation, denoted by ⊕ (i.e. for which 1 ⊕ 1 = 0).

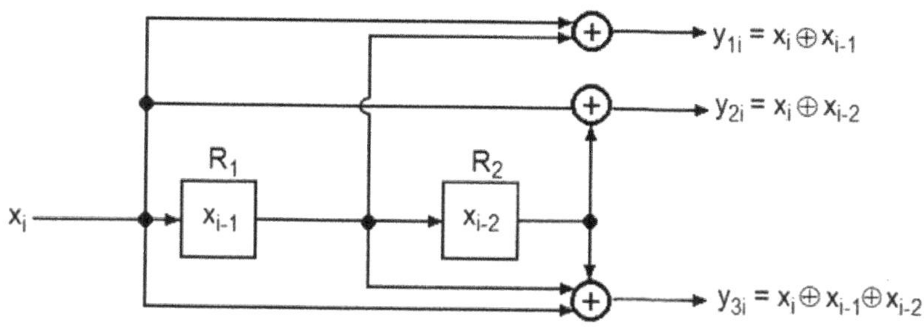

Figure 7.16. *Convolutional code* (3,1)

Thus, each transmitted bit x_i gives rise to a triplet y_{1i}, y_{2i}, y_{3i}. Errors can be detected by calculating the values of x_i, x_{i-1} and x_{i-2} (the latter two already received) as functions of y_{1i}, y_{2i}, y_{3i}, allowing for error checking.

The operation of convolutional coding can be represented in several ways:

a) Representation by *tree diagram* (see Figure 7.17). This representation visualizes the various coding stages from useful information to the codeword. It is well suited for convolutional codes, where nodes represent encoder states and branches correspond to different possible transitions based on input and output bits.

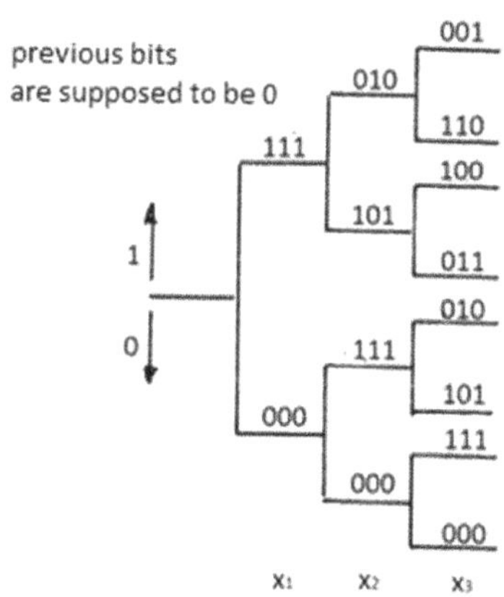

Figure 7.17. *Tree diagram*

b) Representation by *state diagram* (see Figure 7.18). This also visualizes encoder states and explicitly shows transitions between states based on input and output bits. Such a representation is particularly suitable for convolutional codes.

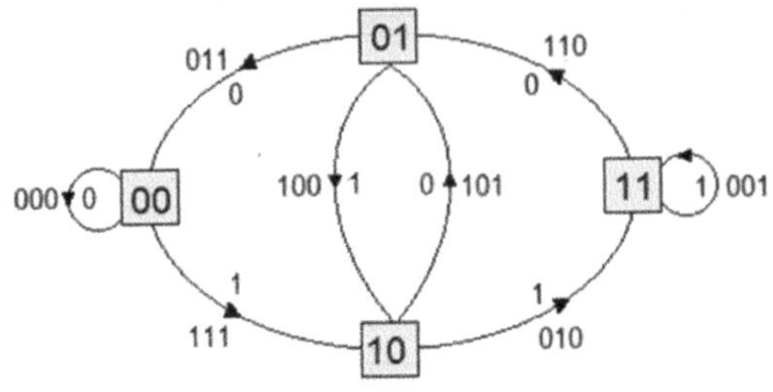

Figure 7.18. *State diagram*

In Figure 7.18, the nodes correspond to the states of registers R_1 and R_2 (the bits x_{i-1} and x_{i-2}). The arcs correspond to the output states $y_1y_2y_3$ depending on whether a 0 or a 1 is input.

c) Representation by *lattice diagram* (see Figure 7.19). This diagram is especially suitable for block type codes. It provides a visual representation of all the processing steps performed on each block and the various error correction procedures. Nodes within the lattice represent distinct processing states, branches indicate potential transitions based on input and output bits, and paths in the lattice illustrate the specific processing steps applied to the blocks.

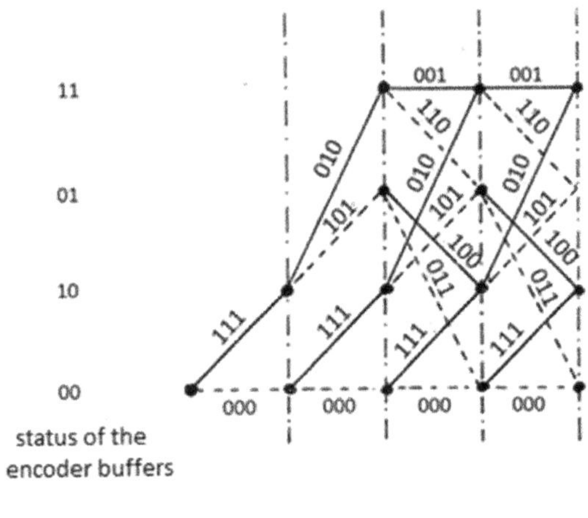

Figure 7.19. *Lattice diagram*

8

Encryption Systems

CONCEPTS COVERED IN THIS CHAPTER.–

Information protection is a concern for all users of binary information. With the rise of website hacking, maintaining information confidentiality has become crucial. This chapter explains the main methods of data encryption.

Following a review of the standard methods of substitution and transposition, two contemporary systems will be detailed: the DES (Data Encryption Standard) system with a secret key and the RSA (Rivest–Shamir–Adleman) system with a public key, highlighting the use of prime numbers in the latter.

The method of elliptic curves will be introduced, building on Chapter 6 on arithmetic.

Finally, the potential of the quantum computer will be discussed, along with an overview of encryption methods based on quantum mechanics.

References: [BEL 94, GHE 08, GOM 13, INA 99, LEH 19, LOV 91, SIN 99].

8.1. Substitution and transposition

Encryption consists of transforming plain text into indecipherable text (cipher text) using an algorithm and an encryption key. The opposite operation is *decryption*, which recovers the plain text from the cipher text using the corresponding decryption key. Generally speaking, two keys are needed in a symmetric encryption system, one for encryption (encryption key) and one for decryption (decryption key), as shown in Figure 8.1.

Mathematics for Digital Science 1,
by Gérard-Michel COCHARD and Mhand HIFI.

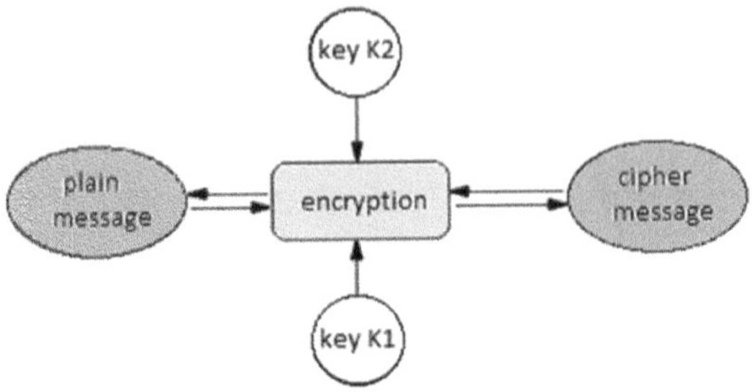

Figure 8.1. *Schema of an encryption system*

If the key K1 is equal to the key K2 (i.e. K1 = K2), the encryption system is called *classic* or *single key*. This means that a single key is used for both encryption and decryption of data.

On the contrary, if the key K1 is different from the key K2 (i.e. $K1 \neq K2$), the encryption system is called *a key pair*. In this case, one key (K1) is used for data encryption and another separate key (K2) is used for decryption.

EXAMPLE 8.1.– Construction of a key: choice of a particular word (see Figure 8.2.)

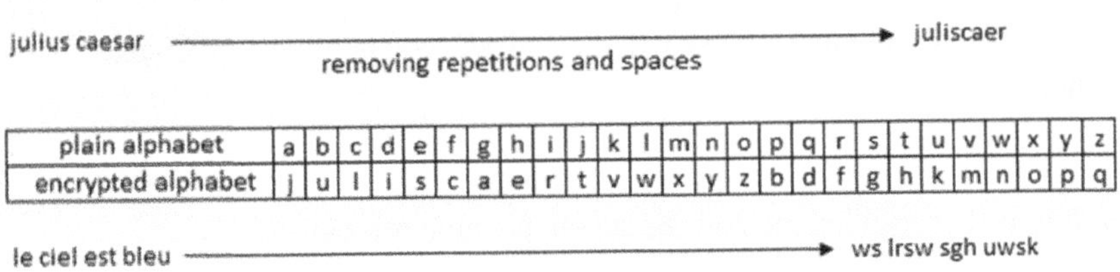

plain alphabet	a	b	c	d	e	f	g	h	i	j	k	l	m	n	o	p	q	r	s	t	u	v	w	x	y	z
encrypted alphabet	j	u	l	i	s	c	a	e	r	t	v	w	x	y	z	b	d	f	g	h	k	m	n	o	p	q

Figure 8.2. *Encryption table*

As shown in Figure 8.2, the cipher alphabet starts with the key "juliscaer", followed by the remaining unused letters in alphabetical order starting from "r". Using this cipher alphabet, the French sentence "*le ciel est bleu*"[1] can be encrypted.

1 The sky is blue.

There are two main basic techniques in the field of encryption: the substitution method and the transposition method.

– *The substitution method*: involves replacing each character or group of characters in plain text with another character or group of characters according to a predefined rule. For example, in monoalphabetic substitution encryption, each letter of the plain text is replaced by a corresponding letter of the alphabet according to a fixed correspondence. This method can be done symmetrically with a single key for encryption and decryption, or asymmetrically with two separate keys.

– *The transposition method*: involves reorganizing the order of the characters in the plain text without modifying them. For example, in transposition encryption, letters in the plain text are rearranged according to a specific rule, sometimes using a matrix or key to determine the order of permutation. This method does not change the characters themselves but makes the cipher text more difficult to understand without knowledge of the transposition method used.

8.2. Substitution methods

8.2.1. *Replacement of a symbol by a symbol*

Encrypting a text involves replacing each symbol (or character in the plain text) with another symbol (or character) according to a specific rule.

Caesar's code, also known as shift encryption, is a historical encryption method attributed to Julius Caesar. According to the tradition reported by Suetonius, such a method involves shifting each letter of the plain text three positions to the right in the alphabet.

VENI VIDI VICI

↓ ↓ ↓

YHQL YLGL YLFL

Figure 8.3. *Caesar's code*

In a "modern" version of shift encryption, the ASCII code of a character is adjusted by a fixed "offset", also referred to as a key. The resulting value is then reduced modulo the size of the alphabet.

EXAMPLE 8.2.– Adding 3 to the ASCII code (see Figure 8.4) to encrypt the French word "*ordinateur*"[2]

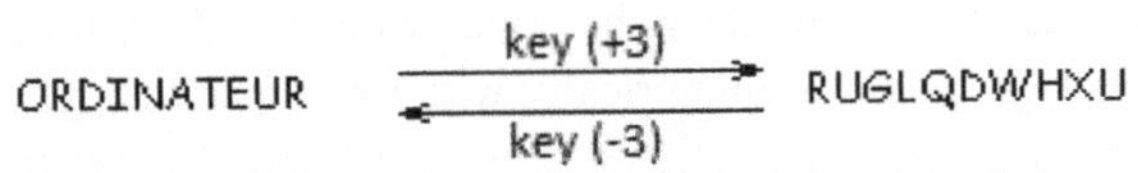

Figure 8.4. *ASCII code offset*

An important cryptographic tool is frequency analysis. By analyzing a large number of texts, the frequency of appearance of different letters is determined. For French texts, the results are shown in Figure 8.5.

letter	%	letter	%
A	9.42	N	7.15
B	1.02	O	5.14
C	2.64	P	2.86
D	3.39	Q	1.06
E	15.87	R	6.46
F	0.95	S	7.90
G	1.04	T	7.26
H	0.77	U	6.24
I	8.41	V	2.15
J	0.89	W	0.00
K	0.00	X	0.30
L	5.34	Y	0.24
M	3.24	Z	0.32

→

letter	%	letter	%
E	15.87	C	2.64
A	9.42	V	2.15
I	8.41	Q	1.06
S	7.90	G	1.04
T	7.26	B	1.02
N	7.15	F	0.95
R	6.46	J	0.89
U	6.24	H	0.77
L	5.34	Z	0.32
O	5.14	X	0.30
D	3.39	Y	0.24
M	3.24	K	0.00
P	2.86	W	0.00

Figure 8.5. *Frequency analysis for French texts*

2 Computer in English.

EXAMPLE 8.3.– Let the following French text be decoded (see Figure 8.6)

```
SGOZXK IUXHKGA YAX AT
GXHXK VKXINK ZKTGOZ
JGTY YUT HKI AT LXUSGMK
SGOZXK XKTGXJ VGX R
UJKAX GRRKINK RAO ZOTZ G
VKA VXKY IK RGTMGMK
KZ HUTPUAX SUTYOKAX RK
IUXHKGA WAK BUAY KZKY
PURO WAK BUAY SK YKSHRKF
HKGA YGTY SKTZOX YO BUZXK
XGSGMK YK XGVVUXZK G
BUZXK VRASGMK BUAY KZKY
RK VNKTOD JKY NUZKY  JK
IKY HUOY  G IKY SUZY RK
IUXHKGA TK YK YKTZ VGY
JK PUOK KZ VUAX SUTZXKX
YG HKRRK BUOD OR UABXK
AT RGXMK HKI RGOYYK
ZUSHKX YG VXUOK RK XKTGXJ
Y KT YGOYOZ KZ JOZ SUT HUT
SUTYOKAX GVVXKTKF WAK
ZUAZ LRGZZKAX BOZ GAD
JKVKTJY JK IKRAO WAO R
KIUAZK IKZZK RKIUT BGAZ
HOKT AT LXUSGMK YGTY
JUAZK RK IUXHKGA NUTZKAD
KZ IUTLAY PAXG SGOY AT
VKA ZGXJ WA UT TK R E
VXKTJXGOZ VRAY
```

Figure 8.6. *The text to decipher*

Count the letters in this text (see Figure 8.7). By using statistics derived from the frequency of letters in French texts, a correspondence can be established between the letters of the alphabet and their probability of appearance. Figure 8.8 displays a list of letters from the French alphabet, arranged according to their average probability of appearance.

K	G	X	A	U	Y	Z	T	O	R	V	S	I	H	J	B	M	N	W	D	L	P	F	E	C	Q
90	41	41	40	38	38	36	34	26	24	17	16	16	15	14	9	7	5	5	4	4	4	2	1	0	0

Figure 8.7. *Number of appearances of letters in the encrypted text*

K	G	X	A	U	Y	Z	T	O	R	V	S	I	H	J	B	M	N	W	D	L	P	F	E	C	Q
E	A	I	S	T	N	R	U	L	O	D	M	P	C	V	Q	G	B	F	J	H	Z	X	Y	K	W

Figure 8.8. *Probable encryption*

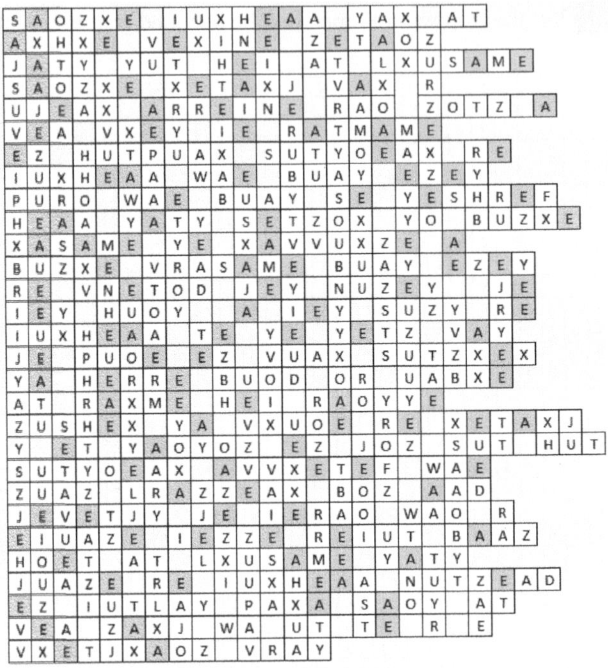

Figure 8.9. *First replacements*

Start by replacing K and G with E and A, resulting in Figure 8.9.

Some words contain doubled letters such as ARREINE. In French, RR can correspond to CC, FF, LL, MM, NN, PP, RR, SS, TT, etc. Since L is statistically closest to R, consider replacing R with L (if this does not work, other options can be explored!).

This substitution results in LE[3], a common article in French, appearing multiple times, indicating a promising direction.

Similarly, for the word RAOYYE, which becomes LAOYYE after substitution, we could consider Y → N, but no fitting word comes to mind. The same applies to Y → T or Y → R which also do not seem appropriate. However, Y → S would

3 THE in English.

yield LAOSSE, suggesting the word LAISSE[4] (any other suggestion?), which implies O → I.

In statistics, S corresponds to M.

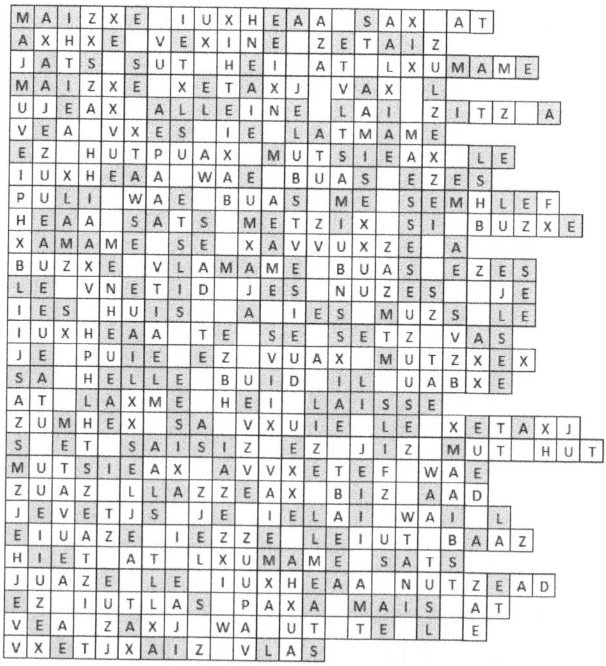

Figure 8.10. *Deciphering attempts*

The text of Figure 8.10 is obtained similarly. The first word is MAIZXE with the unknown letters Z and X. A dictionary search can yield the candidate MAITRE[5], which implies Z → T and X → R, correspondences close to statistics.

4 LEAVE in English.
5 MASTER in English.

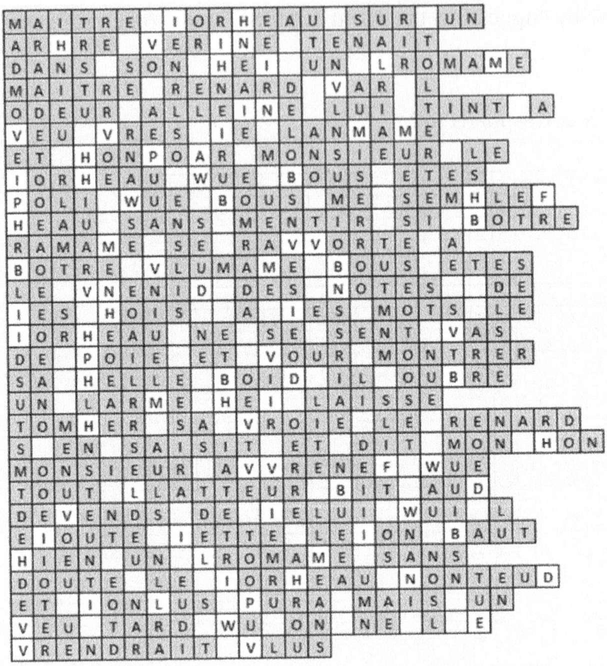

Figure 8.11. *Text almost deciphered*

A word such as METTIR is probably MENTIR[6]; hence, T → N, also close in statistics. Likewise, the word RENARJ is certainly RENARD[7]; hence, J → D. Also, the word MUNSIEAR should be MONSIEUR[8]; hence, U → O and A → U. The text of Figure 8.11 is obtained similarly.

From there, it is no longer very difficult to decipher the text:

IOHEAU must be CORBEAU[9]; therefore, I → C and H → B. From this, EIOUTE → ECOUTE[10] is deduced.

6 LIE in English.
7 FOX in English.
8 MISTER in English.
9 CROW in English.
10 LISTEN in English.

LROMAME must be FROMAGE[11]; hence, L → F and M → G. From this, LANMAME → LANGAGE[12] is deduced.

VLUMAME must be PLUMAGE[13], from which V → P. Therefore, VROY → PROIE[14], etc.

Finally, the text to decipher is the fable "*Le corbeau et le renard*"[15] by Jean de La Fontaine (see Figure 8.12). See the appendix for the translation.

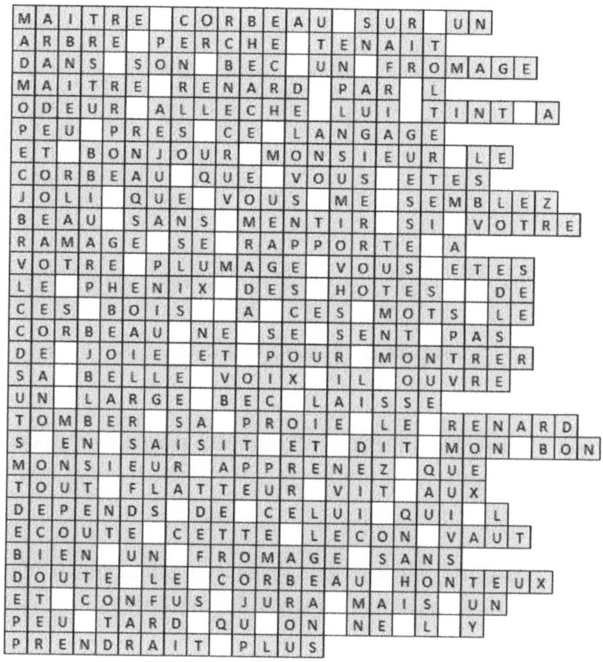

Figure 8.12. *Plain text*

11 CHEESE in English.

12 LANGUAGE in English.

13 PLUMAGE in English.

14 PREY in English.

15 The Crow and the Fox in English.

plain	A	B	C	D	E	F	G	H	I	J	K	L	M	N	O	P	Q	R	S	T	U	V	W	X	Y	Z
1	B	C	D	E	F	G	H	I	J	K	L	M	N	O	P	Q	R	S	T	U	V	W	X	Y	Z	A
2	C	D	E	F	G	H	I	J	K	L	M	N	O	P	Q	R	S	T	U	V	W	X	Y	Z	A	B
3	D	E	F	G	H	I	J	K	L	M	N	O	P	Q	R	S	T	U	V	W	X	Y	Z	A	B	C
4	E	F	G	H	I	J	K	L	M	N	O	P	Q	R	S	T	U	V	W	X	Y	Z	A	B	C	D
5	F	G	H	I	J	K	L	M	N	O	P	Q	R	S	T	U	V	W	X	Y	Z	A	B	C	D	E
6	G	H	I	J	K	L	M	N	O	P	Q	R	S	T	U	V	W	X	Y	Z	A	B	C	D	E	F
7	H	I	J	K	L	M	N	O	P	Q	R	S	T	U	V	W	X	Y	Z	A	B	C	D	E	F	G
8	I	J	K	L	M	N	O	P	Q	R	S	T	U	V	W	X	Y	Z	A	B	C	D	E	F	G	H
9	J	K	L	M	N	O	P	Q	R	S	T	U	V	W	X	Y	Z	A	B	C	D	E	F	G	H	I
10	K	L	M	N	O	P	Q	R	S	T	U	V	W	X	Y	Z	A	B	C	D	E	F	G	H	I	J
11	L	M	N	O	P	Q	R	S	T	U	V	W	X	Y	Z	A	B	C	D	E	F	G	H	I	J	K
12	M	N	O	P	Q	R	S	T	U	V	W	X	Y	Z	A	B	C	D	E	F	G	H	I	J	K	L
13	N	O	P	Q	R	S	T	U	V	W	X	Y	Z	A	B	C	D	E	F	G	H	I	J	K	L	M
14	O	P	Q	R	S	T	U	V	W	X	Y	Z	A	B	C	D	E	F	G	H	I	J	K	L	M	N
15	P	Q	R	S	T	U	V	W	X	Y	Z	A	B	C	D	E	F	G	H	I	J	K	L	M	N	O
16	Q	R	S	T	U	V	W	X	Y	Z	A	B	C	D	E	F	G	H	I	J	K	L	M	N	O	P
17	R	S	T	U	V	W	X	Y	Z	A	B	C	D	E	F	G	H	I	J	K	L	M	N	O	P	Q
18	S	T	U	V	W	X	Y	Z	A	B	C	D	E	F	G	H	I	J	K	L	M	N	O	P	Q	R
19	T	U	V	W	X	Y	Z	A	B	C	D	E	F	G	H	I	J	K	L	M	N	O	P	Q	R	S
20	U	V	W	X	Y	Z	A	B	C	D	E	F	G	H	I	J	K	L	M	N	O	P	Q	R	S	T
21	V	W	X	Y	Z	A	B	C	D	E	F	G	H	I	J	K	L	M	N	O	P	Q	R	S	T	U
22	W	X	Y	Z	A	B	C	D	E	F	G	H	I	J	K	L	M	N	O	P	Q	R	S	T	U	V
23	X	Y	Z	A	B	C	D	E	F	G	H	I	J	K	L	M	N	O	P	Q	R	S	T	U	V	W
24	Y	Z	A	B	C	D	E	F	G	H	I	J	K	L	M	N	O	P	Q	R	S	T	U	V	W	X
25	Z	A	B	C	D	E	F	G	H	I	J	K	L	M	N	O	P	Q	R	S	T	U	V	W	X	Y
26	A	B	C	D	E	F	G	H	I	J	K	L	M	N	O	P	Q	R	S	T	U	V	W	X	Y	Z

Figure 8.13. *Vigenère code*

Vigenère code (Blaise de Vigenère, *Traité des chiffres ou secrètes manières d'écrire*[16], 1586) is a famous substitution type of cipher. It is defined by the matrix in Figure 8.13.

Using the key "MIAGE" and the text to encrypt, the French text "*Le ciel est bleu, la mer est verte*"[17] is arranged as shown in Figure 8.14.

M	I	A	G	E	M	I	A	G	E	M	I	A	G	E	M	I	A	G	E	M	I	A	G	E	M
L	E	C	I	E	L	E	S	T	B	L	E	U	L	A	M	E	R	E	S	T	V	E	R	T	E

Figure 8.14. *Correspondence table*

The gray lines in Figure 8.15 correspond to the letters of the key (second column) and define the substitution.

16 Literally, Treatise on secret writing.

17 The sky is blue, the sea is green.

plain	A	B	C	D	E	F	G	H	I	J	K	L	M	N	O	P	Q	R	S	T	U	V	W	X	Y	Z
1	B	C	D	E	F	G	H	I	J	K	L	M	N	O	P	Q	R	S	T	U	V	W	X	Y	Z	A
2	C	D	E	F	G	H	I	J	K	L	M	N	O	P	Q	R	S	T	U	V	W	X	Y	Z	A	B
3	D	E	F	G	H	I	J	K	L	M	N	O	P	Q	R	S	T	U	V	W	X	Y	Z	A	B	C
4	E	F	G	H	I	J	K	L	M	N	O	P	Q	R	S	T	U	V	W	X	Y	Z	A	B	C	D
5	F	G	H	I	J	K	L	M	N	O	P	Q	R	S	T	U	V	W	X	Y	Z	A	B	C	D	E
6	G	H	I	J	K	L	M	N	O	P	Q	R	S	T	U	V	W	X	Y	Z	A	B	C	D	E	F
7	H	I	J	K	L	M	N	O	P	Q	R	S	T	U	V	W	X	Y	Z	A	B	C	D	E	F	G
8	I	J	K	L	M	N	O	P	Q	R	S	T	U	V	W	X	Y	Z	A	B	C	D	E	F	G	H
9	J	K	L	M	N	O	P	Q	R	S	T	U	V	W	X	Y	Z	A	B	C	D	E	F	G	H	I
10	K	L	M	N	O	P	Q	R	S	T	U	V	W	X	Y	Z	A	B	C	D	E	F	G	H	I	J
11	L	M	N	O	P	Q	R	S	T	U	V	W	X	Y	Z	A	B	C	D	E	F	G	H	I	J	K
12	M	N	O	P	Q	R	S	T	U	V	W	X	Y	Z	A	B	C	D	E	F	G	H	I	J	K	L
13	N	O	P	Q	R	S	T	U	V	W	X	Y	Z	A	B	C	D	E	F	G	H	I	J	K	L	M
14	O	P	Q	R	S	T	U	V	W	X	Y	Z	A	B	C	D	E	F	G	H	I	J	K	L	M	N
15	P	Q	R	S	T	U	V	W	X	Y	Z	A	B	C	D	E	F	G	H	I	J	K	L	M	N	O
16	Q	R	S	T	U	V	W	X	Y	Z	A	B	C	D	E	F	G	H	I	J	K	L	M	N	O	P
17	R	S	T	U	V	W	X	Y	Z	A	B	C	D	E	F	G	H	I	J	K	L	M	N	O	P	Q
18	S	T	U	V	W	X	Y	Z	A	B	C	D	E	F	G	H	I	J	K	L	M	N	O	P	Q	R
19	T	U	V	W	X	Y	Z	A	B	C	D	E	F	G	H	I	J	K	L	M	N	O	P	Q	R	S
20	U	V	W	X	Y	Z	A	B	C	D	E	F	G	H	I	J	K	L	M	N	O	P	Q	R	S	T
21	V	W	X	Y	Z	A	B	C	D	E	F	G	H	I	J	K	L	M	N	O	P	Q	R	S	T	U
22	W	X	Y	Z	A	B	C	D	E	F	G	H	I	J	K	L	M	N	O	P	Q	R	S	T	U	V
23	X	Y	Z	A	B	C	D	E	F	G	H	I	J	K	L	M	N	O	P	Q	R	S	T	U	V	W
24	Y	Z	A	B	C	D	E	F	G	H	I	J	K	L	M	N	O	P	Q	R	S	T	U	V	W	X
25	Z	A	B	C	D	E	F	G	H	I	J	K	L	M	N	O	P	Q	R	S	T	U	V	W	X	Y
26	A	B	C	D	E	F	G	H	I	J	K	L	M	N	O	P	Q	R	S	T	U	V	W	X	Y	Z

Figure 8.15. *Vigenère table*

Therefore, the ciphertext in Figure 8.16 is obtained.

key	M	I	A	G	E	M	I	A	G	E	M	I	A	G	E	M	I	A	G	E	M	I	A	G	E	M
plain text	L	E	C	I	E	L	E	S	T	B	L	E	U	L	A	M	E	R	E	S	T	V	E	R	T	E
cipher text	X	M	C	O	I	X	M	S	Z	F	X	M	U	R	E	Y	M	R	K	W	F	D	E	X	X	Q

Figure 8.16. *Text encryption*

Note that this method solves the problem of frequency analysis.

Another interesting encryption method worth noting is based on modulo calculation, known as the Hill cipher. This technique uses a key represented by a 2×2 matrix and segments the message into blocks of two letters. If the total number of letters is odd, the last block is completed with the "@" symbol.

To illustrate this method, here is an example.

EXAMPLE 8.4.– Consider the key matrix K as well as its inverse K^{-1}, ensuring that the determinant of K is non-zero, which guarantees the existence of the inverse:

$$\mathbf{K} = \begin{bmatrix} 1 & 3 \\ 2 & 7 \end{bmatrix} \quad \mathrm{K}^{-1} = \begin{bmatrix} 7 & -3 \\ -2 & 1 \end{bmatrix}$$

Implement the following correspondence between the letters of the alphabet and numbers, as shown in Figure 8.17.

A	B	C	D	E	F	G	H	I	J	K	L	M	N	O	P	Q	R	S	T	U	V	W	X	Y	Z	@
0	1	2	3	4	5	6	7	8	9	10	11	12	13	14	15	16	17	18	19	20	21	22	23	24	25	26

Figure 8.17. *Letter–number matching*

To encrypt the word "INFORMATION", apply the correspondence shown in Figure 8.18.

I	N	F	O	R	M	A	T	I	O	N
8	13	5	14	17	12	0	19	8	14	13

Figure 8.18. *Matching for key*

Divide the message into pairs of two numbers: (8, 13), (5, 14), (17, 12), (0, 19), (8, 14), (13, 26).

Consider these pairs of numbers as the components of a vector and apply the matrix **K** to these components to perform the encryption:

$$\text{"IN"} \rightarrow \begin{bmatrix} 1 & 3 \\ 2 & 7 \end{bmatrix} \begin{bmatrix} 8 \\ 13 \end{bmatrix} = \begin{bmatrix} 47 \\ 107 \end{bmatrix} \equiv \begin{bmatrix} 20 \\ 26 \end{bmatrix} [27] \rightarrow \text{"U@"}$$

$$\text{"FO"} \rightarrow \begin{bmatrix} 1 & 3 \\ 2 & 7 \end{bmatrix} \begin{bmatrix} 5 \\ 14 \end{bmatrix} = \begin{bmatrix} 47 \\ 108 \end{bmatrix} \equiv \begin{bmatrix} 20 \\ 0 \end{bmatrix} [27] \rightarrow \text{"UA"}$$

$$\text{"RM"} \rightarrow \begin{bmatrix} 1 & 3 \\ 2 & 7 \end{bmatrix} \begin{bmatrix} 17 \\ 12 \end{bmatrix} = \begin{bmatrix} 53 \\ 118 \end{bmatrix} \equiv \begin{bmatrix} 26 \\ 10 \end{bmatrix} [27] \rightarrow \text{"@K"}$$

$$\text{"AT"} \rightarrow \begin{bmatrix} 1 & 3 \\ 2 & 7 \end{bmatrix} \begin{bmatrix} 0 \\ 19 \end{bmatrix} = \begin{bmatrix} 57 \\ 133 \end{bmatrix} \equiv \begin{bmatrix} 3 \\ 25 \end{bmatrix} [27] \rightarrow \text{"DZ"}$$

$$\text{"IO"} \rightarrow \begin{bmatrix} 1 & 3 \\ 2 & 7 \end{bmatrix} \begin{bmatrix} 8 \\ 14 \end{bmatrix} = \begin{bmatrix} 50 \\ 114 \end{bmatrix} \equiv \begin{bmatrix} 23 \\ 6 \end{bmatrix} [27] \rightarrow \text{"XG"}$$

$$\text{"N@"} \rightarrow \begin{bmatrix} 1 & 3 \\ 2 & 7 \end{bmatrix} \begin{bmatrix} 13 \\ 26 \end{bmatrix} = \begin{bmatrix} 91 \\ 208 \end{bmatrix} \equiv \begin{bmatrix} 10 \\ 19 \end{bmatrix} [27] \rightarrow \text{"KT"}$$

The matrix K^{-1} allows the text to be deciphered and the initial message to be found.

Note that there are many variations of this method. For example, instead of grouping letters into pairs, they can be grouped into packs of n letters, requiring a corresponding key ($n \times n$).

8.2.2. *The code: replacement of a word by a word*

Instead of simply replacing one symbol with another symbol, it is also possible to take a more holistic approach by replacing one word with another word. Thus, while the first method is known as *cipher*, the second method is called *code* (or *coding*).

To implement the coding method, it is necessary to create a correspondence table beforehand.

EXAMPLE 8.5.– Either code the following French text:

IL ÉTAIT UNE FOIS UNE MARCHANDE DE FOIE QUI VENDAIT DU FOIE DANS LA VILLE DE FOIX. ELLE ME DIT, MA FOI, C'EST LA PREMIÈRE FOIS QUE JE VENDS DU FOIE DANS LA VILLE DE FOIX[18].

A practical illustration of the lookup table can be found in Figure 8.19, which shows the mappings used in the code. By excluding punctuation marks and spaces, the message is simplified as follows:

B2F6I9VWI9TUGHD4MN11KLD412Z0EFGHXYH8E510C3VWABOPZ0QRVWG7A 111KLD412Z0EFGHXY.

words	code
C	AB
DANS	12
DE	GH
DIT	10
DU	KL
ELLE	H8
EST	OP
ÉTAIT	F6
FOI	RS
FOIE	D4
FOIS	VW
FOIX	XY

words	code
IL	B2
JE	A1
LA	Z0
MA	C3
MARCHANDE	TU
ME	E5
PREMIERE	QR
QUE	G7
QUI	MN
UNE	I9
VENDS	IJ
VENDAIT	11
VILLE	EF

Figure 8.19. *The code*

The encryption key is then a kind of dictionary.

---

18 "ONCE UPON A TIME THERE WAS A LIVER MERCHANT WHO SOLD LIVER IN THE TOWN OF FOIX. SHE TOLD ME, MY FAITH, THIS IS THE FIRST TIME THAT I AM SELLING LIVER IN THE TOWN OF FOIX".

8.3. Transposition methods

The text to be encrypted is divided into blocks of fixed size (e.g. eight characters). Suppose the key K1 = 71265843, which specifies that the first character of the block is placed at position 7 in the encrypted block, the second at position 1, etc. Following this approach, encryption proceeds for the following French text:

VIVE L'INFORMATIQUE[19] becomes IVI'EVLFOITMRNAUE Q

The decryption key is K2 = 23875416.

The process can be further refined, as in the "Lucifer" method developed by IBM and based on the Feistel structure. To illustrate this method, consider the following example.

EXAMPLE 8.6.– To code the letter J in extended ASCII (see Figure 8.20).

0	1	0	0	1	0	1	0

Figure 8.20. *ASCII code of J*

The Lucifer method involves several phases (see Figure 8.21). These operations are then repeated several times (see Figure 8.22).

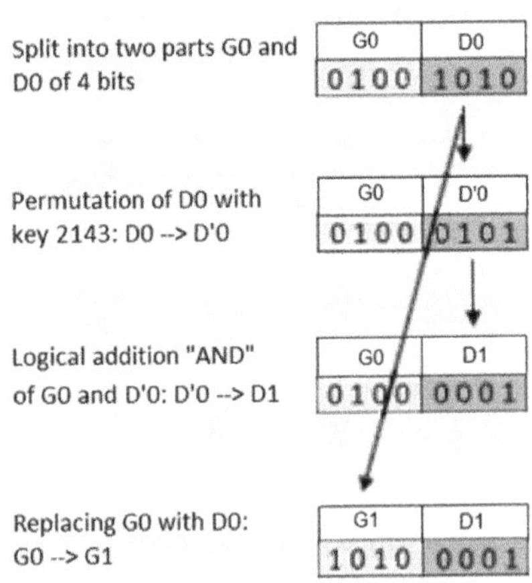

Figure 8.21. *Basic operation*

19 LONG LIVE COMPUTER SCIENCE in English.

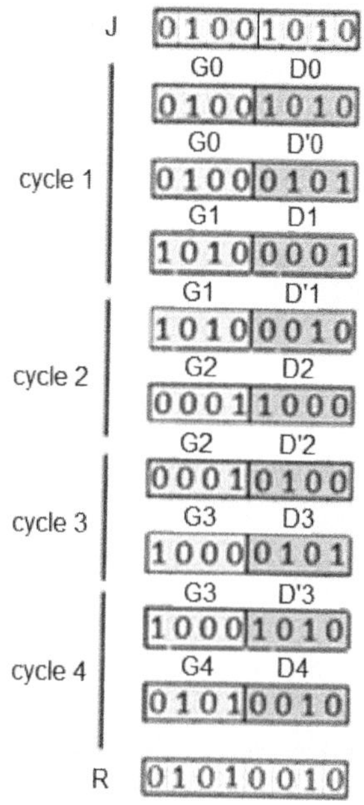

Figure 8.22. *Iterations*

The process is reversible given the permutation key.

8.4. Asymmetric systems

Asymmetric systems use a key pair consisting of a public key and a private key. However, securely transmitting these keys between involved parties poses a challenge.

A solution to this problem is provided by the Bell–Hellman algorithm, developed in 1976. To illustrate its operation, consider two correspondents, A and B, as shown in Figure 8.23.

A	B
A chooses any number kept secret: 3	B chooses any number kept secret: 6

Figure 8.23. *Communication between A and B*

Suppose, as shown in Figure 8.24, that correspondents A and B have knowledge of a function $f(x) \equiv 7^x\ [11]$, which forms part of the public key. Obviously, a potential intruder may also possess knowledge of this function.

A	B
$f(3) \equiv 7^3\ [11] \equiv 343\ [11] = 2$ A sends 2 to B A receives 4 A calculates $4^3\ [11] \equiv 64\ [11] = 9$	$f(6) \equiv 7^6\ [11] \equiv 117649\ [11] = 4$ B sends 4 to A B receives 2 B calculates $2^6\ [11] \equiv 64\ [11] = 9$

Figure 8.24. *Communication between A and B (continued)*

A and B now hold secret key 9.

An intruder possessing the values 2 and 4 faces a significant challenge in solving the equations $7^y \equiv 2\ [11]$ and $7^z \equiv 4\ [11]$ to find $y = 3$ and $z = 6$. Reversing the "modulo" function is indeed extremely difficult.

8.5. DES – secret key system

Classical methods combine substitution and transposition. Two well-known methods are DES (Data Encryption Standard), standardized by the National Bureau of Standards since 1977, and RSA (Rivest–Shamir–Adleman), which will be discussed later.

DES is a standardized method where text is divided into 64-bit blocks and a 64-bit key is used. The encryption process is shown in Figure 8.25.

The encryption (encoding) key K generates, as discovered later, 16 secondary keys used in the encryption process. IP and IP^{-1} are permutation operations defined by the tables shown in Figure 8.26.

The IP operation repositions the 58th bit as the first bit, the 50th bit as the second, etc. Similarly, the IP^{-1} inverse operation ensures that the first bit of the result corresponds to the 40th bit, the second bit of the result corresponds to the 8th, etc.

The function f is defined according to the schema in Figure 8.27.

The function $f(R, K)$ involves several essential operations: operation E, which converts a 32-bit block into a 48-bit block; operations S1–S8, which transform 6-bit groups into 4-bit groups according to Figure 8.29; and the permutation P, which acts on a 32-bit block. These operations are detailed in Figure 8.28.

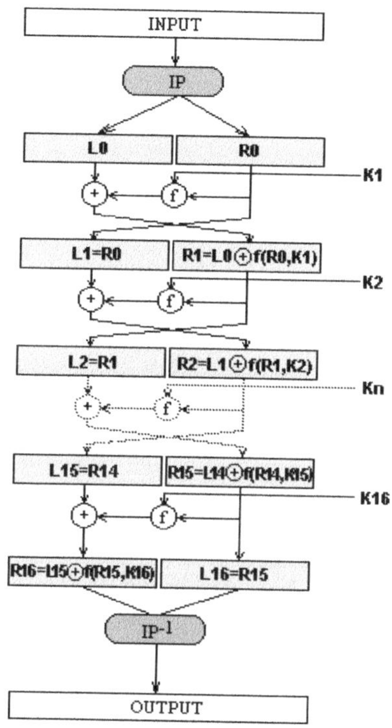

Figure 8.25. *Encryption process*

IP

58	50	42	34	26	18	10	2
60	52	44	36	28	20	12	4
62	54	46	38	30	22	14	6
64	56	48	40	32	24	16	8
57	49	41	33	25	17	9	1
59	51	43	35	27	19	11	3
61	53	45	37	29	21	13	5
63	55	47	39	31	23	15	7

IP^{-1}

40	8	48	16	56	24	64	32
39	7	47	15	55	23	63	31
38	6	46	14	54	22	62	30
37	5	45	13	53	21	61	29
36	4	44	12	52	20	60	28
35	3	43	11	51	19	59	27
34	2	42	10	50	18	58	26
33	1	41	9	49	17	57	25

Figure 8.26. *IP and IP^{-1} swap (permutation operation)*

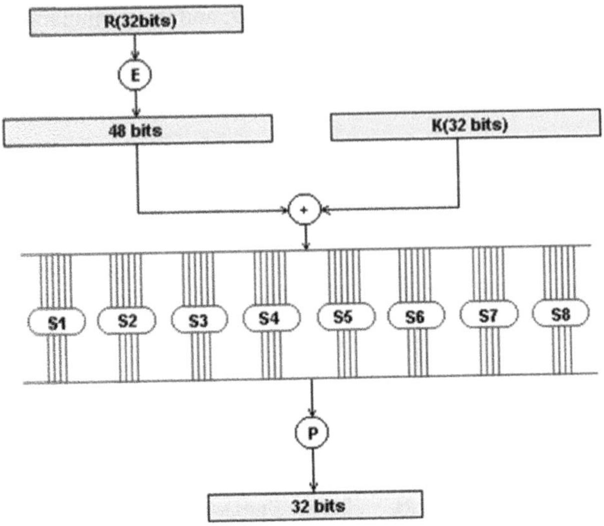

Figure 8.27. *Definition of function f*

operation E					
32	1	2	3	4	5
4	5	6	7	8	9
8	9	10	11	12	13
12	13	14	15	16	17
16	17	18	19	20	21
20	21	22	23	24	25
24	25	26	27	28	29
28	29	30	31	32	1

operation P			
16	7	20	21
29	12	28	17
1	15	23	26
5	18	31	10
2	8	24	14
32	27	3	9
19	13	30	6
22	11	4	25

Figure 8.28. *Operations E and P*

The first bit of E(R) corresponds to bit 32, the second bit corresponds to bit 1, etc. As for the P permutation, the first bit corresponds to bit 16, the second bit to bit 7, etc.

To illustrate the operation of the S-boxes:

– Take an input sextet of 101100 for S1.

– Separate the first bit (1) and the last bit (0).

– Concatenate them to get 10 (equivalent to 2 in decimal).

– The remaining four bits, 0110, represent the decimal number 6.

S1

14	4	13	1	2	15	11	8	3	10	6	12	5	9	0	7
0	15	7	4	14	2	13	1	10	6	12	11	9	5	3	8
4	1	14	8	13	6	2	11	15	12	9	7	3	10	5	0
15	12	8	2	4	9	1	7	5	11	3	14	10	0	6	13

S2

15	1	8	14	6	11	3	4	9	7	2	13	12	0	5	10
3	13	4	7	15	2	8	14	12	0	1	10	6	9	11	5
0	14	7	11	10	4	13	1	5	8	12	6	9	3	2	15
13	8	10	1	3	15	4	2	11	6	7	12	0	5	14	9

S3

10	0	9	14	6	3	15	5	1	13	12	7	11	4	2	8
13	7	0	9	3	4	6	10	2	8	5	14	12	11	15	1
13	6	4	9	8	15	3	0	11	1	2	12	5	10	14	7
1	10	13	0	6	9	8	7	4	15	14	3	11	5	2	12

S4

7	13	14	3	0	6	9	10	1	2	8	5	11	12	4	15
13	8	11	5	6	15	0	3	4	7	2	12	1	10	14	9
10	6	9	0	12	11	7	13	15	1	3	14	5	2	8	4
3	15	0	6	10	1	13	8	9	4	5	11	12	7	2	14

S5

2	12	4	1	7	10	11	6	8	5	3	15	13	0	14	9
14	11	2	12	4	7	13	1	5	0	15	10	3	9	8	6
4	2	1	11	10	13	7	8	15	9	12	5	6	3	0	14
11	8	12	7	1	14	2	13	6	15	0	9	10	4	5	3

S6

12	1	10	15	9	2	6	8	0	13	3	4	14	7	5	11
10	15	4	2	7	12	9	5	6	1	13	14	0	11	3	8
9	14	15	5	2	8	12	3	7	0	4	10	1	13	11	6
4	3	2	12	9	5	15	10	1	14	1	7	6	0	8	13

S7

4	11	2	14	15	0	8	13	3	12	9	7	5	10	6	1
13	0	11	7	4	9	1	10	14	3	5	12	2	15	0	6
1	4	11	13	12	3	7	14	10	15	6	8	0	5	9	2
6	11	13	8	1	4	10	7	9	5	0	15	14	2	3	12

S8

13	2	8	4	6	15	11	1	10	9	3	14	5	0	12	7
1	15	13	8	10	3	7	4	12	5	6	11	0	14	9	2
7	11	4	1	9	12	14	2	0	6	10	13	15	3	5	8
2	1	14	7	4	10	8	13	15	12	9	0	3	5	6	11

Figure 8.29. *Operations S1–S8*

Referring to the intersection of row 2 and column 6 of matrix S1 (rows and columns being numbered starting from 0), the value obtained is 2, represented as 0010 in 4 bits.

Consequently, the sextet 101100 transforms into 0010, as shown in Figure 8.30.

101100 ⇨ 1 0110 0 ⇨ 10 0110 ⇨ 2 6 ⇨ S1(2, 6) = 2 ⇨ 0010

Figure 8.30. *Transformation of sextet 101100*

The keys are generated on 48 bit keys from the starting key K by following the process described below, involving operations PC-1 (initial permutation) and PC-2 (final permutation) according to the schema in Figure 8.31 (known as Permuted Choice: PC).

Detailed definitions of PC-1 and PC-2 are provided in Figure 8.32.

– For PC-1: the bits of the key (KEY) are numbered from 1 to 64. The bits of C0 correspond to bits 57, 49, 41, 33, … of KEY, while the bits of D0 correspond to bits 63, 55, 47, 39, … of KEY.

– For PC-2: the first bit of each Ki key is the 14th bit of CiDi, and the second bit of Ki is the 17th bit of CiDi.

– For left shifts: the shifts are circular, meaning the outgoing bit on the left is reintroduced to the right. The number of shifts varies for each key, as shown in the table in Figure 8.33.

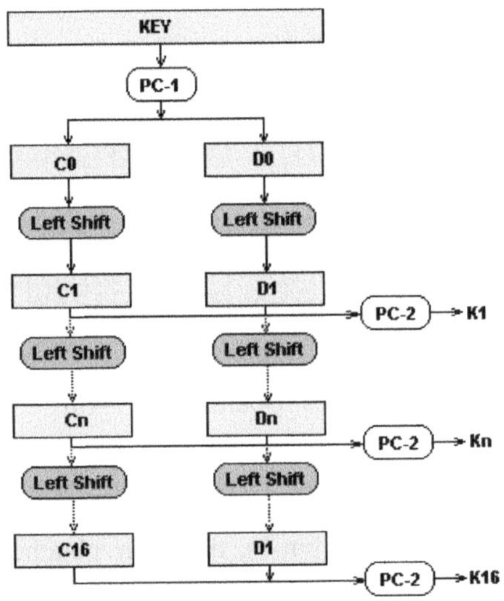

Figure 8.31. *PC-1 and PC-2*

PC-1

57	49	41	33	17	9
1	58	50	42	26	18
10	2	59	51	35	27
19	11	3	60	44	36
63	55	47	39	23	15
7	62	54	46	30	22
14	6	61	53	37	29
21	13	5	28	12	4

PC-2

14	17	11	24	1	5
3	28	15	6	21	10
23	19	12	4	26	8
16	7	27	20	13	2
41	52	31	37	47	55
30	40	51	45	33	48
44	49	39	56	34	53
46	42	50	36	29	32

Figure 8.32. *Definition of PC-1 and PC-2*

key number	1	2	3	4	5	6	7	8	9	10	11	12	13	14	15	16
number of offsets	1	1	2	2	2	2	2	2	1	2	2	2	2	2	2	1

Figure 8.33. *Number of offsets based on key number*

Criticism of DES includes its key length, which is 64 bits (effectively 56 bits due to 8 bits being used for parity). This relatively short length allows attackers to potentially guess the key by trying successive combinations of the 56-bit. While this process may take time, advancements in computer performance make it increasingly feasible. For instance, a demonstration by John Gilmore in January 1999 showed that it was possible to find a DES key in 22 hours and 15 minutes using a machine costing $250,000. Due to these vulnerabilities, DES is being phased out in favor of other systems such as IDEA (International Data Encryption Algorithm), which use 64-bit blocks and 128-bit keys, offering significantly enhanced security.

8.6. RSA – public key system

The principle of RSA encryption relies on a double key (or dual-key) system: an encryption key, known as the public key and shared openly, and a decryption key, kept confidential.

Consider the scenario of two correspondents, Alice and Bob, who wish to exchange messages securely (that cannot be deciphered by a hacker). Alice possesses a key pair consisting of a public key and a private key. She shares her public key with Bob while retaining her private key.

Bob receives Alice's public key and uses it to encrypt his message. He then sends the encrypted message to Alice. Alice, possessing the corresponding private key, decrypts the message.

The encryption system, based on asymmetric cryptography, was developed by Ron Rivest, Adi Shamir and Leonard Adleman in 1977.

To understand the method used, consider the following facts:

1) There exists a quick algorithm for determining whether a large integer is prime. This algorithm is based on Fermat's little theorem, which states that if p is a prime number, there exists a number n less than p ($n < p$) such that $n^{p-1} \equiv 1[p]$. For a number with 130 digits, this verification can be performed in a few minutes on a modern machine.

2) Conversely, there is no efficient algorithm for determining the prime factors of a large non-prime integer. For instance, factoring a product of two prime numbers, each with 63 digits, would require an estimated calculation time of approximately 40,000,000,000,000,000 years on a contemporary computer (as of 1999).

The RSA encryption method involves the following steps:

a) Randomly choosing two distinct large prime numbers, p and q, each of about 100 digits. Calculating the product $r = p \times q$.

b) Randomly choosing a large integer e that is coprime with $(p-1) \times (q-1)$. This integer e becomes the encryption key.

c) Determining the decryption key d such that $d.e \equiv 1\ [(p-1)(q-1)]$.

d) Making the values of r (the modulus) and e (the public encryption exponent) public, but keeping the value of d (the private decryption exponent) secret.

e) Encrypting a message P by calculating $C \equiv P^e[r]$.

f) Decrypting an encrypted message C by calculating $P \equiv C^d[r]$.

EXAMPLE 8.7.– (deliberately simple) $p = 3$, $q = 5$, $r = 15$

$(p - 1).(q - 1) = 8$, $e = 11$ (prime number greater than p and q)

$d.11 \equiv 1\ [8]$, hence $d = 3$ by performing the “modulo” division

Publication of $r = 15$ and $e = 11$

$P = 13$; encryption: $C \equiv 13^{11}\ [15] \equiv 1\ 792\ 160\ 394\ 037\ [15] = 7$

Decryption: $P \equiv 7^3\ [15] \equiv 343\ [15] = 13$

An interesting application of the public key system is *message signing*. Consider the example where Alice (A) sends an RSA encrypted message to Bob (B).

– CPA: public key of A for encryption (used by others to encrypt messages to B).

– CPB: public key of B for encryption (used by others to encrypt messages to B).

– CSA: secret key of A for decryption (used by A to decrypt messages).

– CSB: secret key of B for decryption (used by B to decrypt messages).

When sending a message P from A to B, it is important that B can verify the authenticity of the sender, ensuring that the message truly originates from A. Here is how this verification process works:

1) Instead of directly encrypting P with CPB (public key of B), A first applies her secret key CSA to P, and then encrypts the result with CPB: C = CPB(CSA(P)).

2) B receives the encrypted message C. He then applies his secret key CSB to C, followed by applying the public key CPA of A to verify the authenticity: CPA(CSB(C)) = CPA(CSB(CPB(CSA(P)))) = CPA(CSA(P)) = P.

Therefore, B can be confirmed that the message originated from A (see Figure 8.34).

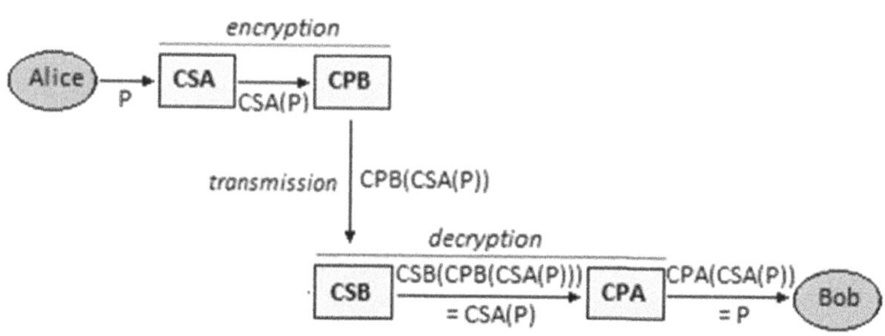

Figure 8.34. *Message signature*

8.7. Cryptography with elliptic curves

Traditionally, when Alice (A) and Bob (B) wish to communicate securely over a telecommunications network, they use a system involving public and private keys.

In this system, there exists a public key that is accessible to all users on the network, and also a private key known only by Alice and Bob.

The public key is used to encrypt messages before transmission over the network. If an interceptor were to capture the message, they would be unable to decrypt it without the corresponding private key.

Conversely, the private key is used by Alice and Bob to decrypt the received messages. Only Alice and Bob possess this key and can therefore read the messages intended for them.

In their setup:

– Alice and Bob agree upon an "abstract" elliptic curve Γ:

$$y^2 \equiv x^3 + ax + b\ [p],$$

where x and y are integers modulo p, and p being a prime number.

– They select a specific point $\mathrm{P}(x_0, y_0)$ that belongs to this curve Γ.

The elliptic curve Γ and the point P constitute their public key, which can be potentially be intercepted by a spy.

To establish a common secret key, Alice chooses an integer n_A and Bob chooses an integer n_B. However, they do not communicate these integers to each other directly to protect them against interception.

Alice sends Bob the point $(n_A \cdot \mathrm{P})$, where P is the agreed-upon point on the curve Γ. Similarly, Bob sends Alice point $(n_B \cdot \mathrm{P})$. Since $n_B(n_A.\mathrm{P}) = n_A(n_B.\mathrm{P}) = (n_A n_B).P$, Alice and Bob can independently calculate the common secret key $(n_A \cdot nB) \cdot \mathrm{P}$. This shared secret key is also a point on the elliptic curve Γ.

When Alice wants to transmit a "point" message M, which can be represented by a point on the elliptic curve, she sends to Bob the point $M + (n_A \cdot n_B) \cdot \mathrm{P}$. Bob can then subtract $(n_A \cdot n_B) \cdot \mathrm{P}$ from the message received:

$$M + (n_A n_B)\mathrm{P} - n_B(n_A \mathrm{P}) = M.$$

Thus, Bob can decrypt the message.

A spy (Eve) can intercept the points $n_A\mathrm{P}$ and $n_B P$, but it is difficult for her to deduce the common secret key $(n_A n_B)P$ needed for decryption, because she knows neither n_A nor n_B. This is due to the difficulty of solving the *discrete logarithm problem*, which involves finding k in the equation $P' = kP$.

Although an algorithm for solving this problem is known, it becomes impractical when the numbers used are very large. Therefore, it is better to choose a large p to strengthen system security.

8.8. Quantum cryptography

To decrypt messages encrypted with advanced encryption methods such as RSA, very powerful computers are required. Even with today's computers, based on the Von Neumann architecture, deciphering encrypted messages remains a complex challenge. The Japanese supercomputer Fugaku considered the most powerful computer in the world as of 2021 (see: https://en.wikipedia.org), displays an impressive performance of 442 PetaFlops. This means it can perform 442.10^{15} floating point operations per second (peta = 1 million trillion). Although this computing power is extremely high, it alone would not be sufficient to quickly decrypt messages encrypted using methods such as RSA.

The new generation of computers will undoubtedly be that of quantum computers. In a "classic" computer, calculations are done on bits (0 or 1). On the contrary, the quantum computer uses qubits, which are quantum elements capable of existing in a superposition of states 0 and 1. Consequently, the quantum computer, which processes qubits and therefore simultaneously handles states 0 and 1, offers potentially much higher computing power than conventional computers. The concept of a quantum computer is not new; it was first formulated by David Deutsch in 1985. Since then, research and development in this area has progressed significantly, with major advances in manipulating qubits, correcting quantum errors, and making working prototypes of quantum computers.

Consider the following example (S. Singh) to illustrate the concept. The goal is to find a number x such that x^2 and x^3 together contain all the integers from 0 to 9. Imagine taking the numbers from 0 to 127 as the search range. A classical computer, working bitwise, must examine each number from 0000000 (0 in decimal) to 1111111 (127 in decimal) to see if it meets the search criterion. It will therefore have to perform 128 operations to explore all the numbers, which will take a "certain" time. The quantum computer, on the contrary, will examine all 128 possibilities in a single operation and give the answer: 69 (since $69^2 = 4761$ and $69^3 = 328509$, and all the numbers 0, 1, 2, 3, 4, 5, 6, 7, 8, 9 are present).

In the case of decrypting a message encrypted with RSA, finding the prime factors of a large number is an essential step. For a classical computer, it would take about 10^{200} operations to find the prime factors of a 10,000-bit number. In contrast, a quantum computer could perform this task in just 10^{10} operations.

In 1994, Peter Shor proposed a revolutionary algorithm that allows a quantum computer to solve the factorization problem in polynomial time. This means that a quantum computer could effectively decipher messages encrypted with RSA by finding the prime factors of the modulo used. Similarly, in 1996, Lov Grover proposed a quantum algorithm for searching for a particular element in an unsorted list. This opens up the possibility of quickly finding a message encrypted with DES by searching the space of possible keys. It is important to note that although these theoretical advances were made several decades ago, the practical realization of a quantum computer capable of running these algorithms remains a major technological challenge. Current quantum computers have limitations in their ability to manipulate and correct qubits, and they are still far from reaching a scale large enough to compete with classical computers in real-world applications.

With the advent of the *quantum computer*, the use of *quantum cryptography* will become essential to ensure secure communications. The foundations of quantum cryptography date back to the 1960s, where innovative concepts were introduced, such as quantum currency proposed by Simon Wiesner. Quantum currency is a concrete example of the application of quantum cryptography. It is based on the use of photon polarization, which makes it possible to create tamper-proof banknotes (see Figure 8.35). The idea is that each quantum currency banknote is associated with a specific polarization of photons, which can be detected and verified by quantum protocols.

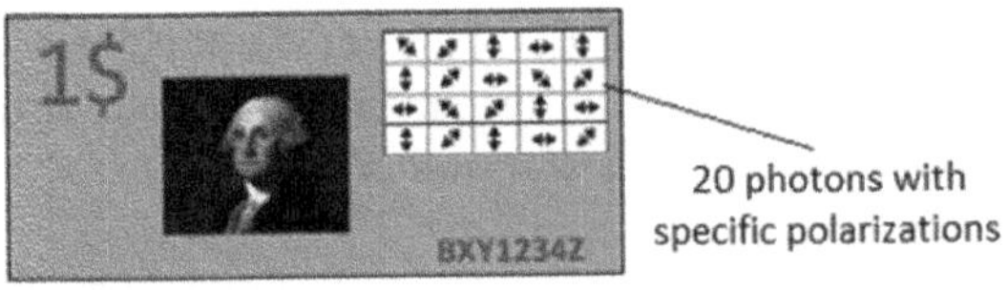

Figure 8.35. *$1 banknote according to Wiesner*

A photon, as a particle of light, possesses both a corpuscular and a wave nature. From the wave point of view, the propagation of a photon is associated with the propagation of an electromagnetic wave. The electric and magnetic fields of this wave lie in a plane perpendicular to the direction of propagation (as shown in Figure 8.36).

The polarization of a photon refers to the orientation of the electric field **E** (or magnetic field **B**, since the two are perpendicular to each other) during its propagation. A photon is said to be polarized when the direction of its electric (or magnetic) field remains constant throughout its path. On the contrary, a non-polarized photon would have a field direction that varies during propagation.

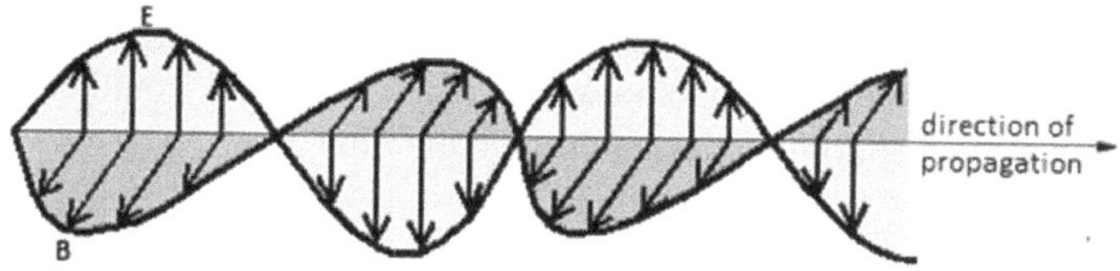

Figure 8.36. *Polarized photon*

Now imagine photons polarized in four different orientations, for example, **E** at 0°, 45°, 90° and 135°. Using a filter in the form of a slit placed at 0°, the results shown in Figure 8.37 are obtained, depending on the initial polarization of the photons. The fraction of photons that pass through the filter is proportional to $cos^2(\theta)$, where θ is the angle between the polarization of the incident photon and the direction of the slit.

If the filter indicates that nothing passes, this means that the incident photons are polarized at 90°. However, in other cases, the filter does not allow determination of the exact polarization of the incident photons (as shown in Figure 8.37).

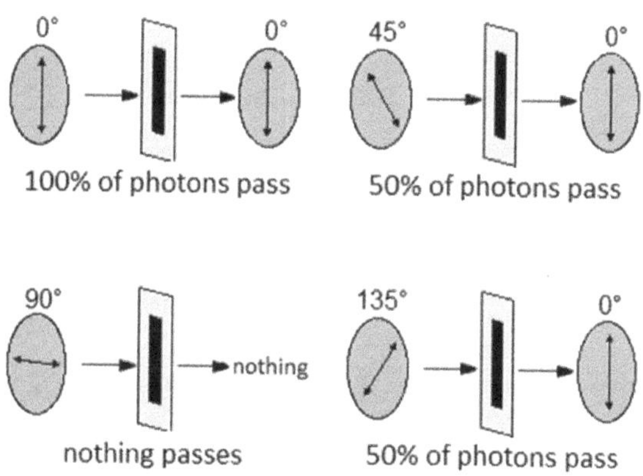

Figure 8.37. *Different scenarios*

The banknote proposed by Wiesner included 20 "photon traps" designed to preserve the polarization of the photon placed there. Each photon was polarized in one of four different orientations, giving a total of 4^{20} possibilities, or about 10^{12} possibilities. Thus, there were numerous variations of different banknotes. The bank was aware of the correspondence between the serial number of the note, for example, BXY1234Z, and the polarization configuration of the photons it contained. By using specific filters adjusted for each photon, the bank could determine whether a banknote was counterfeit. For example, if for photon n, the filter was set to 0° and the predicted polarization of the photon was also at 0°, the bank could conclude that the banknote was fake if nothing passed through the filter.

Charles Bennett (who knew Wiesner) and Gilles Brassard took Wiesner's idea and developed a method of quantum cryptography known as BB84. This method exploits the transmission of polarized photons to ensure communication security. When transmitting polarized photons, two coding schemas can be used:

– The " + " diagram, where the photons are polarized at 0°(1) and at 90°(0) (see Figure 8.38).

Figure 8.38. " + " *schema*

– The " × " diagram, where photons are polarized at 45°(1) and 135°(0) (see Figure 8.39).

Figure 8.39. " × " *schema*

Suppose Alice wants to send the message "10010111" to Bob using specific coding schemes for each bit (see Figure 8.40).

message to send	1	0	0	1	0	1	1	1
used diagrams	+	x	+	x	x	x	+	+
transmitted message	↕	⤡	↔	⤢	⤡	⤢	↕	↕

Figure 8.40. *Coding schemas*

Bob has a filter that he can orient in one of four possible positions (0°, 45°, 90°, 135°). When each bit of the message arrives, Bob randomly chooses a polarization for his filter.

When a photon arrives, three situations are possible:

1) Alice and Bob chose the same polarization. In this case, the sent bit and the received bit are the same, and Bob can determine the value of the bit unambiguously.

2) Alice and Bob chose perpendicular polarizations. In this situation, Bob does not detect any photons. He deduces that the photon sent is polarized perpendicular to the direction of his filter.

3) Alice and Bob chose polarizations that are 45° apart (in either direction). In this case, Bob does receive a photon, but he cannot determine the original polarization with certainty.

Imagine that Bob has two detectors (filters), one following the " + " schema and the other following the " × " schema. With these two detectors, he can receive the correct message. However, a photon is indivisible; it cannot be split in two to pass through each detector separately.

If Alice sends Bob the sequence of schemas used (the secret key) for each sent photon polarization, Bob can then receive the message in clear text. However, a spy (Eve, in the literature) who knows the sequence of patterns used by Alice for each bit will also be able to intercept it. The entire issue thus revolves around the secure transmission of the secret key, a persistent challenge in the key distribution. Of course, methods such as RSA could be used, but with the advent of quantum computers, RSA will become obsolete.

Bennett and Brassard's method is based on three preparatory steps to secure the transmission of quantum keys:

Step 1: Alice sends a random sequence of 0s and 1s using photons with arbitrary patterns.

Step 2: Bob randomly applies filters following the "+" and " × " schemas, which can result in correct or incorrect interpretations. The possible outcomes are detailed in the table in Figure 8.41.

Step 3: after transmitting a certain number of bits, Alice communicates with Bob over an open channel, typically by telephone, to reveal the sequence of patterns she used. However, she does not reveal the actual bits transmitted (see Figure 8.42). This step allows Bob to verify the accuracy of his interpretations.

The result is shown in Figure 8.43.

Alice diagrams	sent bits	Alice polarizations	Bob detector	detected polarizations	received bits	diagnosis
+	1	↕	+	↕	1	YES
			x	⤢	1	YES
				⤡	0	NO
	0	↔	+	↔	0	YES
			x	⤢	1	NO
				⤡	0	YES
x	1	⤢	+	↕	1	YES
				↔	0	NO
			x	⤢	1	YES
	0	⤡	+	↕	1	NO
				↔	0	YES
			x	⤡	0	YES

Figure 8.41. *Table of possible options*

Alice diagrams	sent bits	Alice polarizations	Bob detector	detected polarizations	received bits	diagnosis
[+]	1	↕	[+]	↕	1	[YES]
			x	⤢	1	YES
				⤡	0	NO
	0	↔	[+]	↔	0	[YES]
			x	⤢	1	NO
				⤡	0	YES
[x]	1	⤢	+	↕	1	YES
				↔	0	NO
			[x]	⤢	1	[YES]
	0	⤡	+	↕	1	NO
				↔	0	YES
			[x]	⤡	0	[YES]

Figure 8.42. *Various situations*

The last line in Figure 8.43 provides the list of "good" bits, which in this case is 10011110. The rest of the bits are rejected. This list, which has not been openly communicated, forms the key for the next message. A spy (Eve) attempting to intercept the list of bits sent by Alice would mimic Bob's actions but would not achieve the same outcome. Due to using different filters, the spy cannot determine which bits were discarded and therefore cannot obtain the key.

Alice sent bits	1	0	0	1	0	1	1	1	0	1	0	1	0	0	1	1	0
Alice diagram	+	x	x	x	+	x	+	+	+	x	x	+	+	x	+	+	x
Bob diagram	+	x	x	+	x	+	+	x	x	x	+	+	x	+	x	+	x
Bob detection	1	0	0	1	1	0	1	1	1	1	1	1	0	1	0	1	0
retained bits	1	0	0				1			1		1				1	0

Figure 8.43. *Synthesis*

Basically, interception by a spy involves examining the transmitted photon, drawing a conclusion and then sending it back to Bob. However, a photon that has been observed is inevitably altered!

Let us therefore assume the sequence of bits intended to establish the key proceeds without interference. When the actual message is subsequently sent, if a spy attempts to intercept it, they will alter the polarization of the photon. Bob will detect this alteration as an error. This would inform Bob that the transmission has been intercepted or “listened to”.

To summarize the approach of transmitting a message with polarized photons, it involves five steps. The first three steps are for defining the key. The fourth step involves checking for any interception by a spy; in such cases, error statistics are performed, and if the error rate is less than 10%, the “line” is considered secure. The fifth step is to transmit the message with the key, contingent upon the fourth step indicating no risk of interception.

Real-world transmission experiments are still underway to improve communication range. While exact details are often confidential, it is likely that transmission distances exceeding 100 km have been achieved today. One of the main primary challenges is preventing alteration of the electromagnetic wave, including photon polarization, which can result from collisions with molecules and other interferences. The use of an optical fibers presents a solution to avoid these issues. Additionally, recent research has explored advancements using entangled photons. However, further details on these developments are currently unavailable.

To return to the quantum formulation, a state of a system is represented by a vector in a Hilbert space, which is a vector space. This vector can be written as a linear combination of basis vectors. Taking the states $|\updownarrow\rangle(1)$ and $|\leftrightarrow\rangle(2)$ of the "+" schema as basic states, their vector representation is shown in Figure 8.44, where α and β are complex coefficients describing the amplitude and phase of each state component.

$$|\psi\rangle = \alpha|\updownarrow\rangle + \beta|\leftrightarrow\rangle$$

Figure 8.44. *Vector representation*

Using Dirac's notation $|.\rangle$ for a vector (denoted as "ket"), a vector in space represents a qubit. Any vector (and therefore qubit) can be expressed as a superposition of basic states.

The other states in the "× " schema can be expressed in terms of the basis vectors as shown in Figure 8.45.

$$|\swarrow\!\!\nearrow\rangle = \frac{1}{\sqrt{2}}[|\updownarrow\rangle + |\leftrightarrow\rangle] \quad \text{and} \quad |\nwarrow\!\!\searrow\rangle = \frac{1}{\sqrt{2}}[|\updownarrow\rangle - |\leftrightarrow\rangle]$$

Figure 8.45. *Other schema states*

The squared coefficients represent the probabilities of the occurrence of a state. The factor $\frac{1}{2}$ for the basic states indicates that 50% of the diagonally polarized photons pass through the vertical or horizontal filters.

8.9. Appendix: The Crow and the Fox

A poetic English version by Edward Marsh of Jean de La Fontaine's French original.

A Crow sat perched upon an oak,
And in his beak he held a cheese.
A Fox snuffed up the savory breeze,
And thus in honeyed accent spoke:
"O Prince of Crows, such grace of mien
Has never in these parts been seen.
If but your song be half as good,
You are the Phoenix of the wood!"
The Crow, beside himself with pleasure,
And eager to display his voice,
Opened his beak, and dropped his treasure.
The Fox was on it in a trice.
"Learn, sir," said he, "that flatterers live
On those who swallow what they say.
A cheese is not too much to give
For such a piece of sound advice."
The Crow, ashamed to have been such easy prey
Swore, but too late, he shouldn't catch him twice.

9

Probabilities

CONCEPTS COVERED IN THIS CHAPTER.–

Fortune tellers make predictions, while weather services make forecasts. Both try to anticipate the future with varying degrees of success, as chance imposes its own law.

The evolution of populations, stock prices, climatology and other phenomena are analyzed using models for their forecasts. These mathematical models generally include many variables, making the use of the computer essential for forecast calculations. These models assume the occurrence of events with certain probabilities derived from other probabilities. Therefore, the combined calculation of these different probabilities proves to be quite useful.

The purpose of this chapter is to review the essential foundations of probability, first in a practical way and then with more mathematical formalism. The two sources of calculating probabilities, namely enumeration and statistics, will be discussed.

All these concepts will then be applied to the evolution of processes represented by graphs of transitions between states. In particular, the properties of Markov chains will be exposed.

References: [BOI 68, COC 19, ENG 76, MAS 76, POC 71].

9.1. Chance

The Merriam Webster dictionary provides several meanings to the word "chance" including:

Mathematics for Digital Science 1,
by Gérard-Michel COCHARD and Mhand HIFI.

1) The fortuitous or incalculable element in existence.

2) Something that happens unpredictably without discernible human intention or observable cause.

The first definition is actually twofold:

a) It introduces the notion of causality according, suggesting that an inexplicable event would nevertheless be due to an unknown cause. Knowing the cause and the mechanisms that result from it could (theoretically) explain the inexplicable. For instance, if a pot of flowers falls from a window due to a gust of wind and hits a passerby on the head, this event might be called fortuitous; however, it is entirely explainable. An observer witnessing the scene can see that the accident is inevitable. If the problem were not so complex, the laws of mechanics could explain the accident. Therefore, "inexplicable" does not mean coincidence.

b) It also suggests that the cause is a "power". Could it be God? Consider the famous anecdote of Napoleon asking Pierre-Simon de Laplace why he does not mention the existence of God in his work *Mécanique céleste (Celestial Mechanics)*. Laplace, fiercely deterministic, replied, "*Sire, I didn't need that assumption*."

The second definition brings up the fundamental question of the chapter: can random events be predicted? The question lies at the origin of probability theory. It emerged from games of chance. Chevalier de Méré's bet serves as a representative example from the time when probability theory was developed by Blaise Pascal and Pierre de Fermat. Chevalier de Méré noted that when rolling a die four times, there was a significant chance of getting 6. He deduced that it must be even more likely to get a double 6 by rolling 2 dice 24 times. Pascal showed that his reasoning was incorrect. Indeed, in the first case, the probability of obtaining a 6 is 0.517, while in the second case, the probability of obtaining a double 6 is only 0.491. The example highlights the field of calculating probabilities, which defines the concept of forecasting: it does not provide a certain result, but the chances of obtaining a given result.

9.2. Counting and probabilities

In the study of probability, it is common to refer to well-known examples such as the throwing of a 6-sided die, the drawing of colored balls from a jar, or the drawing of cards from a deck of cards. These examples are easy to imagine and provide a better understanding of the basic concepts. However, it is important to highlight that the concept of probability is not limited to chance games. Probabilities are also used in many areas of daily life, such as meteorology, economics and public health. Thus, in addition to classic examples, more concrete and varied examples will be presented to illustrate the scope of probabilities.

EXAMPLE 9.1.– Consider the roll of a six-sided die and examine the chances of getting a 5 as the result. Assuming that the die is perfect (i.e. fair, with each face having an equal probability of appearing), the chances of rolling a 5 are the same as those of obtaining any other number between 1 and 6. Each possible result is therefore equiprobable.

Since there are six possible outcomes and only one desired outcome, there is a one-in-six chance of getting a 5. This can be expressed as a ratio, where the probability of getting a given outcome is equal to the number of desired results (here 1) divided by the number of possible results (here 6). Thus, the probability of rolling a 5 is as follows: $p(outcome = 5) = \frac{1}{6}$.

Furthermore, if interested in outcome a, anything that is not a is called a contrary outcome. Therefore, the probability of obtaining either the outcome a or the contrary outcome is 1, which can be expressed by the formula: $p(outcome = a) + p(outcome \neq a) = 1$.

Finally, note that the sum of the probabilities of all possible outcomes is equal to 1 (since there are 6 outcomes and each has a probability of $\frac{1}{6}$: $(6 \times \frac{1}{6})$). It is certain that one of the possible outcomes will occur.

EXAMPLE 9.2.– Consider a jar filled with identical balls of different colors. Suppose there are 3 black balls and 2 white balls. If a ball is drawn at random, what are the chances of drawing a black ball?

This problem is more complex than previous examples. However, the probability of drawing a black ball is higher compared to drawing a white ball, due to the greater number of black balls. Assuming each ball is distinct, there are 5 possible results, 3 of which correspond to drawing a black ball. Therefore, the probability of drawing a black ball is given by the ratio $\frac{3}{5}$, so $p(\text{drawing a black ball}) = \frac{3}{5}$.

Similarly, the probability of drawing a white ball can be expressed as $\frac{2}{5}$: the $p(\text{drawing of a white ball}) = \frac{2}{5}$. This probability is calculated as the ratio of the number of favorable results (here, 3 for black balls and 2 for white balls) to the number of possible results (here, 5 balls).

Note that, unlike the previous example, the probabilities are not equal: drawing a black ball has a probability of $\frac{3}{5}$, while drawing a white ball has a probability of $\frac{2}{5}$. Additionally, the sum of the probabilities for all possible results is $\frac{3}{5}+\frac{2}{5}=1$.

EXAMPLE 9.3.– Consider a card game scenario where 2 cards are drawn from a deck of 52 cards (with 13 cards in each suit). To estimate the chances of drawing 2 aces, start by calculating the total number of possible results.

First, estimate all possible results by designating the two cards as X and Y. For the first card drawn (X), there are 52 possibilities, and for the second card (Y), there are 51 remaining possibilities. Since the roles of X and Y are interchangeable, the total number of results, without counting any result twice, is $52 \times \frac{51}{2} = 1{,}326$ possible results.

There are 4 aces in the deck. For the first ace (X), there are 4 possibilities, and for the second ace (Y), there are 3 possibilities. Since the roles X and Y are interchangeable, the number of expected results is $4 \times \frac{3}{2} = 6$. This can be verified by listing all possible combinations:

$$♣♥, ♣♦, ♣♠, ♥♦, ♥♠, ♦♠$$

Therefore, there are 6 favorable outcomes of 1,326 possible results of drawing 2 aces. This can be expressed as the probability:

$$p(drawing\ of\ 2\ aces) = \frac{6}{1{,}326} = 0.00452\ ...$$

Now consider the case where no aces are obtained in the draw of the two cards. For the first drawn, there are 48 possibilities, and for the second card, there are 47 remaining possibilities. Since the order of drawing does not matter, we use the combination formula, giving:

$$\text{Number of favorable outcomes} = 48 \times \frac{47}{2} = 1{,}128.$$

So, there are 1,128 possible results where neither card is an ace. The total number of possible results remains 1,326, as previously calculated. Thus, the probability of drawing no aces in 2 cards is:

$$p\text{ (drawing 2 cards which are not aces)} = \frac{1{,}128}{1{,}326} = 0.85006.$$

There is a third possible result: drawing exactly one ace among the two cards. If the first card drawn X is an ace, there are 4 possibilities for X, and for the second card Y, there are 51 - 3 = 48 possibilities (since Y cannot be an ace). Since X and Y are not interchangeable in this case, the number of such outcomes is 4 x 48 = 192. Thus, the probability of drawing exactly one ace in 2 cards is:

$$p\text{ (drawing 2 cards with exactly one ace)} = \frac{192}{1{,}326} = 0.14479.$$

The three possible results do not have the same probability. They are not equally probable. Note that 6 + 1,128 + 192 = 1,326, so that the sum of the probabilities for all possible expected results is 1.

In summary, the probability calculation method illustrated in the three previous examples involves:

1) Determining the number of possible results: identify the total number of possible outcomes/results (also referred to as possible cases) and the number of favorable outcomes (also referred to as favorable cases). The probability is then calculated by dividing the number of favorable outcomes by the number of possible outcomes. This approach follows Pierre-Simon de Laplace's definition of probability, which assumes that all possible outcomes are equally likely and can be counted.

2) Probability constraints: the number of favorable outcomes can never exceed the number of possible outcomes, which ensures that the probability is always a positive number between 0 and 1, inclusive.

3) Sum of probabilities: the sum of the probabilities of all possible outcomes is equal to 1. Consequently, the sum of the probability of a specific outcome and the probability of its complement (i.e. the contrary outcome) also equals 1.

For now, considerations are based on the previous examples. A more precise, mathematical definition of probability and its properties will be provided later. In this case, the focus is on counting methods, which are necessary for determining probabilities. The concepts of permutation, arrangement and combination discussed in Chapter 3 can be employed to count favorable or possible events.

Recall the corresponding results:

- The number of permutations of n objects: $P_n = n!$
- The number of arrangements of k objects taken from n objects: $A_n^k = \frac{n!}{(n-k)!}$.
- The number of combinations of k objects taken from n objects: $C_n^k = \frac{n!}{k!(n-k)!}$.

EXAMPLE 9.4.– In a race with 20 horses, to determine the number of trifectas in order (i.e. classifying three horses as first, second and third), use:

$$A_{20}^3 = \frac{20!}{17!} = 6{,}840.$$

There is only one favorable case (the winning arrangement), so the probability is:

$$p = \frac{1}{6{,}840} = 0.000146.$$

For a trifecta in "disorder", where the order of arrival does not matter, the numbers of trifectas that can be formed from 20 horses is given by the number of combinations of 3 elements taken from 20, i.e.:

$$C_{20}^3 = \frac{20!}{3!17!} = 1{,}140.$$

There is only one favorable case (the winning combination), so the probability is:

$$p = \frac{1}{1{,}140} = 0.000877.$$

EXAMPLE 9.5.– A jar contains 3 white balls and 2 black balls. The only difference between the balls is their color. Drawing one ball allows for the calculation of the probability of drawing a black ball or a white ball.

The number of possible outcomes is 5 (given by $C_5^1 = 5$). There are 3 favorable outcomes for drawing a white ball (since there are $C_3^1 = 3$ white balls in the jar) and 2 favorable outcomes for drawing a black ball (since there are 2 black balls in the jar). Thus, the probability of drawing a white ball is:

$$p(\text{white}) = \frac{\text{number of favorable cases}}{\text{number of possible cases}} = \frac{3}{5}$$

and that the probability of drawing a black ball is:

$$p(\text{black}) = \frac{\text{number of favorable cases}}{\text{number of possible cases}} = \frac{2}{5}$$

Verification shows that $p(\text{white}) + p(\text{black}) = 1$. This confirms that the sum of the probabilities for all possible outcomes equals 1, as only one ball can be drawn from the jar.

EXAMPLE 9.6.– Consider a deck of 52 cards (13 cards of each color/suit). Calculate the following probabilities for drawing 2 cards: drawing 2 aces, drawing 2 cards that are not aces and drawing 1 ace and 1 card that is not an ace.

The number of possible outcomes is given by: $C_{52}^{2} = \frac{52!}{50!2!} = 1{,}326$.

For drawing 2 aces, the number of favorable outcomes is: $C_{4}^{2} = \frac{4!}{2!2!} = 6$. Thus, the probability of drawing 2 aces is: $p(ace, ace) = \frac{6}{1{,}326}$.

For drawing 2 cards that are not aces, the number of favorable outcomes is: $C_{48}^{2} = \frac{48!}{46!2!} = 1{,}128$. Thus, the probability of drawing 2 cards that are not aces is: $p(noace, noace) = \frac{1{,}128}{1{,}326}$.

For drawing 1 ace and 1 card that is not an ace, the number of favorable outcomes is: $C_{4}^{1}.C_{48}^{1} = \frac{4!48!}{3!47!1!1!} = 192$. Thus, the probability of drawing 1 ace and 1 card that is not an ace is: $p(ace, noace) = \frac{192}{1{,}326}$.

The approach, based on empirical examples, now requires a more formal definition of the concept of probability.

9.3. Events and probabilities

Consider a non-empty set E with arbitrary elements. The *parts* (subsets) of E are the sets that can be formed by selecting some elements of E and leaving others. Both the empty part ∅ (containing no elements) and E itself (containing all elements of E) are also parts (subsets) of E.

EXAMPLE 9.7.– Let the set E be a, b, c, d, where a, b, c and d are the elements of E. All parts of E are:

Part with 0 elements: $\emptyset = \{\}$
Parts with 1 element: a, b, c, d
Parts with 2 elements: $a, b, a, c, a, d, b, c, b, d, c, d$
Parts with 3 elements: $a, b, c, a, b, d, a, c, d, b, c, d$

Part with 4 elements: $E = a, b, c, d$

Thus, there are $2^4 = 16$ subsets in total, including all of the above categories.

The set of all parts of a set E is denoted as $\mathcal{P}(E)$. Note that the number of elements in $\mathcal{P}(E)$ is equal to 2^n if E has n elements.

EXAMPLE 9.8.– In example 9.7, the different parts of E can be considered as combinations of k elements taken from the 4 elements of E. The number of parts of each size is given by the binomial coefficients:

The number of parts with 0 elements is $C_4^0 = 1$.
The number of 1-element parts is $C_4^1 = 4$.
The number of 2-element parts is $C_4^2 = 6$.
The number of 3-element parts is $C_4^3 = 4$.
The number of 4-element parts $C_4^4 = 1$.

The total number of parts is equal to $C_4^0 + C_4^1 + C_4^2 + C_4^3 + C_4^4 = 16$.

This result can also be confirmed using the property of the binomial coefficient that the total number of parts of a set with n elements are 2^n. For $n = 4$, $2^4 = 16$.

Now consider the elements of $\mathcal{P}(E)$ as *events*, and define a map p from $\mathcal{P}(E)$ to $\mathbb{R}$ (the set of real numbers) that satisfies the following axioms:

AXIOMS.–

Axiom 1: $\forall A \in \mathcal{P}(E)\ p(A) \geq 0$
Axiom 2: $p(E) = 1$
Axiom 3: if $A \cap B = \emptyset$, then $p(A \cup B) = p(A) + p(B)$

This map p is called a *probability* measure on $\mathcal{P}(E)$.

A little vocabulary to clarify the terms used:

– E is the certain event.

– $\emptyset$ is the impossible event.

– a, b, c, d are *elementary* events (which cannot be decomposed into simpler events).

– If $A \cap B = \emptyset$, then A and B are said to be *incompatible* (or mutually exclusive) events.

– If $\bar{A}$ is the complement of A, then A and $\bar{A}$ are said to be *complementary* (or *contrary*) events.

EXAMPLE 9.9.– Let us return to example 9.7 and consider the set $E = \{a,b,c,d\}$. The parts (or subsets) of E are: $\emptyset$, $\{a\},\{b\},\{c\},\{d\},\{a,b\}$, $\{a,c\}$, $\{a,d\}$, $\{b,c\}$, $\{b,d\}$, $\{c,d\}$, $\{a,b,c\}$, $\{a,b,d\}$, $\{a,c,d\}$, $\{b,c,d\}, E$.

Elementary events: $\{a\},\{b\},\{c\},\{d\},\emptyset$ are elementary events because they cannot be decomposed into simpler elements.

Incompatible events: Two events are incompatible (or mutually exclusive) if they have no elements in common. The pairs of incompatible events are: $\{a\}$ and $\emptyset$, $\{a\}$ and $\{b\}$, $\{a\}$ and $\{c]$, $\{a\}$ and $\{d\}$, $\{b\}$ and $\{c\}$, $\{b\}$ and $\{d\}$, $\{c\}$ and $\{d\}$, $\{a\}$ and $\{b,c\}$, $\{a\}$ and $\{c,d\}$, $\{a\}$ and $\{b,d\}$, $\{b\}$ and $\{c\}$, $\{b\}$ and $\{d\}$, $\{b\}$ and $\{a,c\}$, $\{b\}$ and $\{a,d\}$, $\{b\}$ and $\{a,c,d\}$, $\{b\}$ and $\emptyset$, $\{c\}$ and $\{d\}$, $\{c\}$ and $\{a,b\}$, $\{c\}$ and $\{a,d\}$, $\{c\}$ and $\{b,d\}$, $\{c\}$ and $\{a,b,d\}$, $\{c\}$ and $\emptyset$, $\{d\}$ and $\{a,b\}$, $\{d\}$ and $\{a,c\}$, $\{d\}$ and $\{b,c\}$, $\{d\}$ and $\{a,b,c\}$, $\{d\}$ and $\emptyset$, $\{a,b\}$ and $\{b,c\}$, $\{a,b\}$ and $\emptyset$, $\{a,c\}$ and $\{b,d\}$, $\{a,c\}$ and $\emptyset$, $\{a,d\}$ and $\{b,c\}$, $\{a,d\}$ and $\emptyset$, $\{a,b,c\}$ and $\emptyset$, $\{a,b,d\}$ and $\emptyset$, $\{a,c,d\}$ and $\emptyset$, $\{b,c,d\}$ and $\emptyset$, E and $\emptyset$

Contrary events: Two events are contrary if one is the complement of the other. The pairs of contrary events are: $\{a\}$ *and* $\{b,c,d\}$, $\{b\}$ *and* $\{a,c,d\}$, $\{c\}$ *and* $\{a,b,d\}$, $\{d\}$ *and* $\{a,b,c\}$, $\{a,b\}$ *and* $\{c,d\}$, $\{a,c\}$ *and* $\{b,d\}$, $\{a,d\}$ *and* $\{b,c\}$, E *and* $\emptyset$.

All these pairs of events A and B satisfy $B = \overline{A}$ (and vice versa).

Contrary and incompatible events: all pairs of events listed in the previous list are both contrary and incompatible. It should be noted that contrary events are always incompatible: if A and B are contrary, then $B = \overline{A}$ and since $A \cap B = \emptyset$, A and B are incompatible. However, the converse is not necessarily true. For example, $\{a\}$ and b, c are incompatible but are not contrary.

EXAMPLE 9.10.– Let us denote drawing a card x from a deck of cards by $\{x\}$. The elements of E are cards such as {ace of hearts}, {jack of diamonds}, {6 of clubs}, etc.

The event "*drawing a heart*" is defined as:

{hearts} = {ace of hearts} ∪ {2 of hearts} ∪ {3 of hearts} ∪ .... ∪ {10 of hearts} ∪ {jack of hearts} ∪ {queen of hearts} ∪ {king of hearts} = {{ace of hearts}, {2 of hearts}, {3 of hearts}, ..., {10 of hearts), {jack of hearts}, {queen of hearts}, {king of hearts}} and is a part of *E*.

The events "*drawing a heart*" and "*drawing a club*" are two incompatible events because: {heart} ∩ {club} = ∅

The events "*drawing a heart*" and "*drawing a club* or *a diamond* or *a spade*" are complementary events because: $\overline{\{hearts\}}$= {clubs} ∪ {diamonds} ∪ {spades}

Let us now list the properties of probability. From the definition axioms, the following conclusions can be easily drawn, leaving the proofs to the reader:

$$\text{P1: if } A \subset B, \text{then } p(A) < p(B)$$
$$\text{P2: } p(A) \leq 1 \;\; \forall A \in \mathcal{P}(E)$$
$$\text{P3: } p(\bar{A}) = 1 - p(A) \;\; \forall A \in \mathcal{P}(E)$$
$$\text{P4: } p(\emptyset) = 0$$
$$\text{P5: } p(A \cup B) = p(A) + p(B) - p(A \cap B) \quad \forall A, B \in \mathcal{P}(E)$$

EXAMPLE 9.8.– Let us return to example 9.10, where {*x*} denotes drawing a card *x* from a deck of cards. Since the events {"hearts"}, {"clubs"}, {"diamonds"} and {"spades"} are pairwise disjoint, the following holds:

p(draw a heart) = *p*({"heart"}) = 1 – *p*({"club"} ∪ {"diamonds"} ∪ {"spade"}) = 1 – *p*({"club"}) – *p*({"diamonds"}) – *p*({"spades"})

Since drawing any card is certain: *p*(draw any card) = 1

To find the probability of drawing a heart or a 10, use the principle of inclusion and exclusion: *p*(draw a heart or a 10) = *p*({"heart"}) + *p*({"10"}) - *p*({"10"}{"heart"})

Since these events are disjoint:

p({"hearts"}) = *p*({ace of hearts} ∪ {2 of hearts} ∪ {3 of hearts} ∪ … ∪ {10 of hearts} ∪ {jack of hearts} ∪ {queen of hearts} ∪ {king of hearts}) = *p*({ace of hearts}) + *p*({2 of hearts}) + *p* ({3 of hearts}) + … + *p*({10 of hearts}) + *p*({jack of hearts}) }) + *p*({queen of hearts}) + *p*({king of hearts})

Similarly, the probability of drawing a 10 is:

$p(\{\text{“10”}\}) = p(\{\text{10 of hearts}\} \cup \{\text{10 of clubs}\} \cup \{\text{10 of diamonds}\} \cup \{\text{10 of spades}\}) = p(\{\text{10 of hearts}\}) + p(\{\text{10 of clubs}\}) + p(\{\text{10 of diamonds}\}) + p(\{\text{10 of spades}\})$

Thus, $p(\text{draw a heart or a 10}) = p(\{\text{ace of hearts}\}) + p(\{\text{2 of hearts}\}) + p(\{\text{3 of hearts}\}) + \ldots\ldots + p(\{\text{jack of hearts}\}) + p(\{\text{queen of hearts}\}) + p(\{\text{king of hearts}\}) + p(\{\text{10 of hearts}\}) + p(\{\text{10 of clubs}\}) + p(\{\text{10 of diamonds}\}) + p(\{\text{10 of spades}\})$.

In general, the concept of *equiprobability* is useful for calculating probabilities. Elementary events are considered equiprobable if they have the same probability.

EXAMPLE 9.12.– Consider the experiment of rolling a die. Assuming the die is fair, all outcomes are equally probable. Therefore:

$$p(1) = p(2) = p(3) = p(4) = p(5) = p(6) = \frac{1}{6}.$$

EXAMPLE 9.13.– Consider drawing a card from a deck of cards. Assuming the deck is standard and all cards are equally likely, each card has the same probability of being drawn:

$p(\{\text{6 of clubs}\}) = p(\{\text{3 of diamonds}\}) = p(\{\text{jack of hearts}\}) = p(\{\text{queen of spades}\}) = \ldots$

In the case of n elementary equiprobable events X_i, and X_j, where $p(X_i) = p(X_j)$ for all i and j, the following holds:

$$p(X_1) + p(X_2) + \cdots + p(X_n) = p(X_1 \cup X_2 \cup \ldots \cup X_n) = p(E) = 1$$

Therefore:

$$p(X_i) = \frac{1}{n}$$

It follows that for any event $A \in \mathcal{P}(E)$, which can be expressed as $A = X_1 \cup X_2 \cup \ldots \cup X_k$, where X_i are elementary events, the probability is:

$$p(A) = p(X_1) + p(X_2) + \cdots + p(X_k) = \frac{k}{n}$$

This relationship is commonly expressed as follows: "The probability of an event is the ratio of the number of favorable outcomes to the number of possible outcomes."

However, this definition is only valid for equiprobable events. This usual definition has already been used previously. Now, let us consider some examples to illustrate this concept.

EXAMPLE 9.14.– In the case of rolling a die, the elementary events are equiprobable. Therefore: $p(1) = p(2) = p(3) = p(4) = p(5) = p(6) = \frac{1}{6}$.

Thus, the probability of rolling an odd number is:

$$p(\{odd\}) = p(\{1,3,5\}) = p(\{1\}) + p(\{3\}) + p(\{5\}) = \frac{1}{6} + \frac{1}{6} + \frac{1}{6} = \frac{3}{6} = \frac{1}{2}$$

Note that this usual definition of probability as a ratio of favorable outcomes to possible outcomes is valid only for equiprobable events.

EXAMPLE 9.15.– Drawing a card from a deck of cards, where all draws are equally likely (assuming a fair deck), the probability of drawing a heart is calculated as follows:

$p(\{\text{"hearts"}\}) = p(\{\text{ace of hearts}\}) + p(\{\text{King of hearts}\}) + p(\{\text{Queen of hearts}\}) + \ldots + p(\{\text{2 of hearts}\})$

Since each card in the deck has a probability of $\frac{1}{52}$:

$$p(\{\text{"hearts"}\}) = \frac{1}{52} + \frac{1}{52} + \ldots + \frac{1}{52} = \frac{13}{52} = \frac{1}{4} = 0.25$$

A significant consequence of the equiprobability schema is that the sum of the probabilities of all possible outcomes of an experiment equals 1. Let $A, B, \ldots.$ be the incompatible (mutually exclusive) events resulting from the experiment, and let $n(A), n(B), \ldots$ be their respective (cardinal) numbers of outcomes (elements). If $n = n(A) + n(B) + \cdots$ represents the total number of outcomes in the sample space E, then:

$$p(A) + p(B) + \cdots .. = \frac{n(A)}{n} + \frac{n(B)}{n} + \cdots \ldots = \frac{n}{n} = 1$$

The basic principle of equiprobability for elementary events can be applied in two ways:

1) By counting possible cases and favorable cases: this involves counting the number of possible cases and the number of favorable cases. Combinatorial analysis formulas can then be used to calculate the probability of each event.

2) By using the law of large numbers: this law provides an approximate value of the probability through statistical observations. It states that as the number of observations increases, the relative frequency of an event approaches its theoretical probability.

It is important to note that the law of large numbers only provides an estimate of probability. For rare events, a large number of observations are required to obtain a reliable estimate.

9.4. Statistics and probabilities

The previous approach to probability, based on counting possible and favorable events, is not always feasible.

ICT equipment/service	population	frequency
website equipment	126 850	0,28
description of goods or services, price lists	94 244	0,21
ordering or reservation service for products and/or services (virtual shopping basket)	33 237	0,07
ability to personalize or design goods or services online	14 293	0,03
online order tracking	20 255	0,05
personalized content	21 574	0,05
links allowing access to company pages from social networks	53 964	0,12
job offers or online applications	41 524	0,09
development of a mobile site adapted to portable devices	31 626	0,07
development of a mobile site adapted to portable devices allowing orders to be received	11 312	0,03
total	**448 879**	**1,00**

Figure 9.1. *INSEE 2016 survey*

For example, consider the INSEE[1] survey conducted in 2016 on the use of information and communication technologies and e-commerce in businesses. An extract from this survey is shown in Figure 9.1. The survey covered a sample of

1 *Institut National de la Statistique et des Etudes Economiques* (French National Institute for Statistics and Economic Studies).

12,500 active companies or individual businesses, each employing at least 10 people (employees and non-employees), located in mainland France or the overseas departments. INSEE specifies that this sample is representative of approximately 191,000 companies.

The last column of the extract includes the statistical frequency, which is the ratio of the number of companies meeting a specific criterion to the total number of companies meeting at least one criterion. Since this is a survey, the results are estimates intended to reflect the actual situation.

A valuable theorem can be used here in Bernoulli's theorem, known as the *weak law of large numbers*. This law can be expressed as follows:

Consider a population of n elements. Let f be the statistical frequency and p the probability. For any positive number ε (as small as desired), the weak law of large numbers states that the probability $P(|f - p| < \varepsilon)$ approaches 1 as the number n of the population tends towards infinity, i.e. $P(|f - p| < \varepsilon) = 1$.

In other words, the probability that the statistical frequency closely approximates the true probability increases as the number of elements becomes very large. This implies that the statistical frequency can be considered a good approximation of the probability when n is very large. In the given example, although the number of company responses is not infinite, it is sufficiently large. Therefore, an approximate value of the probability can be obtained using the statistical frequency. Consequently, the probability of a company having a website is approximately 0.2826. This result is consistent with the primary definition of probability.

However, it does not guarantee an exact value, but only an approximate one. While the number of company responses to the survey is "relatively" large, larger samples are possible. Additionally, the sample is assumed to be representative. This situation illustrates how another approach to the concept of probability arises from statistical results.

EXAMPLE 9.16.– According to an (imaginary) survey conducted on 2,000,000 French people, 34% have a nose length greater than 6 cm. Thus, the approximate probability that a French person has a nose length greater than 6 cm can be estimated as $p \approx 0.34$.

EXAMPLE 9.17.– In a factory, out of 1,200 mechanical parts, 60 are found to be defective. Thus, the probability of a part being defective can be estimated as: $p = \frac{60}{1{,}200} = 0.05$.

Note that this method provides us a rough estimate of the probability rather than an exact value, as the sample size is finite. This implies that the estimated probability can be affected by bias and random variations. Therefore, it is important to interpret the results with caution and consider the margins of error associated with the estimate.

9.5. Compound probabilities

Given a set E and the set of events $\mathcal{P}(E)$, let $A, B \in \mathcal{P}(E)$ and suppose that $p(A)$ is non-zero. Define the conditional probability of B given A as:

$$p(B|A) = \frac{p(A \cap B)}{p(A)}$$

This defines a new probability, denoted as $p(B|A)$ or $p(BifA)$ or $p(BafterA)$ or $p_A(B)$. This is known as *conditional probability* of B given A. It can be verified that the axioms of probability (1, 2 and 3) are satisfied for this conditional probability. The relationship can also be expressed as:

$$p(A \cap B) = p(A).p(B|A)$$

This expresses the idea that to have both A and B occur, A must first occur with probability $p(A)$, and then B must occur given that A has already occurred, which can affect the calculation of $p(B|A)$. Thus, the probabilities multiply in this way.

Similarly, the expression can be written as: $p(A \cap B) = p(B).p(B)$, so that

$$p(A \cap B) = p(A).p(B|A) = p(B).p(A|B)$$

From this, Bayes' theorem is derived:

$$p(A|B) = \frac{p(B|A)p(A)}{p(B)}$$

EXAMPLE 9.18.– Take the example of the jar containing black and white balls. Suppose a ball is drawn, without putting it back in the jar, then a second is drawn. The probability of drawing two black balls is calculated as follows:

Apply the previous relationship: p(2 black balls) = p(1 black ball) × p(1 black ball | 1 black ball) = $(\frac{3}{5}) \times (\frac{2}{4}) = 0.3$.

Now, consider the same scenario, but this time the drawn ball is put back into the jar after each draw. The probability of drawing two black balls in this case is: $p(2 \text{ black balls}) = p(1 \text{ black ball}) \times p(1 \text{ black ball} \mid 1 \text{ black ball}) = \left(\frac{3}{5}\right) \times \left(\frac{3}{5}\right) = \frac{9}{25} = 0.36$.

EXAMPLE 9.19.– From studies on accidents, the following results are given:

– The probability of having an accident if the driver is sober is 0.001:

$$p_{sober}(\text{accident}) = 0.001.$$

– The probability of having an accident if the driver is drunk is 0.02:

$$p_{drunk}(\text{accident}) = 0.02.$$

– The probability of having a drunk driver is 0.01:

$$p(\text{drunk}) = 0.01.$$

To find the overall probability of having an accident, p(accident), consider the following scenarios (Figure 9.2 shows the cases in the form of a tree):

drunk ∩ accident

drunk ∩ no accident

sober ∩ accident

sober ∩ no accident

Define D for drunk, S for sober, A for accident and N for no accident. The given probabilities are:

$$p(D) = 0.01, \text{ from which } p(S) = 1 - p(D) = 0.99$$

$$p(A|S) = 0.001$$

$$p(A|D) = 0.02$$

Calculate the probabilities of each scenario involving an accident:

– Drunk and accident:

$$p(D \cap A) = p(A|D) \times p(D) = 0.002 \times 0.01 = 0.0002$$

– Sober and accident:

$$p(S \cap A) = p(A|S) \times p(S) = 0.001 \times 0.99 = 0.00099$$

Sum the probabilities of the scenarios involving an accident to find the overall probability of having an accident:

$$p(A) = p(D \cap A) + p(S \cap A) = 0.0002 + 0.00099 = 0.00119$$

Therefore, the probability of having an accident is approximately 0.00119.

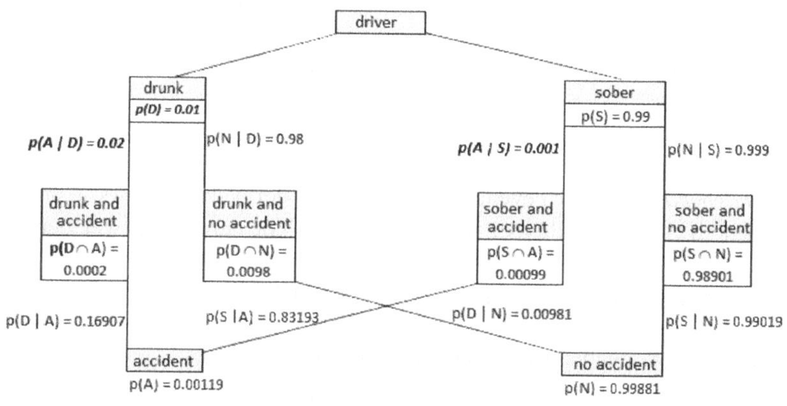

Figure 9.2. *Case tree*

The other probabilities can be calculated using the same methods:

$$p(N|D) = 1 - p(A|D) = 1 - 0.02 = 0.98$$
$$p(N|S) = 1 - p(A|S) = 1 - 0.001 = 0.999$$

Now, calculate the probabilities of each scenario involving no accident:

– Sober and no accident:

$$p(S \cap N) = p(N|S) \times p(S) = 0.999 \times 0.99 = 0.98901$$

– Drunk and no accident:

$$p(D \cap N) = p(N|D) \times p(D) = 0.98 \times 0.01 = 0.0098$$

Sum the probabilities of the scenarios involving no accident to find the overall probability of having no accident:

$$p(N) = p(S \cap N) + p(D \cap N) = 0.98901 + 0.0098 = 0.99881$$

Verify that the sum of the probabilities of having an accident and not having an accident is 1:

$$p(A) + p(N) = 1$$

Finally, calculate the conditional probabilities:

– Probability that a driver is sober given that an accident has occurred:

$$p(S|A) = p(S \cap A)/p(A) = 0.00099/0.00119 = 0.83193$$

– Probability that a driver is sober given that no accident has occurred:

$$p(S|N) = p(S \cap N)/p(N) = 0.98901/0.99881 = 0.99019$$

The *independence* of two events A and B can be defined as follows: A and B are independent if the probability of the event B does not depend on whether or not event A occurs. In other words, A and B are independent if the conditional probability of B given A is equal to the probability of B:

$$A \text{ and } B \text{ are independent if and only if } p(B|A) = p(B)$$

This can also be expressed by:

$$p(A \cap B) = p(A).p(B)$$

If this equality holds, then A and B are independent events (as in the case of the second part of example 9.18).

EXAMPLE 9.20.– Chevalier de Méré's bet involves comparing two probabilities: obtaining at least one double 6 by rolling two dice 24 times versus obtaining at least one 6 by rolling a single die four times. To do this, calculate the probabilities of these two events.

a) The probability of rolling at least one 6 in 4 rolls of a single die

First, examine the contrary event, which is the appearance of a number that is not a 6. In a simple roll, there is a 5 in 6 chance of not getting a 6. The corresponding probability is $p(\text{no } 6) = 5/6$.

The successive rolls of a die are independent. For 4 rolls, the probability of not rolling a 6 in any of the rolls is: $Prob(\text{not } 6) = \left(\frac{5}{6}\right)\left(\frac{5}{6}\right)\left(\frac{5}{6}\right)\left(\frac{5}{6}\right) = \left(\frac{5}{6}\right)^4$ Consequently, the probability of rolling at least one 6 in 4 rolls is:

$$Prob(6) = 1 - Prob(\text{not } 6) = 1 - \left(\frac{5}{6}\right)^4 = 0.517.$$

b) Probability of rolling at least one double six in 24 rolls of two dice

The probability of not rolling a double 6 with two dice in a single throw is $\frac{35}{36}$. For 24 independent throws of two dice, the probability of not getting a double 6 in any of the throws is:

$$Prob(\text{not } 66) = \left(\frac{35}{36}\right)^{24}.$$

Consequently, the probability of getting at least one double 6 in 24 throws is:

$$Prob(66) = 1 - Prob(\text{not } 66) = 1 - \left(\frac{35}{36}\right)^{24} = 0.491\ldots$$

Thus, Chevalier de Méré assertion was incorrect.

9.6. Graphs, states, transitions

It is relevant (and more explicit) to use graphs to represent composite probabilities, which will be of great use later. Suppose that the system being studied starts in state 0 and can transition to discrete states numbered 1, 2, 3..., as shown in Figure 9.3.

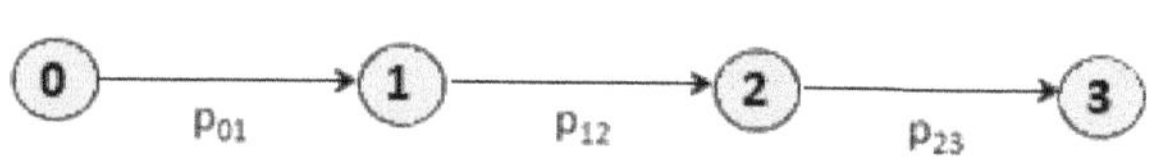

Figure 9.3. *Evolution of a system by transitions between states*

Transitions from one state to another are events whose probability is indicated by the arrow connecting the two states. The notation p_{ij} represents the probability of moving from state i to state j, and these probabilities are commonly called *transition probabilities*. In the case of the graph presented previously, the only way to go from state 0 to state 3 is to first go through state 1, then through state 2, before arriving at state 3. Therefore,

$$p_{03} = p_{01} \times p_{12} \times p_{23}$$

If the previous graph summarizes all the possibilities of transitions between states, it is clear that $p_{01} = 1,\ p_{12} = 1,\ p_{23} = 1$. Thus, it is certain to go from state 0 to state 3.

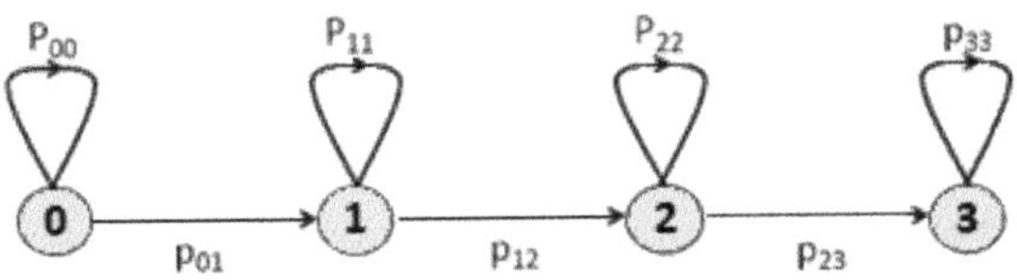

Figure 9.4. *Evolution of a system*

When the system is in state 0, it can either move to state 1 with probability p_{01} or stay in state 0 with probability p_{00}. It is important to take into account these possibilities of remaining in the same state because it can happen for a certain period of time. The graph in Figure 9.3, augmented with transitions of the type "$state\ x \Rightarrow state\ x$", is presented in Figure 9.4. There are therefore only two possible eventualities, which implies that $p_{00} + p_{01} = 1$. Similarly, $p_{11} + p_{12} = 1$ and $p_{22} + p_{23} = 1$. Finally, in state 3, the system can only remain in this state, which means that the transition probability p_{33} must be equal to 1.

Consider the graph in Figure 9.5, representing a system that can evolve between several states, where all possible transitions are represented.

Taking state 0 as an example, the possible evolutions are: $0 \rightarrow 0$, $0 \rightarrow 1$ and $0 \rightarrow 2$. Therefore, the probabilities must satisfy: $p_{00} + p_{01} + p_{02} = 1$. Similarly, for other states, the probabilities are as follows:

For state 1: $p_{11} + p_{10} + p_{12} + p_{13} = 1$.

For state 2: $p_{22} + p_{20} + p_{23} = 1$.

For state 3: $p_{33} + p_{32} = 1$.

These equations ensure that the sum of probabilities for all possible transitions from a given state is equal to 1.

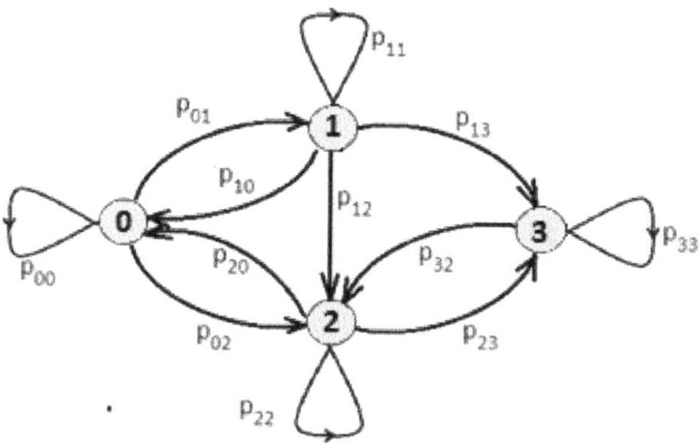

Figure 9.5. *Various transitions between states*

Now suppose the goal is to determine the probability that the system transitions from state 0 to state 3. This can be a complex task due to the infinite numbers of possible paths leading from state 0 to state 3. The direct paths include:

– 0-1-3;

– 0-1-2-3;

– 0-2-3.

Less direct paths can include "on the spot" transitions (such as 0-0, 1-1, 2-2, 3-3) or cycles (such as 0-1-2-0 or 0-1-3-2-0) repeated a certain number of times. It is possible to determine the corresponding probability for each defined path.

For example:

– For the direct path 0-1-2-3, the probability is $p_{01} \cdot p_{12} \cdot p_{23}$.

– For the path 0-1-2-2-3, the probability is $p_{01} \cdot p_{12} \cdot p_{22} \cdot p_{23}$.

These observations lead to the following two rules:

– Multiplication rule: the probability of a sequence of independent transitions (traversing the path) is the product of the probabilities of each transition in the sequence (along the path), as shown in Figure 9.6.

– Summation rule: the total probability of reaching a state j from a state i can be found by summing the probabilities of all distinct paths that lead from i to j (see Figure 9.7).

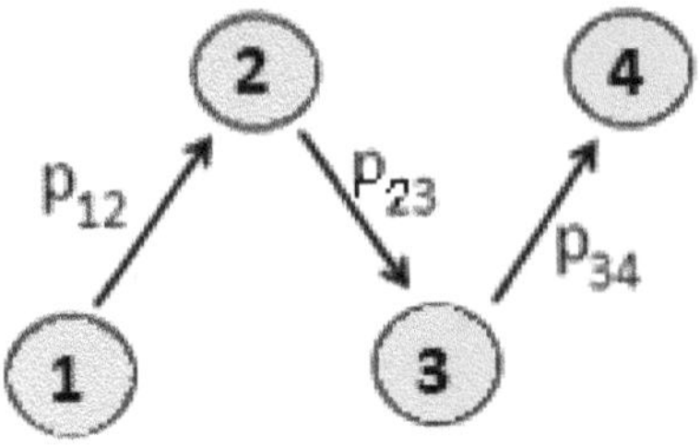

Figure 9.6. $p_{12}.p_{23}.p_{34}$

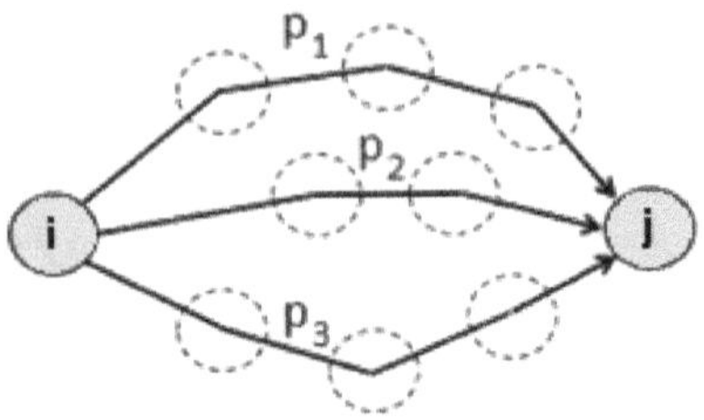

Figure 9.7. $p_1 + p_2 + p_3$

EXAMPLE 9.21.– In a jar, there are 6 identical balls, each bearing a letter A, N, A, N, A, S.

A ball is drawn, then a second one (without replacing the first). The question is: what is the probability that the second ball drawn has the letter A?

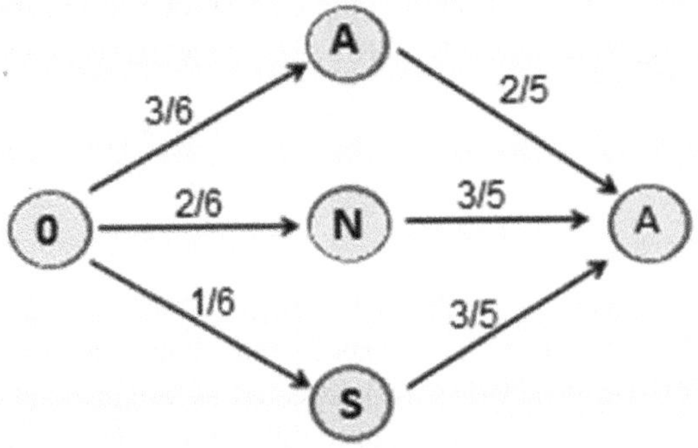

Figure 9.8. *Transition probabilities*

To analyze this, consider the system initially in state 0 (no ball is drawn). After drawing the first ball, the system can be in several states represented by the letter drawn: A, N or S. The final state of interest is drawing the letter A on the second draw. This evolution can be represented by the graph in Figure 9.8, which indicates the transition probabilities. By applying the previous rules, the probability can be obtained as follows:

$$p = \frac{3}{6}.\frac{2}{5} + \frac{2}{6}.\frac{3}{5} + \frac{1}{6}.\frac{3}{5} = \frac{1}{2}$$

9.7. Markov chains

9.7.1. *Definition*

Consider a changing system (transition from one state to another state). This system is supposed to pass from *defined state* to *defined state*, the number of states being *finite*: $E_1, E_2, E_3, \ldots, E_m$ (a finite sequence of m discrete states). The change of state is random, and the process of change of state is called a *stochastic sequence* (a fancy word to simply say *random*).

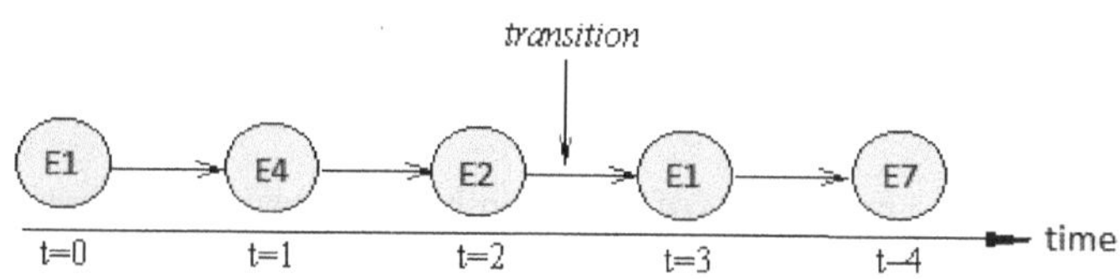

Figure 9.9. *Evolution of a system over time*

The system evolves from state to state through states E_1 to E_m. It can obviously pass through the same state multiple times, as shown in Figure 9.9, which represents the evolution of a system over time.

States are characterized by the values of random variables $X(t)$, one random variable for each discrete value of time. As time is discrete, the random variables can be written with an index reminiscent of time: X_n for $X(t_n)$. For a pair of states i and j at a time $t = t_n$, $p_{ij}(n)$ is the probability that the system being in state i at time $t = t_n$ moves to state j at time $t = t_{n+1}$. This relationship can be explained using compound probabilities which make it possible to calculate the probability of several independent events occurring successively:

$$p_{ij}(n) = p(X_{n+1} = j | X_n = i, X_{n-1} = i_{n-1}, X_{n-2} = i_{n-2}, \ldots, X_0 = i_0),$$

where $p_{ij}(n)$ represents the probability of transition from state i to state j during the duration from $t = t_n$ to $t = t_{n+1}$. It is a conditional probability, meaning the system must first be in state i at time t_n, in state X_{n-1} at time t_{n-1}, etc.

In the following, two simplifying hypotheses will be made:

Hypothesis H1: the system evolves "without memory", meaning that the transition from state i to state j does not depend on the previous states of the system, but only on the present state of the system (specifically, the fact that, at moment t_n, the system is in state i). This hypothesis is called the *Markov* property, and the stochastic process is called *a Markov chain.* Mathematically, this translates into the equality:

$$p(X_{n+1} = j|X_n = i, X_{n-1} = i_{n-1}, X_{n-2} = i_{n-2}, \dots, X_0 = i_0) = p(X_{n+1}j|X_n = i) = p_{ij}(n),$$

where $i_{n-1}, i_{n-2}, i_{n-3}, \dots, i_0$ represent any states.

Hypothesis H2: the transition probabilities are stationary, meaning they do not depend on time. Therefore, the notation can be simplified to:

$$p_{ij}(n) = p_{ij}$$

In other words, the probability of transition between two states does not change over time. It remains the same between time t_n and t_{n+1} as it does between time t_k and t_{k+1}, where t_k is any time. This assumption is called the *homogeneity* property. Therefore, the focus will be solely on homogeneous Markov chains.

EXAMPLE 9.22.– Imagine a consumer who can only buy 3 similar products of different brands *A*, *B* and *C*. The consumer's reactions are of interest, particularly in knowing which brand they purchased and predicting the brand they will prefer during their next purchase.

Assuming these products are consumable and the consumer does not remember the product he bought during his penultimate purchase, the evolution between brands *A*, *B* and *C* (the states) corresponds to a homogeneous Markov chain.

EXAMPLE 9.23.– Consider a rabbit in an enclosure comprising 4 compartments linked together by doors. Each compartment is associated with a possible state of the rabbit's presence. The rabbit moves from one compartment to another by passing through a door, which corresponds to a stage in the process of moving from one state to another. In Figure 9.10, the four compartments are represented by states 1, 2, 3 and 4.

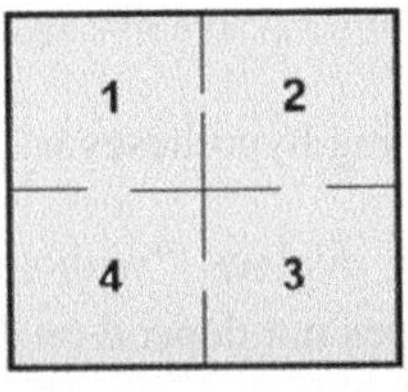

Figure 9.10. *The world of the rabbit*

The rabbit's preferences for certain compartments (based on the presence of food, for example) influence transitions between states. However, the rabbit has no memory and only remembers the compartment it is currently in (its last "stay"). Thus, the process of passage between compartments forms a homogeneous Markov chain.

9.7.2. *Transition matrix*

To describe a homogeneous Markov chain, two types of probability are used:

– State probability vector: the probability $q_k(n)$ represents the chance of finding the system in state E_k at the moment t_n. With m states, there are m probabilities $q_k(n)$ (for $k = 1$ to m). These probabilities are conveniently arranged (and for the rest of the calculations) in a probability vector:

$$Q(n) = [q_1(n)q_2(n)q_3(n)\cdots q_m(n)]$$

The sum of these probabilities must equal 1:

$$\sum_{i=1}^{m} q_i(n) = 1$$

– Transition probability matrix: the transition probability p_{jk} represents the probability of transitioning from state E_j to state E_k in one time step. These probabilities are stored in a matrix known as the *transition matrix* ***T***. In matrix ***T***, the rows correspond to the departure states and the columns correspond to the arrival states.

$$\boldsymbol{T} = \begin{bmatrix} p_{11} & p_{12} & \cdots & p_{1m} \\ p_{21} & p_{22} & \cdots & p_{2m} \\ \cdots & \cdots & \vdots & \cdots \\ p_{m1} & p_{m2} & \cdots & p_{mm} \end{bmatrix}$$

Each row of the matrix ***T*** sums to 1:

$$\sum_{k=1}^{m} p_{jk} = 1$$

This ensures that the sum of the elements (probabilities) for transitioning from any given state to all possible states is 1, which is a key feature of transition matrices.

Such matrices are also referred to as stochastic matrices. Transition matrices have the following remarkable property:

PROPERTY The product of two transition matrices with the same number of rows and columns is also a transition matrix.

In particular, if T is a transition matrix, then T^2, T^3, …., T^n are also transition matrices.

This property holds for transition matrices because they satisfy the following two conditions:

1) The sum of the elements in each row is equal to 1.

2) All elements of the matrix are non-negative (since they represent probabilities).

EXAMPLE 9.24. Consider the following two transition matrices:

$$T = \begin{bmatrix} 0 & 1 \\ \frac{1}{2} & \frac{1}{2} \end{bmatrix} \text{ and } T' = \begin{bmatrix} 0 & 1 \\ 1 & 0 \end{bmatrix}$$

To verify the remarkable property of transition matrices, consider the sum of the elements in a row of the product of the two matrices $T.T'$. It can be observed that the sum of the elements in each row of the resulting matrix is equal to 1 (as shown below).

$$T.T' = \begin{bmatrix} 0 & 1 \\ \frac{1}{2} & \frac{1}{2} \end{bmatrix} \begin{bmatrix} 0 & 1 \\ 1 & 0 \end{bmatrix} = \begin{bmatrix} 1 & 0 \\ \frac{1}{2} & \frac{1}{2} \end{bmatrix}$$

The calculation can also be extended to T^2 and T^3 as follows:

$$T^2 = \begin{bmatrix} 0 & 1 \\ \frac{1}{2} & \frac{1}{2} \end{bmatrix} \begin{bmatrix} 0 & 1 \\ \frac{1}{2} & \frac{1}{2} \end{bmatrix} = \begin{bmatrix} \frac{1}{2} & \frac{1}{2} \\ \frac{1}{4} & \frac{3}{4} \end{bmatrix} \qquad T^3 = \begin{bmatrix} \frac{1}{2} & \frac{1}{2} \\ \frac{1}{4} & \frac{3}{4} \end{bmatrix} \begin{bmatrix} 0 & 1 \\ \frac{1}{2} & \frac{1}{2} \end{bmatrix} = \begin{bmatrix} \frac{1}{4} & \frac{3}{4} \\ \frac{3}{8} & \frac{5}{8} \end{bmatrix}$$

Hence, it can be observed that the remarkable property of transition matrices holds for to T^2 and T^3, as the sum of the elements in each row of the resulting matrices equals 1.

EXAMPLE 9.25. Let us return to example 9.22. Suppose the known transition probabilities are those in Figure 9.11, where the states A, B and C are renamed as 1, 2 and 3.

The diagram provided is incomplete because the sum of the transition probabilities from each state to all others must equal 1. Some transitions are missing. For a complete representation, refer to Figure 9.12.

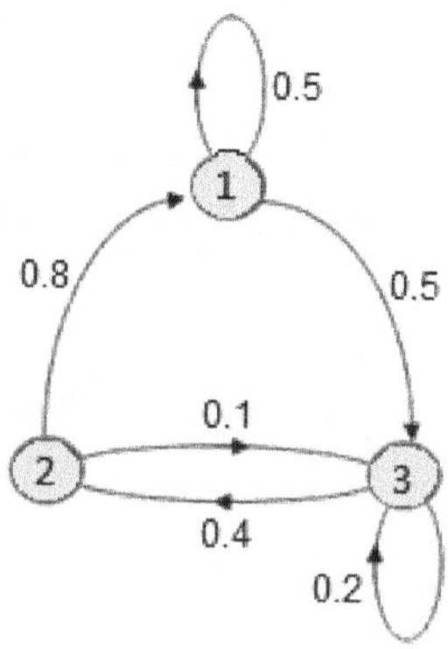

Figure 9.11. *Incomplete graph*

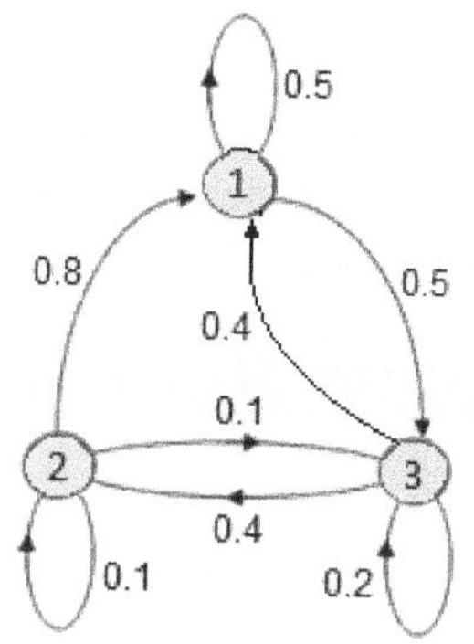

Figure 9.12. *Completed graph*

And the transition matrix is:

$$T = \begin{bmatrix} 0.5 & 0 & 0.5 \\ 0.8 & 0.1 & 0.1 \\ 0.4 & 0.4 & 0.2 \end{bmatrix}$$

It is important to note that there is no probability of transition from state 1 to state 2, meaning that this probability is zero.

EXAMPLE 9.26.– Let us return to example 9.23 and represent the states as a graph with the transition probabilities (see Figure 9.13), assuming that the steps consist of a change of compartment:

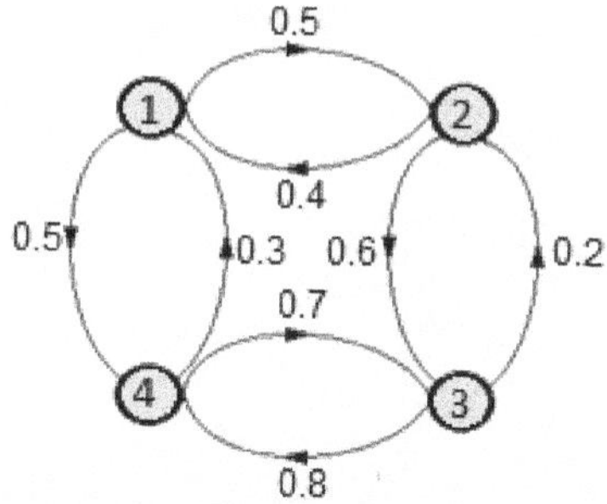

Figure 9.13. *Transition graph*

The transition matrix is then:

$$T = \begin{bmatrix} 0.0 & 0.5 & 0.0 & 0.5 \\ 0.4 & 0.0 & 0.6 & 0.0 \\ 0.0 & 0.2 & 0.0 & 0.8 \\ 0.3 & 0.0 & 0.7 & 0.0 \end{bmatrix}$$

It should be noted that the transition probabilities from state 1 to state 3, from state 2 to state 4, from state 3 to state 1 and from state 4 to state 2 are all zero (see Figure 9.13).

EXAMPLE 9.27.– Consider a system that evolves through five states, numbered from 1 to 5.

The transition graph associated with this system is shown in Figure 9.14.

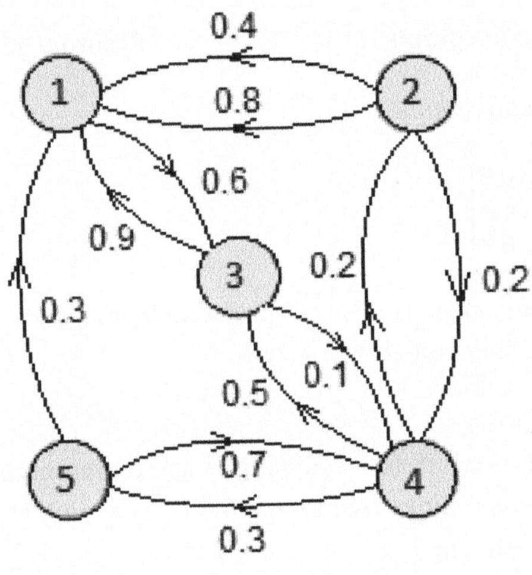

Figure 9.14. *Transition graph*

The transition matrix is as follows:

$$T = \begin{bmatrix} 0.0 & 0.4 & 0.6 & 0.0 & 0.0 \\ 0.8 & 0.0 & 0.0 & 0.2 & 0.0 \\ 0.9 & 0.0 & 0.0 & 0.1 & 0.0 \\ 0.0 & 0.2 & 0.5 & 0.0 & 0.3 \\ 0.3 & 0.0 & 0.0 & 0.7 & 0.0 \end{bmatrix}$$

9.7.3. Evolution rules

Now, let us examine the relationship between the probability vector and the transition matrix. To determine the probability of being in state E_k after n steps, given that the system was in state E_j just after step $n-1$, apply the result to composite probabilities (using hypothesis *H1*):

$$p\text{ (system in } E_k \text{ at } n) = p\text{ (system in } E_j \text{ at } n - 1).p_{jk} \text{ or } q_k(n) = q_j(n - 1).p_{jk,}$$

If any state at time $n-1$ is considered, or if all possibilities of previous states at time $n-1$ are taken into account, the probability of being in state E_k after n steps is obtained by summing all the corresponding probabilities (see Figure 9.15).

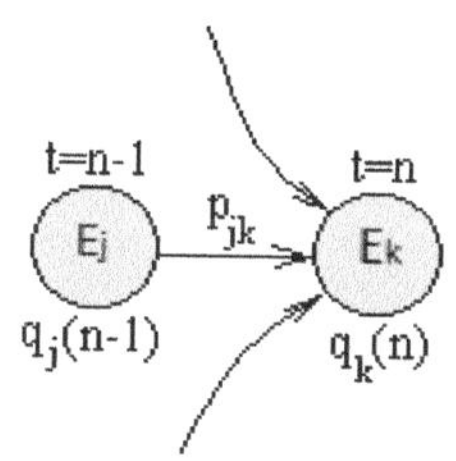

$$q_k(n) = \sum_{j=1}^{m} q_j(n-1).p_{jk}$$

Figure 9.15. *Evolution towards E_k*

This relationship can also be expressed in a matrix form:

$$\boldsymbol{Q}(n) = \boldsymbol{Q}(n - 1).\boldsymbol{T}$$

So, $\boldsymbol{Q}(1) = \boldsymbol{Q}(0).\boldsymbol{T}$, $\boldsymbol{Q}(2) = \boldsymbol{Q}(1).\boldsymbol{T} = \boldsymbol{Q}(0).\boldsymbol{T}^2$, $\boldsymbol{Q}(3) = \boldsymbol{Q}(2).\boldsymbol{T} = \boldsymbol{Q}(0).\boldsymbol{T}^3$, etc.

By induction, it can be deduced that:

$$\boldsymbol{Q}(n) = \boldsymbol{Q}(0).\boldsymbol{T}^n$$

EXAMPLE 9.28.– Let us return to example 9.27 and assume that the system is initially in state E_1 with certainty at time 0. In this case, the initial probability vector is as follows:

$$\boldsymbol{Q}(0) = [1\ 0\ 0\ 0\ 0]$$

To determine the probability vectors at subsequent stages, the relationship previously discussed can be applied:

$$Q(1) = [1\ 0\ 0\ 0\ 0]\begin{bmatrix} 0.0 & 0.4 & 0.6 & 0.0 & 0.0 \\ 0.8 & 0.0 & 0.0 & 0.2 & 0.0 \\ 0.9 & 0.0 & 0.0 & 0.1 & 0.0 \\ 0.0 & 0.2 & 0.5 & 0.0 & 0.3 \\ 0.3 & 0.0 & 0.0 & 0.7 & 0.0 \end{bmatrix} = [0\ 0.4\ 0.6\ 0\ 0]$$

and

$$Q(2) = [0\ 0.4\ 0.6\ 0\ 0]\begin{bmatrix} 0.0 & 0.4 & 0.6 & 0.0 & 0.0 \\ 0.8 & 0.0 & 0.0 & 0.2 & 0.0 \\ 0.9 & 0.0 & 0.0 & 0.1 & 0.0 \\ 0.0 & 0.2 & 0.5 & 0.0 & 0.3 \\ 0.3 & 0.0 & 0.0 & 0.7 & 0.0 \end{bmatrix} = [0.86\ 0\ 0\ 0.14\ 0]]$$

This process allows for the computation of the probability distribution across states at any given step n.

9.7.4. *Ergodicity*

What happens when n tends to infinity in the relation $Q(n) = Q(0).T^n$, i.e. long after the start of the process? If T^n converges to a limit T_0, then the transition matrix T is said to be *ergodic*. In this case, $Q(n)$ approaches a limit vector Q_0. Here are three theorems that help determine the limit behavior of the matrix T, though they are not proved here.

THEOREM 9.1.– If T is a regular transition matrix, the following properties hold:

– T admits a fixed point Q_0, which is a probability vector with strictly positive components.

– $T_0 = \lim T^n$ for n tending towards infinity. The matrix T_0 has rows that are identical and equal to Q_0.

– Q being any probability vector, when n tends to infinity, $\lim Q.T^n = Q_0$.

To understand the theorem, it is necessary to clarify the notions of a *regular matrix* and a *fixed point.*

a) A transition matrix T is considered *regular* if, from a certain power n; all elements of T^n are strictly positive.

EXAMPLE 9.29.– Consider the following transition matrices:

$$T_1 = \begin{bmatrix} 0 & 1 \\ 0.6 & 0.4 \end{bmatrix} \quad T_2 = \begin{bmatrix} 0 & 0.5 & 0.5 \\ 1 & 0 & 0 \\ 0.3 & 0.3 & 0.4 \end{bmatrix} \quad T_3 = \begin{bmatrix} 1 & 0 \\ 0.6 & 0.4 \end{bmatrix}$$

Square them:

$$T_1^2 = \begin{bmatrix} 0.6 & 0.4 \\ 0.2 & 0.8 \end{bmatrix} \quad T_2^2 = \begin{bmatrix} 0.65 & 0.15 & 0.20 \\ 0 & 0.5 & 0.5 \\ 0.42 & 0.27 & 0.31 \end{bmatrix} \quad T_3^2 = \begin{bmatrix} 1 & 0 \\ 0.84 & 0.16 \end{bmatrix}$$

The squared matrix is sufficient to show that T_1 is regular. However, for T_2^2 and T_3^2, zeros remain. Therefore, continue by cubing the matrices:

$$T_2^3 = \begin{bmatrix} 0.21 & 0.385 & 0.405 \\ 0.65 & 0.15 & 0.20 \\ 0.363 & 0.303 & 0.334 \end{bmatrix} \quad T_3^3 = \begin{bmatrix} 1 & 0 \\ 0.936 & 0.064 \end{bmatrix}$$

which is sufficient to prove that T_2 is regular. It remains T_3, but the result of T_3^n will always be of the form:

$$T_3^n = \begin{bmatrix} 1 & 0 \\ \mathrm{a} & \mathrm{b} \end{bmatrix}$$

T_3 is therefore not regular.

The definition of a regular matrix can be applied to matrices that are not transition matrices. However, if a matrix is a transition matrix and it is regular, it is referred to as an *ergodic matrix*.

a) A *fixed point* of a square matrix $\boldsymbol{T}$ is a vector $\boldsymbol{Q_0}$ such that $\boldsymbol{Q_0 T} = \boldsymbol{Q_0}$.

EXAMPLE 9.30.– Let the transition matrix $\boldsymbol{T}$ be defined by:

$$\boldsymbol{T} = \begin{bmatrix} 0 & 1/2 & 1/2 \\ 1 & 0 & 0 \\ 0 & 0 & 1 \end{bmatrix}$$

To find the fixed point $\boldsymbol{Q_0} = [x\ y\ z]$ where $x + y + z = 1$ because $\boldsymbol{Q_0}$ is a probability vector, the following must hold if $\boldsymbol{Q_0}$ is indeed a fixed point:

If $\boldsymbol{Q_0}$ is a fixed point, then the following must hold:

$$[x\ y\ z]\begin{bmatrix} 0 & \frac{1}{2} & \frac{1}{2} \\ 1 & 0 & 0 \\ 0 & 0 & 1 \end{bmatrix} = \begin{bmatrix} y & \frac{x}{2} & \frac{x}{2} + z \end{bmatrix} = [x\ y\ z]$$

Thus, $x = y = 0$ and $z = 1$. The vector $\boldsymbol{Q_0} = [0\ 0\ 1]$ is the fixed point of the matrix $\boldsymbol{T}$.

NOTE.– It is essential to note that theorem 9.1 states that if the matrix $\boldsymbol{T}$ is regular, then it has a fixed point $\boldsymbol{Q_0}$. However, the converse is not true: a matrix $\boldsymbol{T}$ having a fixed point $\boldsymbol{Q_0}$ is not necessarily regular. This is evident from the previous example, where, despite the existence of a fixed point, the transition matrix $\boldsymbol{T}$ was not regular for sufficiently large n:

$$T^n = \begin{bmatrix} 0 & 0 & 1 \\ 0 & 0 & 1 \\ 0 & 0 & 1 \end{bmatrix} \quad \text{for a large n}$$

To illustrate the application of theorem 9.1, consider the following example.

EXAMPLE 9.31.– Consider a motorist who changes cars every year according to the following rules:

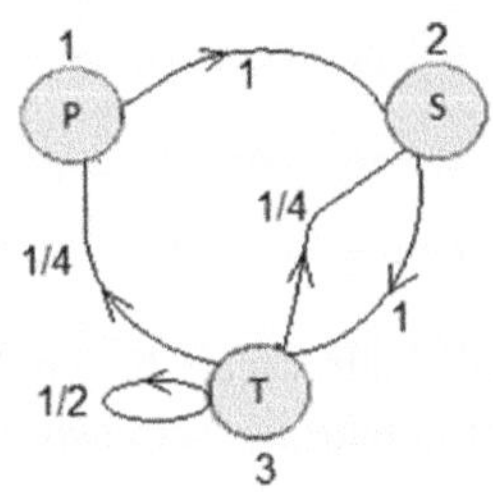

Figure 9.16. *Transition graph*

– When he owns a Pampa, he buys a Savanna.

– When he owns a Savanna, he buys a Tundra.

– When he owns a Tundra, he buys a Tundra 50% of the time, a Pampa 25% of the time and a Savanna 25% of the time.

The car brands are treated as states, and the process is shown in Figure 9.16 (where Pampa = 1, Savanna = 2 and Tundra = 3).

To determine if the transition matrix $\boldsymbol{T}$ is regular, examine the powers of the matrix $\boldsymbol{T}$:

$$\boldsymbol{T}^2 = \begin{bmatrix} 0 & 0 & 1 \\ 0.25 & 0.25 & 0.5 \\ 0.125 & 0.375 & 0.5 \end{bmatrix} \quad and \quad \boldsymbol{T}^3 = \begin{bmatrix} 0.25 & 0.25 & 0.5 \\ 0.125 & 0.375 & 0.5 \\ 0.125 & 0.25 & 0.625 \end{bmatrix}$$

$\boldsymbol{T}$ is therefore very regular. According to theorem 9.1, $\boldsymbol{T}$ has a fixed point $\boldsymbol{Q_0} = [x\ y\ z]$ with $x + y + z = 1$

The equation $\boldsymbol{Q_0T} = \boldsymbol{Q_0}$ translates to:

$$[x \quad y \quad z]\begin{bmatrix} 0 & 1 & 0 \\ 0 & 0 & 1 \\ \frac{1}{4} & \frac{1}{4} & \frac{1}{2} \end{bmatrix} = \begin{bmatrix} \frac{z}{4} & x+\frac{z}{4} & y+\frac{z}{2} \end{bmatrix}$$

$$\boldsymbol{Q_0T} = \boldsymbol{Q_0} \quad \Rightarrow \quad \begin{cases} x = \frac{z}{4} \\ y = x + \frac{z}{4} \\ z = y + \frac{z}{2} \end{cases} \quad \Rightarrow \quad x = \frac{1}{7} \quad y = \frac{2}{7} \quad z = \frac{4}{7}$$

$$\text{So, } \boldsymbol{T}_0 = \begin{bmatrix} \frac{1}{7} & \frac{2}{7} & \frac{4}{7} \\ \frac{1}{7} & \frac{2}{7} & \frac{4}{7} \\ \frac{1}{7} & \frac{2}{7} & \frac{4}{7} \end{bmatrix}$$

In 2023, the motorist had a Savanna. What car will he have in 2043 (assuming the brands still exist)?

2023 corresponds to the date $n = 0$; the probability vector is $\boldsymbol{Q}(0) = [0\ 1\ 0]$.

To find the probability vector for $n = 20$ (the year 2043), calculate $\boldsymbol{Q}(20)$. This can be obtained by multiplying the initial vector $\boldsymbol{Q}(0)$ by the transition matrix $\boldsymbol{T}$ raised to the 20th power:

$$\boldsymbol{Q}(20) = \boldsymbol{Q}(0).\boldsymbol{T}^{20} \cong \boldsymbol{Q}(0).\boldsymbol{T}_0 = [0 \quad 1 \quad 0]\begin{bmatrix} \frac{1}{7} & \frac{2}{7} & \frac{4}{7} \\ \frac{1}{7} & \frac{2}{7} & \frac{4}{7} \\ \frac{1}{7} & \frac{2}{7} & \frac{4}{7} \end{bmatrix} = \begin{bmatrix} \frac{1}{7} & \frac{2}{7} & \frac{4}{7} \end{bmatrix}$$

Here, $n = 20$ is considered sufficient large, so the approximation with $n = \infty$ is used.

He will therefore be four times more likely to own a Tundra than a Pampa and two times more likely to own a Tundra than a Savanna.

THEOREM 9.2.– If the number of states is discrete and finite, and if the transition graph is strongly connected, the conclusions of theorem 9.1 are valid.

A graph is strongly connected if, for every pair of vertices (i, j), there exists a path from i to j and a path from j to i.

Theorem 9.2 extends theorem 9.1 to strongly connected transition graphs. Specifically, a transition graph is strongly connected if every state is reachable from any other state. In other words, no subset of states is isolated from the rest of the graph.

Ergodicity is closely related to the connectivity of the transition graph. Specifically, the connectivity of the transition graph is a necessary condition for the system to be ergodic. This implies that for each state i, it must be possible to reach any other state j from i, either directly or by passing through one or more intermediate states.

EXAMPLE 9.32.– Check on the previous example (motorist changing cars: example 9.31) to see if theorem 9.2 applies.

The number of states is discrete and finite (3 states). Refer to the transition graph in Figure 9.16:

Vertices 1 and 2: $1 \rightarrow 2$ and $2 \rightarrow 3 \rightarrow 1$
Vertices 1 and 3: $1 \rightarrow 2 \rightarrow 3$ and $3 \rightarrow 1$
Vertices 2 and 3: $2 \rightarrow 3$ and $3 \rightarrow 2$

The graph is therefore strongly connected, and thus theorem 9.2 applies: the process is ergodic.

Lastly, a final result on ergodicity will be considered.

THEOREM 9.3.– If the transition matrix $\boldsymbol{T}$ has a simple eigenvalue equal to 1, and all other eigenvalues have absolute values less than 1 (or moduli less than 1 if they are complex), then the conclusions of theorem 9.1 are valid.

In other words, if the matrix $\boldsymbol{T}$ meets the conditions specified in this theorem, the process can be considered ergodic. This theorem is especially useful for verifying the ergodicity of a stochastic process when the conditions of theorem 9.2 are not met.

EXAMPLE 9.33. Consider the case where the driver makes the following purchasing choices based on his current car:

– If he owns a Pampa, he buys a Pampa 50% of the time and a Tundra 50% of the time.

– If he owns a Savanna, he buys a Pampa 75% of the time and a Savanna 25% of the time.

– If he owns a Tundra, he buys a Savanna.

Represent these choices in the form of a transition matrix $\boldsymbol{T}$:

$$\boldsymbol{T} = \begin{bmatrix} 1/2 & 0 & 1/2 \\ 3/4 & 1/4 & 0 \\ 0 & 1 & 0 \end{bmatrix}$$

To find the eigenvalues λ of $\boldsymbol{T}$, solve the equation $\det(\boldsymbol{T} - \lambda\boldsymbol{I}) = 0$, where $\boldsymbol{I}$ is the identity matrix.

Thus:

$$\begin{vmatrix} \frac{1}{2} - \lambda & 0 & \frac{1}{2} \\ \frac{3}{4} & \frac{1}{4} - \lambda & 0 \\ 0 & 1 & -\lambda \end{vmatrix} = -\lambda\left(\frac{1}{2} - \lambda\right)\left(\frac{1}{4} - \lambda\right) + \frac{3}{8} = 0$$

which leads to the equation $8\lambda^3 - 6\lambda^2 + \lambda - 3 = 0$.

In this case, $\lambda = 1$ is immediately identified as a root of the equation. Factoring the equation yields:

$$(\lambda - 1)(8\lambda^2 + 2\lambda + 3) = 0.$$

The quadratic term $8\lambda^2 + 2\lambda + 3$ has roots $-\frac{1}{8} \pm \frac{i}{8}\sqrt{23}$ both of which have a modulus less than 1. According to theorem 9.3, the process is therefore ergodic. Additionally, the fixed point of the transition matrix $\boldsymbol{T}$ is $\boldsymbol{Q_0} = (\frac{6}{13}, \frac{4}{13}, \frac{3}{13})$.

10

Descriptive Statistics

CONCEPTS COVERED IN THIS CHAPTER.–

Using computer resources in the field of statistics is essential, as statistics involves handling large quantities of data to extract relevant information.

This chapter aims to review fundamental statistical concepts, including numerical tables, graphic representations, measures of central tendency and measures of dispersion.

Subsequently, elementary modeling techniques, such as linear adjustment/regression, are presented in detail, along with the concept of correlation.

This modeling is then applied to statistical series that describe the evolution of phenomena over time, specifically chronological series.

References: [BOI 68, GRA 78, MAS 76].

10.1. Statistical description

10.1.1. *A little vocabulary*

The key terms of the statistical vocabulary are defined by the standard NF X05-001. Essential elements of statistical terminology include the following:

Population: this refers to the entire set of elements or subjects that the statistical study concerns. It typically includes a large number of elements. In practice, statistical studies are often conducted on a *sample* drawn from this population.

Mathematics for Digital Science 1,
by Gérard-Michel COCHARD and Mhand HIFI.

Sample: this is a subset of the population selected for conducting a statistical study. The sample is chosen to accurately represent the population, allowing for meaningful analysis and inference.

EXAMPLE 10.1.– The objective is to measure the average diameter of sand grains on a beach. Given the large number of grains, measuring the diameter of each one is impractical. Instead, a sample of 1,000 grains of sand are selected for diameter measurements, allowing for an estimation of the average diameter.

Individual or *statistical unit*: refers to any element within a population or sample. This can include a person, object, event or any entity that is subject to a statistical study.

EXAMPLE 10.2.– In the context of the previous example (example 10.1), each grain of sand serves as a statistical unit rather than an individual.

Character or *statistical variable*: a property or aspect under study within a population. It is essential to differentiate between:

– *Quantitative characters*: these are expressed in numerical terms. Examples include a person's age, the number of employees in a company or an employee's wage. Quantitative characteristics are further divided into two categories:

1) *Measurable characters*: their values represent measurements in a chosen unit. For instance, a person's height in centimeters or an employee's wage in euros are measurable characters. Mathematical operations such as addition can be applied to these values.

2) *Identifiable characters*: their values correspond to a measurement scale but do not permit arithmetic operations. For example, temperature in degree Celsius (°C) is an identifiable character. The sum of two temperatures is not meaningful (e.g. combining 20°C water with 100°C water does not result in 120°C water).

– *Qualitative characteristics*: these are not inherently numerical. To use these characteristics, values (either numerical or non-numerical) are assigned through coding. Examples include marital status (e.g. single = 1, married = 2, widowed = 3, divorced = 4) and occupation codes (e.g. INSEE codes).

Size: the number of elements corresponds to a specific value of a character. For example, in a dataset of 100 families, if only 4 families have exactly 6 children, then the size for the value 6 of the "number of children" character is 4.

Statistical series: a collection of pairs (x, n), where x represents a specific value of the character and n denotes the corresponding number of occurrences of that value.

EXAMPLE 10.3.– Out of 100 families, the distribution of children is as follows:

12 families with 0 children

26 families with 1 child

20 families with 2 children

15 families with 3 children

14 families with 4 children

9 families with 5 children

4 families with 6 children

The corresponding statistical series for this distribution is: (12, 0), (26, 1), (20, 2), (15, 3), (14, 4), (9, 5), (4, 6).

10.1.2. *Tabular presentation and frequency*

As previously mentioned, a statistical series associates each value x_i with its corresponding frequency n_i. This series can be represented in a table with two columns (or two rows), x_i and n_i, making it easier to read. The total number n of observations is the sum of all n_i values, and it is used to determine the statistical frequency for each value x_i:

$$f_i = \frac{n_i}{n}$$

where:

$$\sum_i f_i = 1$$

Frequency is often expressed as a percentage rather than as a decimal fraction (number between 0 and 1).

EXAMPLE 10.4.– Given the classification of 94,641 voters in an electoral constituency of their chosen candidate (with abstentions, blank or invalid votes excluded), the candidates are numbered 1 through 6, representing a qualitative and discreet variable. The corresponding statistical series is as follows:

$$(1, 33{,}521)\ (2, 21{,}775)\ (3, 19{,}496)\ (4, 15{,}752)\ (5, 2{,}812)\ (6, 1{,}285)$$

This statistical series can be effectively represented in Figure 10.1.

candidate	1	2	3	4	5	6
number of votes	33521	21775	19496	15752	2812	1285

Figure 10.1. *Statistical series table*

The character corresponds to the number of votes obtained by each candidate. Since the character is discrete, it is convenient to denote its values by x_i, where i varies from 1 to 6.

The same notation is used for the values of the candidates, noted n_i. The index i numbers the entries in the previous table (for this example, $x_i = i$). For instance,

$$\text{for } x_3 = 3, \quad n_3 = 19{,}496.$$

The total number of votes is the sum of all n_i: $n = n_1 + n_2 + n_3 + n_4 + n_5 + n_6 = 94{,}641$.

The statistical frequency f_i is obtained by dividing each n_i by n and can be added to the table (see Figure 10.2).

candidate	1	2	3	4	5	6	sum
number of votes	33521	21775	19496	15752	2812	1285	94641
frequency	0.354	0.23	0.206	0.166	0.03	0.014	1.000

Figure 10.2. *Effective table and frequency*

With the frequencies, it can be observed that candidate 3 received slightly more than 20% of the votes.

The previous example dealt with a statistical series with a discrete character. When the character is continuous, it is necessary to replace the value of the character with a range of values, known as a *class*. In this case, the value of the character falls within an interval defined by two values, typically represented as $a \leq x_i < b$, which is denoted as $x \in [a, b[$. The difference $(b - a)$ is referred to as the *amplitude* class.

EXAMPLE 10.5.– In the competitive entrance examination for higher education, where the scores obtained by the 280 eligible candidates range from 0 to 20.

Their distribution can be represented using a table, as shown in Figure 10.3.

grades over 20	[6. 8[	[8. 10[	[10. 12[	[12. 14[	[14. 16[	[16. 18[	[18. 20}	sum
size	1	8	29	63	100	70	9	280
frequency	0.0036	0.0286	0.1036	0.2250	0.3571	0.2500	0.0321	1.0000

Figure 10.3. *Effective table and frequency*

When classes have equal amplitude, each class (a, b) can be represented by its midpoint (*center)*, without ambiguity

$$\frac{a+b}{2}$$

This approach simplifies the conversion of a continuous statistical series into a discrete one. Additionally, it allows for the aggregation of classes by merging some of them, thereby simplifying the analysis of the data.

EXAMPLE 10.6.– For instance, in example 10.5, it is possible to consolidate classes of amplitude 2 into larger classes with amplitude 4 (see Figure 10.4) or even amplitude 6 (see Figure 10.5). While this approach simplifies the data presentation, it is important to note that increasing the class amplitude results in a loss of precision.

Thus, there is a trade-off between simplification and the level of detail retained.

grades over 20	[4, 8[	[8, 12[	[12, 16[	[16, 20]	sum
size	1	37	163	79	280
frequency	0.0036	0.1321	0.5821	0.2821	1.0000

Figure 10.4. *Amplitude class 4*

grades over 20	[2, 8[	[8, 14[	[14, 20]	sum
size	1	100	179	280
frequency	0.0036	0.3571	0.6393	1.0000

Figure 10.5. *Amplitude class 6*

EXAMPLE 10.7.– From the general agricultural census of 1955, the distribution of farms in France based on their total surface area is shown in Figure 10.6. This data excludes woodlands and provides a snapshot of farm sizes across the country.

farm surface (ha)	[0,1[	[1, 2[	[2, 5[	[5, 10[	[10, 20[	[20, 50[	[50, 100[	[100,200[	≥ 200	sum
size (in thousands)	149.72	229.57	411.27	470.76	530.05	373.81	74.75	16.70	3.52	2260.16
frequency	0.07	0.10	0.18	0.21	0.23	0.17	0.03	0.01	0.00	1.00

Figure 10.6. *Statistical series with different amplitude classes*

For clarity in presentation, classes with varying amplitudes are often used. Directly representing a class by its center is generally avoided unless accompanied by additional information.

In some cases, it is necessary to determine the cumulative sizes (or cumulative frequencies) up to a specific value of the characteristic being considered. These are calculated as follows:

$$Cumulative\ size = \sum_{i=1}^{p} n_i \qquad Cumulative\ frequency = \sum_{i=1}^{p} f_i$$

EXAMPLE 10.8.– For example 10.7, Figure 10.7 shows both cumulative size and cumulative frequency for the distribution of farms based on their total surface area.

farm surface (ha)	[0,1[	[1, 2[	[2, 5[	[5, 10[	[10, 20[	[20, 50[	[50, 100[	[100,200[	≥ 200	sum
size (in thousands)	149.72	229.57	411.27	470.76	530.05	373.81	74.75	16.70	3.52	2260.16
frequency	0.07	0.10	0.18	0.21	0.23	0.17	0.03	0.01	0.00	1.00
cumulative size (in thousands)	149.72	379.30	790.56	1261.33	1791.37	2165.18	2239.93	2256.63	2260.16	
cumulative frequency	0.07	0.17	0.35	0.56	0.79	0.96	0.99	1.00	1.00	

Figure 10.7. *Number, frequency and increasing cumulative frequency*

The table presented in Figure 10.7 answers the question: "*How many farms are less than 50 hectares, both in terms of number and percentage?*"

For the interval [20, 50[, the table shows: a number of 2,165,183 farms, representing a frequency of 0.9580 or 95.8%. It is important to note that in the final class, the cumulative size figures represent the total size, and the cumulative frequency is equal to 1. While cumulative numbers and frequencies are typically expected to increase, there are cases where they may decrease, as shown in Figure 10.8.

farm area (ha)	[0,1[	[1, 2[	[2, 5[	[5, 10[	[10, 20[	[20, 50[	[50, 100[	[100,200[	≥ 200	sum
size (in thousands)	149.72	229.57	411.27	470.76	530.05	373.81	74.75	16.70	3.52	2260.16
frequency	0.07	0.10	0.18	0.21	0.23	0.17	0.03	0.01	0.00	1.00
cumulative size (in thousands)	2260.16	2110.43	1880.86	1469.59	998.83	468.78	94.97	20.23	3.52	
cumulative frequency	1.00	0.93	0.83	0.65	0.44	0.21	0.04	0.01	0.00	

Figure 10.8. *Number, frequency, decreasing cumulative frequency*

This data indicates that 1,880,858 farms, or 83.22% of the total, have an area of 2 hectares or more.

10.1.3. *Two-dimensional series*

So far, one-dimensional series have been explored, focusing on a single character. However, it is also possible to analyze statistical series with multiple characters, resulting in multidimensional series. A series involving two-character series is known as a two-dimensional series and can be represented in a "crosstab" or matrix. In this setup, each pair of character values (x_i, y_j) corresponds to a frequency n_{ij}.

The sums of the elements in rows and columns create *marginal distributions*, which are one-dimensional series. These marginal distributions enable the breakdown of a two-dimensional statistical series into two separate one-dimensional statistical series.

EXAMPLE 10.9.– Consider the statistical series representing a population of 142 boys from a nursery school, analyzed based on their weight (y) in kg and age (x) in years.

y \ x	[3, 4[	[4, 5[	[5, 6[	marginal distribution
less than 15	19	7	1	27
[15, 20[	32	21	12	65
[20, 25[	3	18	28	49
[25, 30[	0	0	1	1
marginal distribution	54	46	42	142

Figure 10.9. *Two-dimensional series*

The representative table of this series is presented in Figure 10.9.

10.2. Graphical representations

A table provides a structured way to represent a statistical series. However, it often lacks the clarity needed to grasp the overall trends and patterns. To address this, graphical representations are frequently used. These visual tools can effectively convey complex data in an accessible format. The primary types of graphical representations include Cartesian diagrams, which give rise to bar charts and histograms, polar charts, pie charts, figurative charts and scatter plots. This discussion will focus on the most important types of graphical representations.

10.2.1. *Bar charts*

In a system of orthogonal axes, each pair (x_i, n_i) of a statistical series is represented by a point on the graph, where x_i represents a value of a character and n_i represents the corresponding size.

The abscissa x_i and the ordinate n_i are used to place this point.

Frequency f_i can also be used instead of size n_i, resulting in a Cartesian diagram. For a continuous character distributed into classes of equal width, the representation can be simplified by using the center of each class as the character's value.

Additionally, connecting the vertices of the bars with a broken line forms the *polygon of sizes* (or the *polygon of frequencies* if the frequencies are used instead of numbers).

EXAMPLE 10.10.– The results of the voters' votes for the six candidates are shown in Figure 10.10.

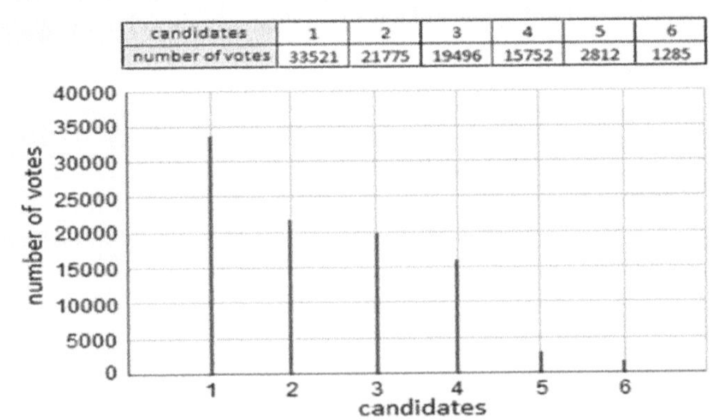

Figure 10.10. *Bar chart. For a color version of this figure, see www.iste.co.uk/cochard/mathematics1.zip*

EXAMPLE 10.11.– Oral grades obtained in a competitive examination for higher education are shown in Figure 10.11.

grades over 20	[6, 8[	[8, 10[	[10, 12[	[12, 14[	[14, 16[	[16, 18[	[18, 20]
centers	7	9	11	13	15	17	19
size	1	8	29	63	100	70	9

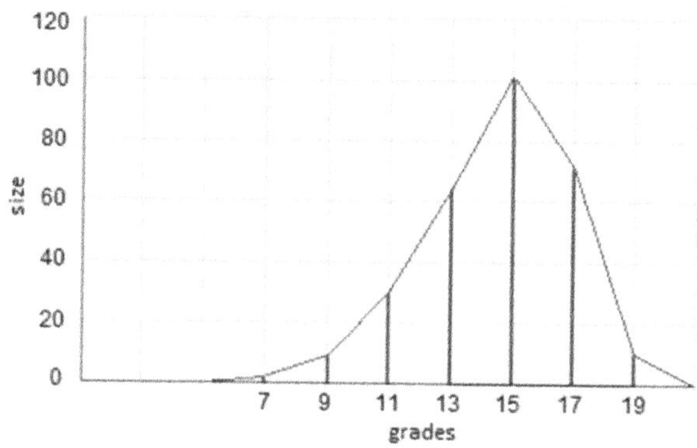

Figure 10.11. *Polygon of sizes. For a color version of this figure, see www.iste.co.uk/cochard/mathematics1.zip*

When the character is continuous, there are an infinite number of points, and the polygon of sizes (or frequencies) becomes a continuous curve known as the curve of sizes (or frequencies), as shown in Figure 10.12

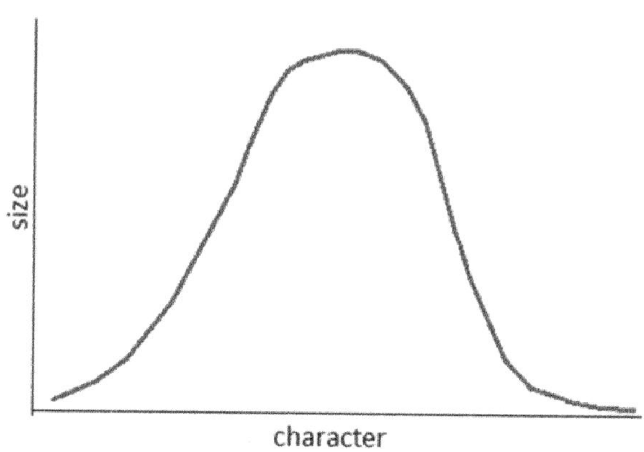

Figure 10.12. *Case of a continuous character. For a color version of this figure, see www.iste.co.uk/cochard/mathematics1.zip*

10.2.2. *Histograms*

Histograms are commonly used to represent continuous characters. Instead of displaying individual values of the character on the x-axis, histograms use classes to group these values together. Consequently, histograms employ rectangles, rather than bars, to visualize data.

EXAMPLE 10.12.– Return to example 10.11. Figure 10.13 shows the histogram representation in addition to the bar representation.

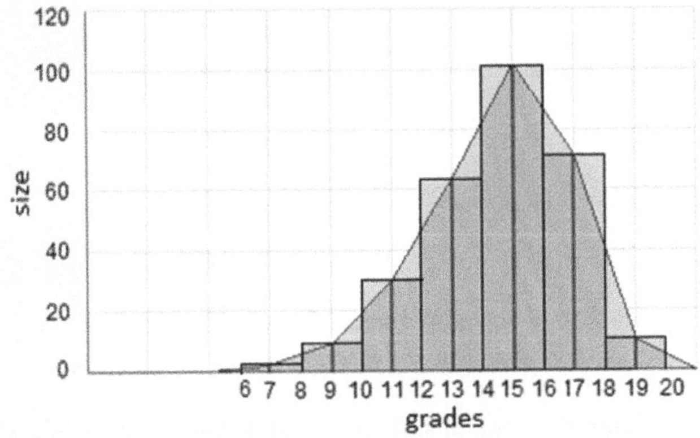

Figure 10.13. *Histogram representation. For a color version of this figure, see www.iste.co.uk/cochard/mathematics1.zip*

In a histogram, the height of each rectangle is proportional to the corresponding size, while the widths of the classes remain equal. Thus, the total area of the histogram reflects the total size.

The polygon of sizes is obtained by connecting the midpoints of the upper edges of the rectangles in a histogram. By convention, this polygon starts and ends at zero, as shown in Figure 10.13. It is evident that the total area of the histogram equals the area enclosed by the polygon of the sizes. Both areas are proportional to the total size.

When the amplitudes are not uniform (equal), it is important to ensure the area of each rectangle in the histogram is proportional to the corresponding size.

EXAMPLE 10.13.– Returning to example 10.11, which involves a distribution with classes of unequal amplitudes (see Figure 10.14), Figure 10.15 shows the application of the area rule. For example, the total area A = A1 + A2 + A3. The value represented by the last class on the histogram must be divided by 3.

grades over 20	[6, 8[	[8, 10[	[10, 14[	[14, 20]
size	1	8	92	179

Figure 10.14. *Unequal classes*

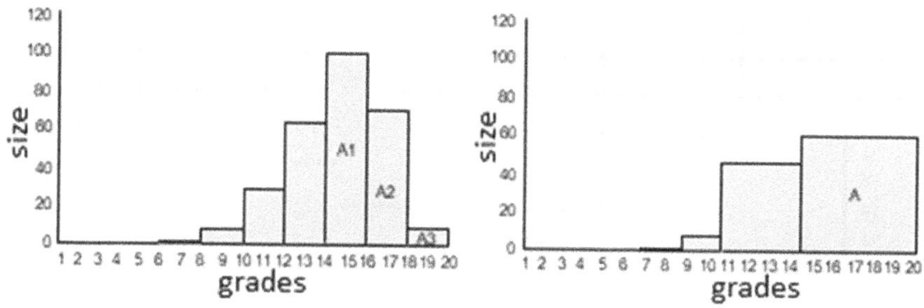

Figure 10.15. *Histograms*

10.2.3. *Cumulative diagrams*

In a Cartesian coordinate system, cumulative sizes and frequencies can be represented in three distinct cases:

– when the character is discrete;

– when the character is continuous and divided into classes;

– when the character is continuous and the classes have zero amplitude.

To illustrate the three cases of representing cumulative sizes and frequencies, let us consider an example.

1) *When the character is discrete*:

EXAMPLE 10.14.– The statistical series shown in Figure 10.16 represents the number of children per woman for a sample of 30 women.

x_i	0	1	2	3	4	5	6	7
n_i	4	3	8	7	4	3	0	1
cumulative size	4	7	15	22	26	29	29	30

Figure 10.16. *Cumulative size*

Plotting the cumulative number against the character values results in a staircase-shaped curve, as shown in Figure 10.17.

This shape arises because, when the character reaches a specific discrete value, the cumulative size increases abruptly to a new level, remaining constant until the next discrete value is reached. This discontinuity in the character values gives the curve its "stepped" appearance.

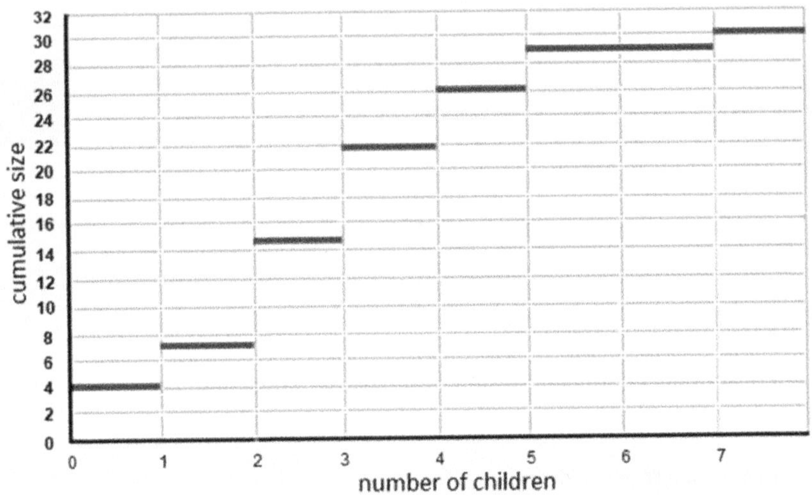

Figure 10.17. *Cumulative chart. For a color version of this figure, see www.iste.co.uk/cochard/mathematics1.zip*

2) *When the character is continuous and divided into classes:*

In general, the cumulative number at the start of a class differs from that at the end. Unless specified otherwise, it is assumed that within a class, the cumulative size varies linearly with the character value. This assumption allows for the construction of a *cumulative polygon* made up of line segments.

EXAMPLE 10.15.– Based on the data from example 10.11, the corresponding cumulative polygon is shown in Figure 10.18.

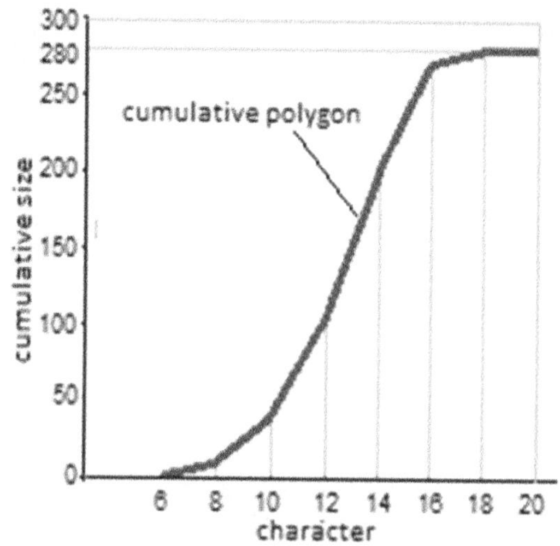

Figure 10.18. *Cumulative polygon. For a color version of this figure, see www.iste.co.uk/cochard/mathematics1.zip*

3) *The character is continuous and the classes have zero amplitude:*

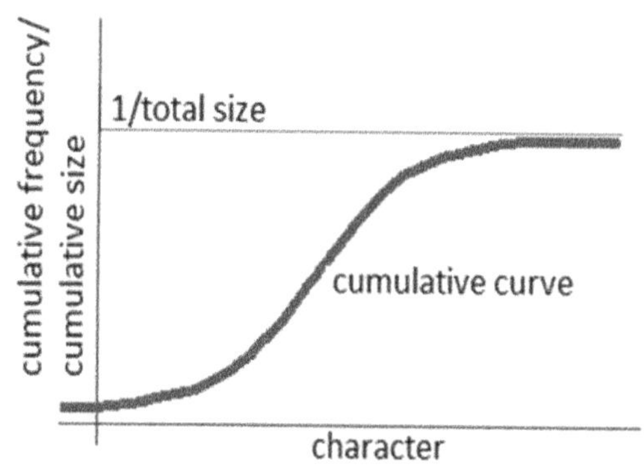

Figure 10.19. *Case of a continuous character. For a color version of this figure, see www.iste.co.uk/cochard/mathematics1.zip*

When the amplitude of the classes tends towards zero in the case of a continuous character, the cumulative polygon converges to a cumulative curve.

The general shape of this curve is shown in Figure 10.19.

10.2.4. *Polar diagrams*

The representation principle is outlined in Figure 10.20:

– The angle θ is proportional to the character.

– The vector radius ρ is proportional to the corresponding size.

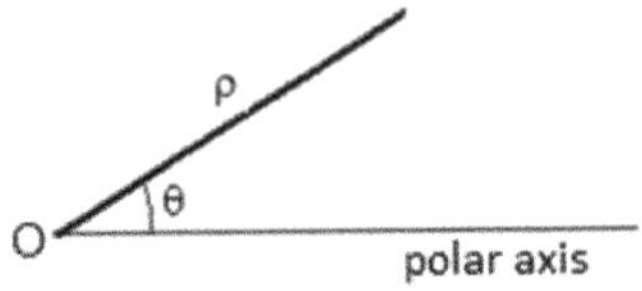

Figure 10.20. *Principle of a polar diagram*

In general, polar diagrams are used for representing periodic characters.

EXAMPLE 10.16.–

Monthly automobile declaration statistics are shown in Figure 10.21 using a polar plot, which facilitates a month-by-month comparison for each year.

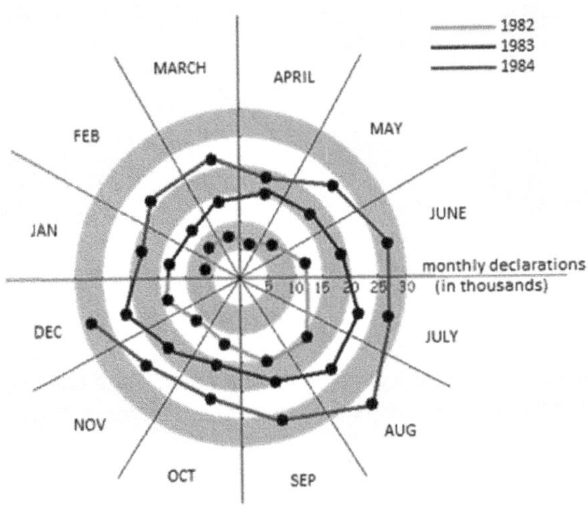

Figure 10.21. *Polar diagram. For a color version of this figure, see www.iste.co.uk/cochard/mathematics1.zip*

10.2.5. *Pie charts*

Pie charts are used to highlight the relative importance of sizes. The diagram resembles a pie, where each sector's central angle is proportional to the value of the corresponding size.

EXAMPLE 10.17.–

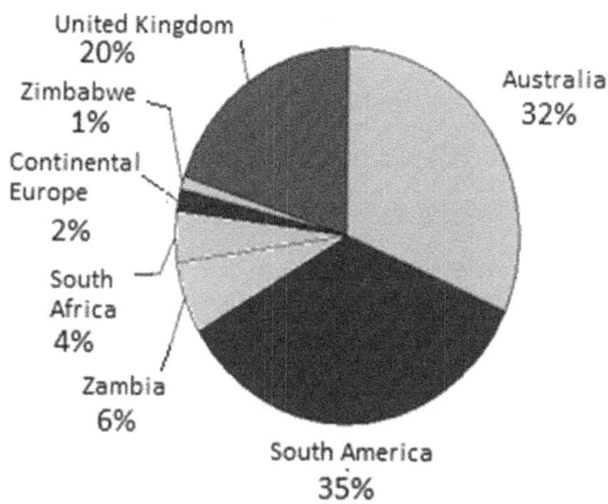

Figure 10.22. *Pie chart. For a color version of this figure, see www.iste.co.uk/cochard/mathematics1.zip*

The geographic distribution of activities for a multinational corporation is shown in Figure 10.22 as a pie chart.

10.2.6. *Figurative diagrams*

A figurative diagram uses illustrative drawings to represent quantities, providing a qualitative understanding of the relative importance of different phenomena. However, these diagrams lack precision on a quantitative level.

EXAMPLE 10.18.– Figure 10.23 is a figurative diagram illustrating the distribution of wool sheep herds by country of breeding in 1960.

Figure 10.23. *Figurative diagram*

10.2.7. *Point clouds*

In the case of a two-dimensional series, each pair (x_i, y_j) of character values can be represented by a point, forming a global point cloud.

When dealing with a continuous character, where not all pairs of values are available but rather a distribution into classes, the approach differs. For each pair of classes (forming a rectangle), a number of points equal to the corresponding size is plotted, distributed homogeneously within the rectangle.

EXAMPLE 10.19.– For the statistical series in example 10.9, the resulting point cloud is shown in Figure 10.24.

To create this visualization, each pair of classes is represented by a number of points corresponding to the size, with these points distributed homogeneously within the rectangle formed by the classes.

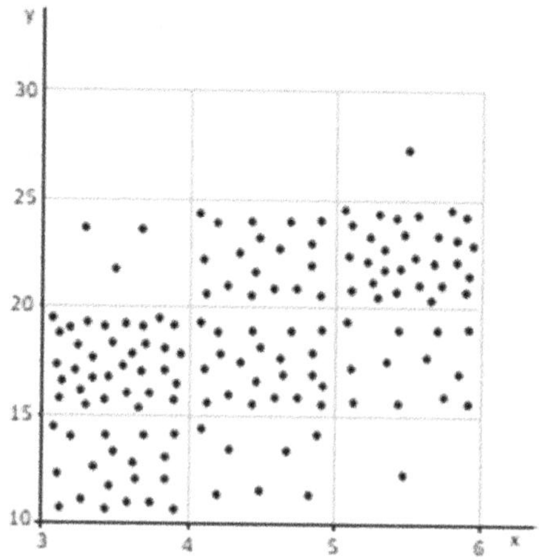

Figure 10.24. *A point cloud*

10.3. Position parameters

When comparing two statistical series, two key types of parameters are employed:

– Position parameters: these include measures such as the mode, median and various averages.

– Dispersion parameters: these cover metrics such as range, mean deviation, variance, standard deviation, coefficient of variation and dispersion index.

This section will focus on positional parameters, with dispersion parameters discussed in the subsequent section.

10.3.1. *Mode*

The *mode*, or *dominant* value, x_0, is the value of the character which corresponds to the largest size.

EXAMPLE 10.20.– Referring to the statistical series from example 10.4, which is a discrete statistical series, the highest count is 33,521, corresponding to the character value 1. Thus, the mode of the series is $x_0 = 1$.

EXAMPLE 10.21.– Referring to the statistical series from example 10.5, which is a continuous statistical series, the highest number is 100, corresponding to the class [14, 16[. This class is called a modal class. By convention, the mode is defined as the center of this class. Therefore, the mode of the series is $x_0 = 15$.

NOTE.– The size of a statistical series can present multiple relative maxima, resulting in several modes (see Figure 10.25). Therefore, the mode is most useful and informative in the case of unimodal series.

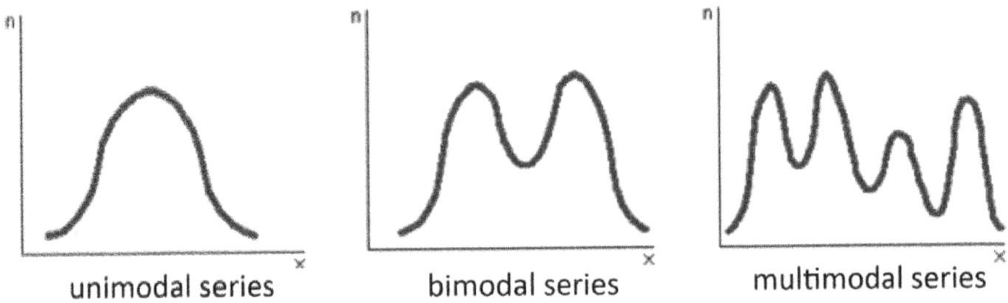

Figure 10.25. *Multiple modes. For a color version of this figure, see www.iste.co.uk/cochard/mathematics1.zip*

NOTE.– In the case of a continuous character, two different distributions into classes can lead to two different mode values.

EXAMPLE 10.22.– In example 10.21, the distribution into classes of unequal amplitudes, as shown in Figure 10.14, results in a mode of $x_0 = 17$ rather than $x_0 = 15$. This occurs because the mode is defined as the class with the highest frequency, and in this case, the class [16, 18[has the maximum number of observations.

10.3.2. *Median*

The *median* is defined as the value of the character that divides the statistical series into two equal parts. It corresponds to the median value x_m where the cumulative frequency reaches $\frac{1}{2}$, as shown in Figure 10.26.

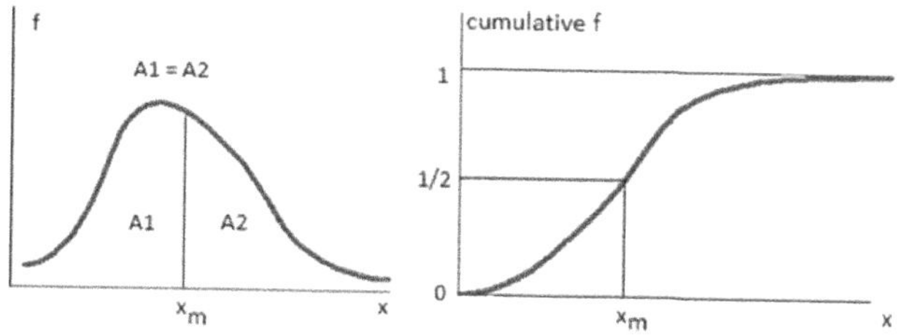

Figure 10.26. *Determining the median. For a color version of this figure, see www.iste.co.uk/cochard/mathematics1.zip*

Now indicate how the median is determined in practice. First, look at the method of determination by calculation. There are two cases to take into account:

i) *The character is discrete*

To calculate the median of a discrete character, follow these steps:

1) Arrange the data: list the character values of the statistical series in ascending (or descending) order.

2) Calculate cumulative frequencies: compute the cumulative frequency by progressively adding the frequencies of each character value.

3) Identify the median: find the value associated with the cumulative frequency that reaches or exceeds half of the total size.

3a) If the total size is odd: the median is the character value for which the cumulative frequency first reaches or exceeds half of the total size.

3b) If the total size is even: the median is the average of the two-character values adjacent to the cumulative frequency that reaches or exceeds half of the total size.

4) Determine the median: the value identified in step 3 is the median of the discrete character.

x_i	1	2	3	4	5	6
n_i	2	7	3	1	7	5

12 12

Figure 10.27. *First case*

x_i	1	2	3	4	5	6
n_i	2	5	8	7	4	2

7+1

14 14

Figure 10.28. *Second case*

Applying these steps to the discrete case:

– In the first case (see Figure 10.27), the median is $x_m = 4$.

– In the second case (see Figure 10.28), there is no distinct median, which is generally true for discrete characters. However, by convention, the median can be referred to as $x_m = 3$ in this case.

ii) *The character is continuous*

To calculate the median of a continuous character:

1) Group the data: organize the data into classes, either with equal or unequal amplitudes, ensuring that each class covers a range of character values.

2) Calculate cumulative frequencies: compute the cumulative frequency by summing the frequencies of all preceding classes.

3) Identify the median class: determine the class, where the cumulative frequency first reaches or exceeds half of the total population.

4) Apply the median formula for continuous classes: use the following formula to calculate the median:

$$\begin{aligned}&\text{Median}\\&= \text{Lower limit of the median class}\\&+ \left(\frac{\text{Median size } - \text{ Cumulative size of the previous class}}{\text{Frequency of the median}}\right)\\&\times \text{Amplitude of the median class}\end{aligned}$$

5) Determine the median: the value obtained from the formula is the median of the continuous character.

EXAMPLE 10.23.– Refer to example 10.8 using the table provided in Figure 10.29:

farm area (ha)	[0,1[	[1, 2[	[2, 5[	[5, 10[	[10, 20[	[20, 50[	[50, 100[	[100,200[	≥ 200	sum
size (in thousands)	149.72	229.57	411.27	470.76	530.05	373.81	74.75	16.70	3.52	2260.16
frequency	0.07	0.10	0.18	0.21	0.23	0.17	0.03	0.01	0.00	1.00
cumulative size (in thousands)	149.72	379.30	790.56	1261.33	1791.37	2165.18	2239.93	2256.63	2260.16	
cumulative frequency	0.07	0.17	0.35	0.56	0.79	0.96	0.99	1.00	1.00	

Figure 10.29. *Statistical table*

The median size is half the total size, which is 1130077.5. By examining the cumulative frequencies, the interval closest to 1130077.5 is the interval [5, 10[. Therefore, the median class (or median interval) is the class [5, 10[.

To find the specific median within this class, use the formula: $x_m = 5 + \Delta x$, where $0 < \Delta x < 5$.

To determine the value of Δx, perform linear interpolation.

Between $x = 5$ and $x = 10$, the cumulative frequency is 470,763.

Between $x = 5$ and $x = 5 + \Delta x$, the cumulative frequency is $1130077.5 - 790{,}562 = 339515.5$.

Therefore, the median x_m is approximately computed as follows:

$$\frac{\Delta x}{10-5} = \frac{339515.5}{470{,}763} \quad \Rightarrow \quad \Delta x \approx 3.61 \quad \text{and} \quad x_m \approx 8.61$$

To determine the median graphically, we can use one of the following methods:

– Increase cumulative polygon: plot the increasing cumulative frequency polygon. The median corresponds to the point where the cumulative frequency reaches half $\left(\frac{1}{2}\right)$ of the total frequency.

– Decrease cumulative polygon: plot the decreasing cumulative frequency polygon. The median corresponds to the point where the cumulative frequency reaches half $\left(\frac{1}{2}\right)$ of the total frequency.

– Intersection of both polygons: plot both the increasing and decreasing cumulative frequency polygons on the same graph. The abscissa (x-coordinate) at the intersection of these two polygons provides the value of the median.

EXAMPLE 10.24.– In this example, 150 workers are classified according to their age, as shown in Figure 10.30.

age (years)	[20, 25[	[25, 30[	[30, 35[	[35, 40[	[40, 45[	[45, 50[	[50, 55[	[55, 60]
size	9	27	36	45	18	9	3	3
frequency	0.06	0.18	0.24	0.3	0.12	0.06	0.02	0.02
increasing cumulative frequency	0.06	0.24	0.48	0.78	0.9	0.96	0.98	1
decreasing cumulative frequency	1	0.94	0.76	0.52	0.22	0.1	0.04	0.02

Figure 10.30. *Statistical table*

Draw the cumulative polygons as shown in Figure 10.31. The intersection of the two polygons provides the median value $x_m = 35.3$.

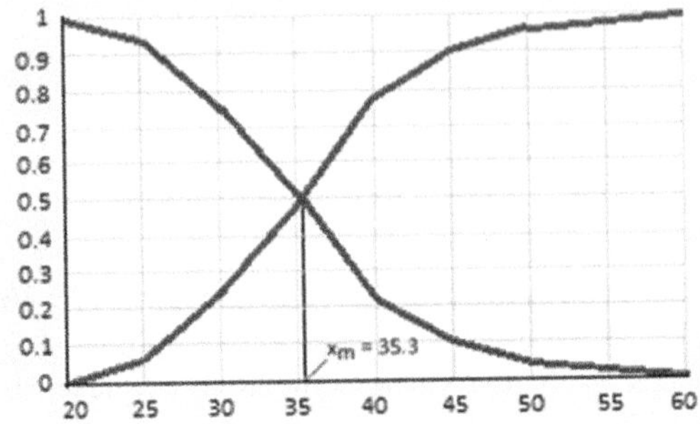

Figure 10.31. *Cumulative polygons. For a color version of this figure, see www.iste.co.uk/cochard/mathematics1.zip*

The notion of median can be generalized by introducing *quartiles*. Quartiles divide the entire population into four equal parts, each representing 25% of the dataset.

– First quartile (Q1): represents the value that separates the lowest 25% of the data from the rest. It corresponds to the median of the lower half of the data.

– Second quartile (Q2): also known as the median, Q2 divides the data into two equal parts, with 50% of the data below and 50% above.

– Third quartile (Q3): separates the highest 25% of the data from the rest. It is equivalent to the median of the upper half of the data.

In Figure 10.32, these quartiles are graphically representation. The values of Q1, Q2 (median) and Q3 are marked on the x-axis. The range between Q1 and Q3 is known as the interquartile range, which contains 50% of the data.

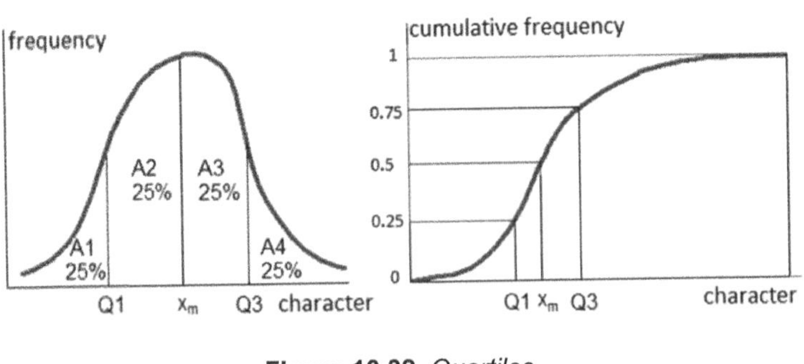

Figure 10.32. *Quartiles*

The total dataset can also be divided into 10 equal parts, known as *deciles*, or into 100 equal parts, known as *centiles*:

– Deciles: divide the data into 10 equal parts, with each part representing 10% of the total data. The corresponding values are called deciles.

– Centiles: divide the data into 100 equal parts, with each part representing 1% of the total data. The corresponding values are called centiles.

10.3.3. *Averages*

There are several types of averages, but the focus here is on the main definitions, starting with the weighted arithmetic average, commonly known as the mean.

The *weighted arithmetic average m* (or *mean*) of a discrete statistical series is defined by the formula:

$$m = \frac{\sum_i n_i x_i}{\sum_i n_i}$$

where n_i represents the frequency of the value x_i of the character and $\sum_i n_i$ represents the total size of the dataset. The mean m in often denoted as $< x >$ or $\bar{x}$. But here it is simply referred to as m.

When frequencies are expressed as proportions (i.e. f_i, where $\sum_i f_i = 1$), the previous formula for the mean becomes:

$$m = \sum_i f_i x_i$$

EXAMPLE 10.25.– Refer to example 10.14 and calculate the frequency f_i using the data provided in Figure 10.33.

x_i	0	1	2	3	4	5	6	7
n_i	4	3	8	7	4	3	0	1
frequency	0.13	0.10	0.27	0.23	0.13	0.10	0.00	0.03

Figure 10.33. *Frequency table*

According to the values in Figure 10.33, the mean is calculated as follows:

$m = 0.13 \times 0 + 0.10 \times 1 + 0.27 \times 2 + 0.23 \times 3 + 0.13 \times 4 + 0.10 \times 5 + 0.00 \times 6 + 0.05 \times 7 = 2.56$

Since the number of children must be an integer, the mean is rounded to 3.

Another method of calculating the mean, which can simplify the process, involves using a provisional average, denoted as μ. The provisional average is typically chosen to be close to the mean; for example, the median $\mu = x_m$ can be used.

Each value x_i can be expressed as the sum of m and the difference between x_i and μ, either: $x_i = \mu + (x_i - \mu)$.

This representation allows each value x_i to be written in terms of the provisional average μ. Consequently, the mean m can be calculated as follows:

$$m = \frac{\sum_i n_i x_i}{\sum_i n_i} = \frac{\sum_i n_i[\mu + (x_i - \mu)]}{\sum_i n_i} = \mu + \frac{\sum_i (x_i - \mu)}{\sum_i n_i}$$

EXAMPLE 10.26.– Apply the provisional average method to example 10.25 by setting $\mu = 2$ (see Figure 10.34)

x_i	0	1	2	3	4	5	6	7	sum
n_i	4	3	8	7	4	3	0	1	30
$x_i - \mu$	-2	-1	0	1	2	3	4	5	
$n_i(x_i - \mu)$	-8	-3	0	7	8	9	0	5	18

Figure 10.34. *Table of operations*

Further simplification, the mean is given as follows:

$$m = 2 + \frac{18}{30} = 2.6$$

Apart from the weighted arithmetic mean, other types of averages are sometimes used. One such average is the *mean square*, which is defined as the square root of the mean of the squares of the character values.

Mean square: the formula for the mean square q is:

$$q = \sqrt{\frac{\sum_i n_i x_i^2}{\sum_i n_i}}$$

Geometric mean:

$$g = \sqrt[n]{\prod_{i=1}^{n} x_i}$$

Harmonic mean: h such that

$$\frac{1}{h} = \frac{\sum_i \frac{n_i}{x_i}}{\sum_i n_i}$$

It can be demonstrated that, among these averages, the following relation holds:

$$h \leq g \leq m \leq q.$$

10.3.4. *Comparison of position parameters*

In general, a frequency curve that is assumed to be unimodal often presents asymmetry, leading to different values for the mode, median and arithmetic mean. These values are distributed in distinct ways, as shown in Figure 10.35.

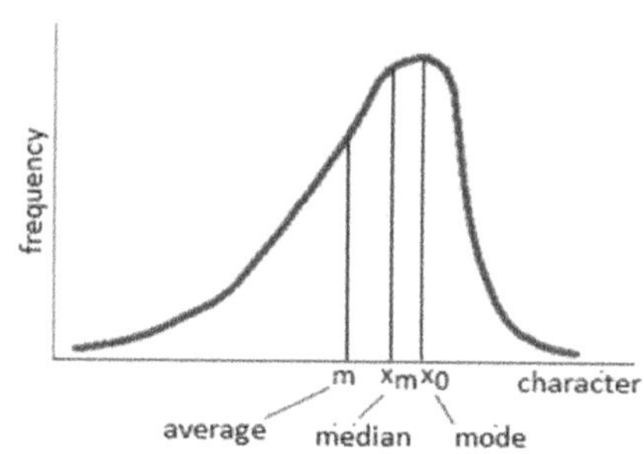

Figure 10.35. *Comparison of position parameters. For a color version of this figure, see www.iste.co.uk/cochard/mathematics1.zip*

When the frequency curve is only slightly asymmetric, the three position parameters approximately satisfy the following relationship, known as the empirical Pearson–Morice formula:

$$x_0 = 3x_m - 2m$$

where x_0 represents the mode, x_m is the median and m is the arithmetic mean.

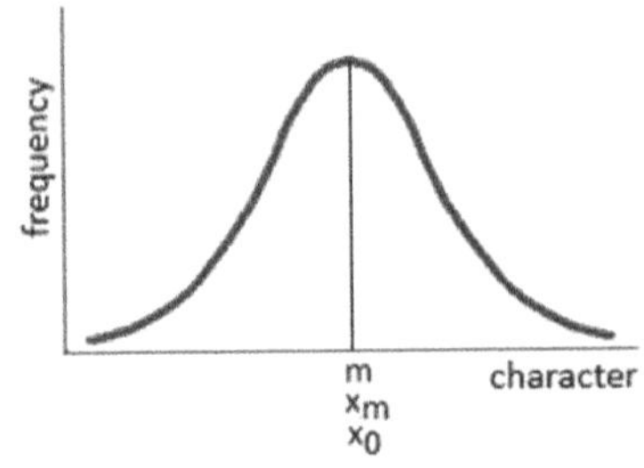

Figure 10.36. *Symmetric statistical distribution. For a color version of this figure, see www.iste.co.uk/cochard/mathematics1.zip*

When the frequency curve is symmetrical (and unimodal), the three position parameters are combined, as shown in Figure 10.36.

10.4. Dispersion parameters

Dispersion parameters measure the spread of a statistical distribution. The main types include the range, the interquartile range, the mean deviation and the standard deviation.

10.4.1. *Range, interquartile range, mean range*

The *extent* of a statistical series is defined as the difference between the extreme values of the character. The difference between the first quartile (Q1) and third quartile (Q3) is known as the *interquartile range*. These definitions are shown in Figure 10.37.

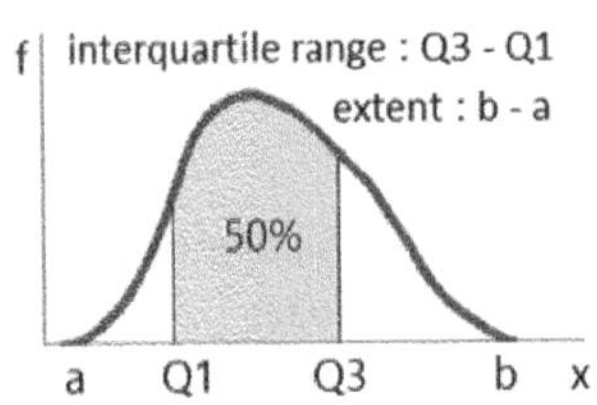

Figure 10.37. *Extent and interquartile range. For a color version of this figure, see www.iste.co.uk/cochard/mathematics1.zip*

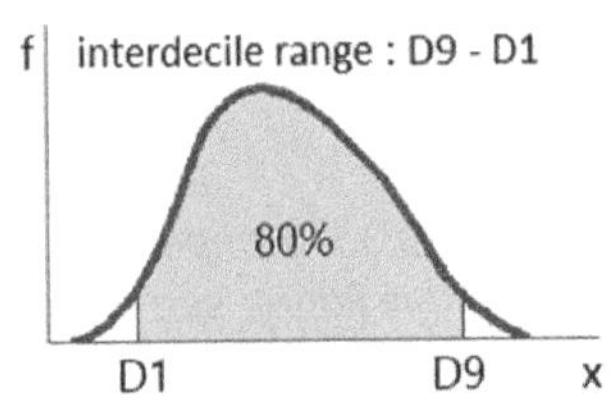

Figure 10.38. *Interdecile range. For a color version of this figure, see www.iste.co.uk/cochard/mathematics1.zip*

The *interdecile range* and the *intercentile range* are defined analogously. The interdecile range is the difference between the 9th and 1st deciles, while the intercentile range is the difference between specified centiles. These concepts are shown in Figure 10.38.

The arithmetic mean of the absolute deviations from the arithmetic mean is called the *mean deviation*. It is calculated using the following formula:

$$e = \frac{\sum_i n_i |x_i - m|}{\sum_i n_i}$$

where n_i represents the frequency of the value x_i and m is the arithmetic mean.

EXAMPLE 10.27.– Consider the statistical series represented in the table in Figure 10.39, which presents the distances of 1,000 localities in Île-de-France from Paris.

The calculated statistics are as follows:

Arithmetic mean: $m = \frac{4{,}240}{1{,}000} = 4.24$,

Range: 50,

First quartile: $Q_1 = 0 + 1 \times \frac{250-0}{353} = 0.71$,

Third quartile: $Q_3 = 2 + 3 \times \frac{750-512}{255} = 4.80$,

Interquartile range: $Q3 - Q1 = 4.80 - 0.71 = 4.09$,

Mean deviation: $e = \frac{388{,}916}{1{,}000} = 3.89$.

distance in km	[0, 1[	[1, 2[	[2, 5[	[5, 10[	[10, 20[	[20, 50]	sum
centers x_i	0.5	1.5	3.5	7.5	15	35	
size n_i	353	159	255	147	59	27	1000
cumulative size	353	512	767	914	973	1000	
$n_i x_i$	176.5	238.5	892.5	1102.5	885	945	4240
$\|x_i - m\|$	3.74	2.74	0.74	3.26	10.76	30.76	
$n_i\|x_i - m\|$	1320.22	435.66	188.7	479.22	634.84	830.52	3889.16

Figure 10.39. *Table for calculating dispersion parameters*

10.4.2. *Variance and standard deviation*

The *variance* is the average of the squared deviations from the mean, calculated as follows:

$$v = \frac{\sum_i n_i(x_i - m)^2}{\sum_i n_i}$$

The *standard deviation* or *quadratic mean deviation* is the square root of the variance:

$$\sigma = \sqrt{v}$$

The variance (and hence the standard deviation) can also be expressed in an alternative, equivalent form using a provisional mean μ. Consider:

$$x_i - m = (x_i - \mu) - (m - \mu)$$

Then:

$$(x_i - m)^2 = (x_i - \mu)^2 + (m - \mu)^2 - 2(x_i - \mu)(m - \mu)$$

Summing over all i:

$$\begin{aligned}\sum_i n_i(x_i - m)^2 \\ &= \sum_i n_i(x_i - \mu)^2 + (m - \mu)^2 \sum_i n_i \\ &- 2(m - \mu)\sum_i n_i(x_i - \mu)\end{aligned}$$

Since,

$$m\sum_i n_i = \sum_i n_i x_i$$

The term $\sum_i n_i(x_i - \mu)$ simplifies to zero:

$$\begin{aligned}\sum_i n_i(x_i - m)^2 &= \sum_i n_i(x_i - \mu)^2 + (m - \mu)^2 \sum_i n_i - 2(m - \mu)^2 \sum_i n_i \\ &= \sum_i n_i(x_i - \mu)^2 - (m - \mu)^2 \sum_i n_i\end{aligned}$$

Thus:

$$v = \frac{\sum_i n_i (x_i - \mu)^2}{\sum_i n_i} - (m - \mu)^2$$

This result is known as Koenig's theorem.

If μ *is chosen* to be 0, the result simplifies to:

$$v = \sigma^2 = q^2 - m^2.$$

where q^2 is the mean square, leading to the Huygens–Koenig result.

EXAMPLE 10.28.– Apply the previous results to example 10.27 using the data provided in the table in Figure 10.40.

distance in km	[0, 1[	[1, 2[	[2, 5[	[5, 10[	[10, 20[	[20, 50]	sum
centers x_i	0.5	1.5	3.5	7.5	15	35	
size n_i	353	159	255	147	59	27	1000
$n_i x_i$	176.5	238.5	892.5	1102.5	885	945	4240
$n_i x_i^2$	88.25	357.75	3123.75	8268.75	13275.00	33075.00	58188.50
$(x_i - m)$	-3.74	-2.74	-0.74	3.26	10.76	30.76	
$(xi - m)^2$	13.99	7.51	0.55	10.63	115.78	946.18	
$n_i(x_i - m)^2$	4937.62	1193.71	139.64	1562.26	6830.88	25546.80	40210.90
$(x_i - \mu)$	-4.5	-3.5	-1.5	2.5	10	30	
$(x_i - \mu)^2$	20.25	12.25	2.25	6.25	100.00	900.00	
$n_i(x_i - \mu)^2$	7148.25	1947.75	573.75	918.75	5900.00	24300.00	40788.50

Figure 10.40. *Table in Figure 10.39 completed*

Applying the definition, the variance is calculated as follows:

$$v = 40210.90/1{,}000 = 40.21$$

Using a provisional mean $\mu = 5$: $v = 40788.50 - (4.24 - 5)2 = 40.21$

With a provisional mean of 0: $v = 58118.50 - (4.24)2 = 40.21$

The standard deviation is:

$$\sigma = \sqrt{40.21} = 6.34$$

10.4.3. *Coefficient of variation and concentration index*

When the scale of a character is expanded or contracted, the dispersion of a statistical series is affected. Specifically, if x_i, ρ,and m are transformed to x'_i, ρ' and $m'$with:

$$x'_i = kx_i, \qquad \rho' = k\rho \quad \text{and} \quad m' = km.$$

Although the standard deviation and the mean are altered by this transformation, the ratio of these two quantities remains unchanged. This ratio is known as *the coefficient of variation*:

$$Coefficient\ of\ variation = \frac{\sigma}{m}$$

The coefficient of variation is invariant under homothetic changes of scale, allowing for comparisons between two statistical series with different scales.

Given a statistical series (x_i, n_i), plotting the cumulative percentage of $n_i x_i$ against the cumulative percentage of n_i yields a curve known as the *concentration curve* or *Lorenz curve*. This curve is typically represented within a square with sides of length 100.

EXAMPLE 10.29.– Modify the table in Figure 10.40 to include the percentages as shown in Figure 10.41.

distance in km	[0, 1[	[1, 2[	[2, 5[	[5, 10[	[10, 20[	[20, 50]
centers x_i	0.5	1.5	3.5	7.5	15	35
size n_i	353	159	255	147	59	27
cumulative size	353	512	767	914	973	1000
cumulative size (%)	35.30	51.20	76.70	91.40	97.30	100.00
$n_i x_i$	176.50	238.50	892.50	1102.50	885.00	945.00
cumulative $n_i x_i$	176.50	415.00	1307.50	2410.00	3295.00	4240.00
cumulative $n_i x_i$ (%)	4.16	9.79	30.84	56.84	77.71	100.00

Figure 10.41. *Percentage table*

Draw the curve of the cumulative $n_i x_i$ percentages as a function of the cumulative n_i percentages (see Figure 10.42.).

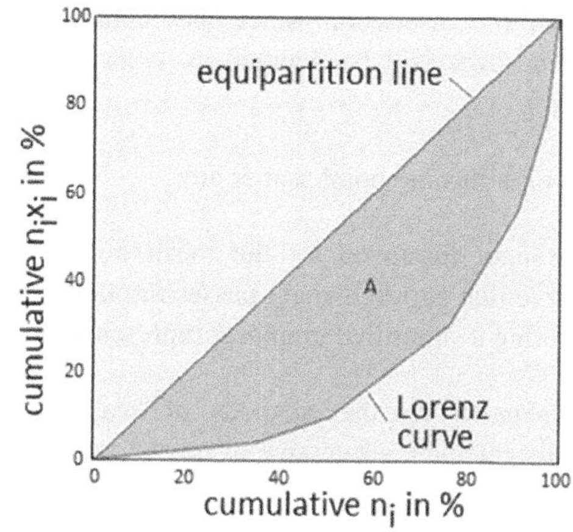

Figure 10.42. *Lorenz curve*

The equipartition line divides the square into two regions of equal areas. When the Lorenz curve deviates from this line, it indicates a higher concentration of data. To measure this concentration, *the concentration index* or *Gini index* is used. This index is defined by the area A between the equipartition line and the Lorenz curve, relative to the area of the half-square, which is 5,000. The Gini index is calculated as follows:

$$Gini\ index = \frac{A}{5{,}000}$$

The Gini index ranges from 0 to 1. A value of 0 indicates perfect equality (0 concentration), meaning that the data is evenly distributed. Conversely, a value of 1 indicates maximum concentration, where the entire area A ($A = 5{,}000$) is between the equipartition line and the Lorenz curve.

10.5. Linear adjustment

10.5.1. *Principle of adjustment*

A statistical series consists of a set of numerical data. To explore relationships between these data, it is often useful to model them in a more concise and understandable way. One common approach is fitting, which involves finding a function $n_i = f(x_i)$ that best represents the data. In practice, the points (x_i, n_i) form a scatter of point, and the goal is to find a curve that best fits this scatter.

In the remainder of the discussion, since the focus is on curves and their graphical representation, sizes will be denoted as y_i instead of n_i, to obtain the desired curve $y_i = f(x_i)$.

The main methods for adjusting point scatter are:

1) *Graphical adjustment* improved by the mean point method: this method involves drawing a curve that approximately passes through the mean points of the grouping intervals, offering a simplified graphical representation of the point scatter.

2) *Mechanical adjustment* by the methods of scaled averages or moving averages: these methods calculate successive averages over subsets of neighboring points, helping to obtain an adjustment curve by smoothing out the fluctuations in the scatter points.

3) *Analytical adjustment*: this method uses the least squares technique to determine the mathematical equation that best represents the relationship between the variables in point scatter. It minimizes the sum of the squares of the differences between the observed values and those predicted by the adjustment curve.

For a graphical adjustment, the principle is relatively simple: it involves drawing a freehand curve that appears to best fit among the scatter points (see Figure 10.43). This method has a significant drawback: the adjustment obtained is imprecise and dependent on the designer's judgment. However, it also offers two important advantages: first, it helps in identifying and potentially eliminating "outlier" points that may be excluded in other adjustment methods; second, it is characterized by its simplicity and low cost, as it requires only basic drawing skills.

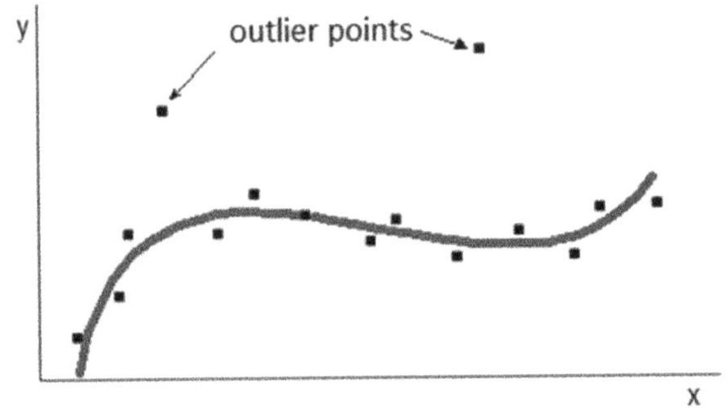

Figure 10.43. *Graphic adjustment. For a color version of this figure, see www.iste.co.uk/cochard/mathematics1.zip*

To obtain a more precise fitting curve, the *average point method* can be used. This method involves connecting successive points so as to form triangles and then drawing a median for each triangle (note that the median, in the geometric sense, is the line segment connecting a vertex of a triangle to the midpoint of the opposite side). Next, the midpoints of these medians are connected (see Figure 10.44). This approach results in a smoother curve that better captures the overall trends in the point scatter.

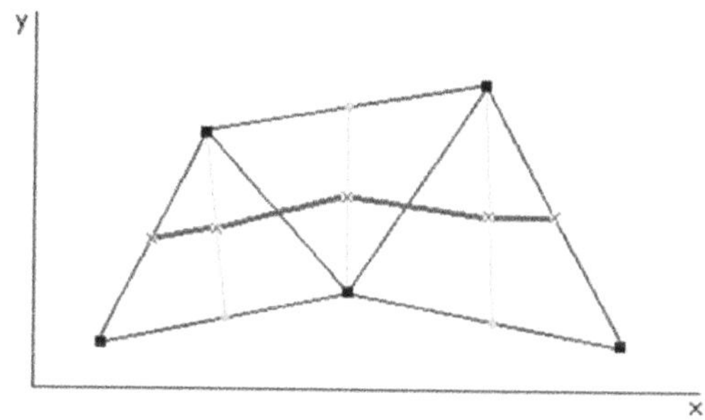

Figure 10.44. *Mean point method. For a color version of this figure, see www.iste.co.uk/cochard/mathematics1.zip*

The principle of mechanical adjustment involves replacing the series of pairs (x_i, y_i) with a new series of pairs (x_i, y_i'), where y_i' are calculated from y_i using a specific method. Two main processes are commonly used:

The scaled average method

The pairs are divided into groups, as shown in Figure 10.45.

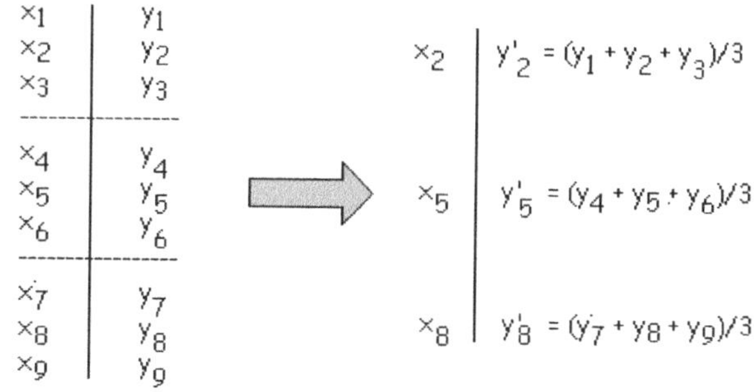

Figure 10.45. *Scaled average method*

The resulting points, which are fewer in number, are then connected with line segments. Another variation involves applying a weighting to the central element. For example, for the point (x_2, y_2') in Figure 10.45, the calculation can be expressed as:

$$y_2' = y_1 + 2y_2 + y_3.$$

However, this method has some drawbacks. First, it reduces the number of points available for adjustment. Additionally, it relies on an arbitrary splitting of points and the arbitrary calculation of y_i'. Another disadvantage is that the point cloud must have a certain structure to be effectively modeled by a curve.

The moving average method

This method uses an integer p, for example, 3, allowing calculation by recurrence:

$$y'_i = \frac{y_{i-1} + y_i + y_{i+1}}{3}$$

Figure 10.46 shows the transformation applied.

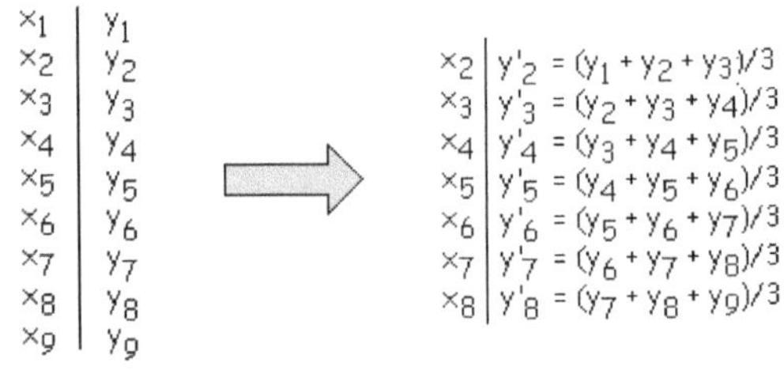

Figure 10.46. *Moving average method*

This method results in a loss of only 2 points, and the operation resembles a form of smoothing.

Graphical and mechanical adjustments provide an approximation of the adjustment curve's shape. Next, a curve of known type, represented by the equation $y = f(x)$, is chosen. This equation includes several parameters for which optimal values need to be determined.

For example, for a line with the equation $y = ax + b$, the parameters to be determined are a and b. For a parabola with the equation $y = ax^2 + bx + c$, the parameters to be determined are a, b and c. Determining these parameters is the goal of *analytical fitting*, with the *least square's method* being the most commonly used technique.

Consider Figure 10.47, which shows a few points and the curve $y = f(x)$ supposed to best represent these points.

For each point (x_i, y_i), the corresponding point on the assumed adjustment curve (x_i, y'_i), where $y'_i = f(x_i)$, is evaluated. The difference in absolute value between the ordinates of these two points is then measured:

$$|y_i - y'_i| = |y_i - f(x_i)|.$$

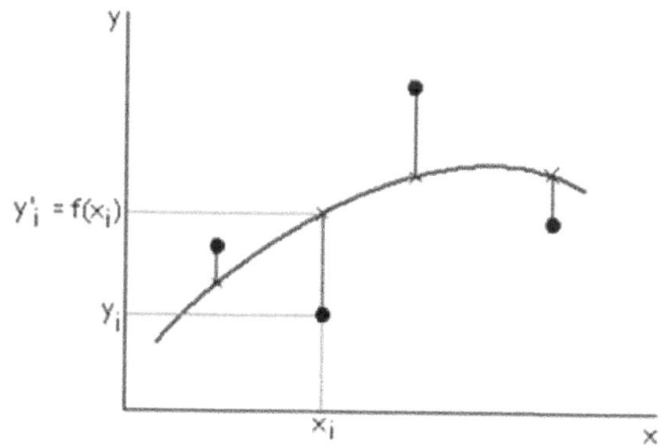

Figure 10.47. *Least squares method. For a color version of this figure, see www.iste.co.uk/cochard/mathematics1.zip*

The following quantity is then formed:

$$M = \sum_{i=1}^{n} (y_i - f(x_i))^2$$

To determine the unknown parameters a, b, c, etc. of the function $f(x)$, M is minimized. This involves solving the system of equations obtained by setting the partial derivatives of M to zero:

$$\frac{\partial M}{\partial a} = 0 \quad \frac{\partial M}{\partial b} = 0 \quad \frac{\partial M}{\partial c} = 0 \quad \ldots$$

This process results in a system of equations that can be complex and requires solving to find the optimal parameters.

10.5.2. *Linear adjustment*

If the shape of the point cloud permits, a linear fit can be attempted using the linear equation $y = ax + b$. This straightforward fitting method is referred to as *linear adjustment*.

For a statistical series consisting of *n* points (x_i, y_i), where the focus is on these points as a collection rather than a series, the arithmetic means of the abscissa and ordinate are calculated as follows:

$$m_x = \frac{1}{n}\sum_{i=1}^{n} x_i \qquad m_y = \frac{1}{n}\sum_{i=1}^{n} y_i$$

As previously described, the objective function is:

$$M = \sum_{i=1}^{n} (y_i - ax_i - b)^2$$

The partial derivatives: to find the optimal values for a and b, set the partial derivatives of M with respect to a and b to zero:

$$\frac{\partial M}{\partial a} = -2\sum_{i} x_i y_i + 2a\sum_{i} {x_i}^2 + 2b\sum_{i} x_i = 0$$

$$\frac{\partial M}{\partial b} = -2\sum_{i} y_i + 2a\sum_{i} x_i + 2nb = 0$$

The system of two linear equations with the unknowns a and b and can be derived from minimizing the sum of squared deviations. Using the averages m_x and m_y, the system of equations is:

$$\left(\sum_{i} x_i^2\right) a + nm_x \times b = \sum_{i} x_i y_i$$

$$nm_x \times a + nb = nm_y$$

Here is a step-by-step approach to solving the system:

1) Sum of squares and cross products:

$$\left(\sum_i x_i^2\right)a + nm_x \times b = \sum_i x_i y_i$$

This equation comes from setting the partial derivative of M with respect to a to zero.

$$nm_x \times a + nb = nm_y$$

This equation comes from setting the partial derivative of M with respect to b to zero.

2) Simplify the system:

Using the averages m_x and m_y, where:

$$m_x = \frac{1}{n}\sum_{i=1}^{n} x_i \qquad m_y = \frac{1}{n}\sum_{i=1}^{n} y_i$$

The system of equations can be written as:

$$\left(\sum_i x_i^2\right)a + nm_x \times b = \sum_i x_i y_i$$

$$nm_x \times a + nb = nm_y$$

3) Matrix form: this can be represented in matrix form

$$\begin{bmatrix} \sum_i x_i^2 & nm_x \\ nm_x & n \end{bmatrix}\begin{bmatrix} a \\ b \end{bmatrix} = \begin{bmatrix} \sum_i x_i y_i \\ nm_y \end{bmatrix}$$

4) Solving for parameters: Calculate the determinants to find a and b,

$$\Delta = \begin{vmatrix} \sum_i x_i^2 & nm_x \\ nm_x & n \end{vmatrix} = \left(\sum_i x_i^2\right) \times n - (n^2 m_x^2)$$

In parallel,

$$\Delta_a = \begin{vmatrix} \sum_i x_i y_i & nm_x \\ nm_y & n \end{vmatrix} = n\sum_i x_i y_i - n^2 m_x m_y$$

and

$$\Delta_b = \begin{vmatrix} \sum_i x_i^2 & \sum_i x_i y_i \\ nm_x & nm_y \end{vmatrix} = nm_y \sum_i x_i^2 - nm_x \sum_i x_i y_i$$

The parameters a and b are then computed as follows:

$$a = \frac{\Delta_a}{\Delta} \quad \text{and} \quad b = \frac{\Delta_b}{\Delta}$$

Thus:

$$a = \frac{\sum_i x_i y_i - nm_x m_y}{\sum_i x_i^2 - nm_x^2} \quad and \quad b = \frac{m_y \sum_i x_i^2 - m_x \sum_i x_i y_i}{\sum_i x_i^2 - nm_x^2}$$

It is implicitly assumed that the denominator $\sum_i x_i^2 - nm_x^2$ is non-zero. This will be further explained.

By working in *centered coordinates*, simpler expression for a and b can be obtained. Define:

$$X_i = x_i - m_x \quad and \quad Y_i = x_i - m_y$$

These new coordinates satisfy the relationships $\sum_i X_i = 0 \; and \; \sum_i Y_i = 0$. Specifically,

$$\sum_i X_i = \sum_i (x_i - m_x) = \sum_i x_i - nm_x = nm_x - nm_x = 0$$

and

$$\sum_i Y_i = \sum_i (y_i - m_y) = \sum_i y_i - nm_y = nm_y - nm_y = 0$$

In the new coordinates, the formulas of a and b become:

$$\sum_i x_i y_i = \sum_i (X_i + m_x)(Y_i + m_y) = \sum_i X_i Y_i + n m_x m_y$$

since $\sum_i X_i = \sum_i Y_i = 0$.

Likewise

$$\sum_i x_i^2 = \sum_i (X_i + m_x)^2 = \sum_i X_i^2 + n m_x^2$$

and

$$\sum_i y_i^2 = \sum_i (Y_i + m_y)^2 = \sum_i Y_i^2 + n m_y^2$$

hence the following simplified expressions for a and b:

$$a = \frac{\sum_i X_i Y_i}{\sum_i X_i^2} \quad \text{and} \quad b = m_y - m_x \frac{\sum_i X_i Y_i}{\sum_i X_i^2} = m_y - m_x a$$

These relationships are simpler, highlighting the advantage of working with cantered coordinates. Additionally, the denominator $\sum_i X_i^2$ is strictly positive (assuming not all X_i are zero simultaneously), which justifies solving the system of equations ($\Delta \neq 0$).

To find the minimum value of M, start with the following expression:

$$\begin{aligned} y_i - a x_i - b &= Y_i + m_y - \frac{\sum_j X_j Y_j}{\sum_j X_j^2}(X_i + m_x) - m_y + m_x \frac{\sum_j X_j Y_j}{\sum_j X_j^2} \\ &= Y_i - \frac{\sum_j X_j Y_j}{\sum_j X_j^2} X_i \end{aligned}$$

Thus:

$$(y_i - a x_i - b)^2 = Y_i^2 + \left(\frac{\sum_j X_j Y_j}{\sum_j X_j^2}\right)^2 X_i^2 - 2 X_i Y_i \frac{\sum_j X_j Y_j}{\sum_j X_j^2}$$

Summing this over all i yields:

$$\sum_i (y_i - ax_i - b)^2 = \sum_i Y_i^2 + \left(\frac{\sum_j X_j Y_j}{\sum_j X_j^2}\right)^2 \sum_i X_i^2 - 2\frac{\sum_j X_j Y_j}{\sum_j X_j^2} \sum_i X_i Y_i$$
$$= \sum_i Y_i^2 - \frac{(\sum_i X_i Y_i)^2}{\sum_i X_i^2}$$

Then, the minimum value of M is:

$$M_{min} = \sum_i (y_i - ax_i - b)^2 = \frac{\sum_i X_i^2 \sum_i Y_i^2 - (\sum_i X_i Y_i)^2}{\sum_i X_i^2}$$

Because $M_{min} \geq 0$, it follows that:

$$(\sum_i X_i^2)(\sum_i Y_i^2) \geq \left(\sum_i X_i Y_i\right)^2$$

This inequality is known as the *Cauchy–Schwarz inequality*.

EXAMPLE 10.30.– Consider the statistical series represented in Figure 10.48, which shows the average hourly wages (in euros) of professional workers in the metal industries of the Paris region from 1950 to 1964 (note that these values are fictitious).

The appearance of the scatterplot suggests that a linear fit may be appropriate.

The calculations required to determine the fitting line are detailed in Figure 10.49, with the result of this fitting shown in Figure 10.50.

The slope a and intercept b of the fitting line are computed as follows:

$$a = \frac{36.5}{168} = 0.22 \quad \text{and} \quad b = 2.69 - 1957 \times 0.22 = -422.49.$$

Thus, the equation of the fitting line is:

$$y = 0.22x - 422.49$$

This linear model allows for interpolation:

– Average hourly wage in 1955: 2.25 euros.

– Average hourly wage in 1966: 4.64 euros.

year	hourly wage
1950	1.15
1952	1.82
1954	1.99
1956	2.36
1958	2.83
1960	3.20
1962	3.75
1964	4.40

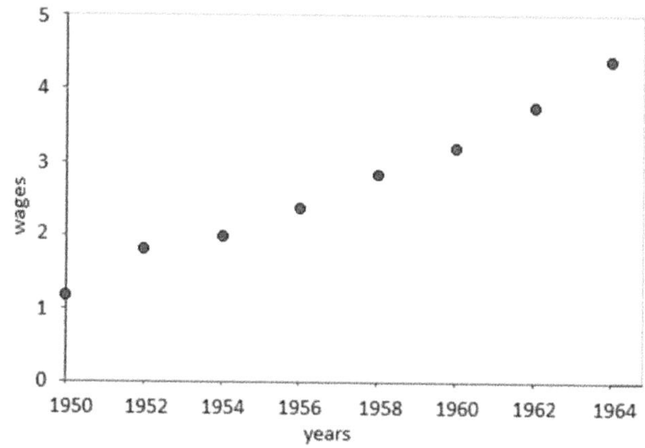

Figure 10.48. *Statistical distribution. For a color version of this figure, see www.iste.co.uk/cochard/mathematics1.zip*

x_i	y_i	X_i	Y_i	X_i^2	X_iY_i
1950	1.15	-7	-1.54	49	10.76
1952	1.82	-5	-0.87	25	4.34
1954	1.99	-3	-0.70	9	2.09
1956	2.36	-1	-0.33	1	0.33
1958	2.83	1	0.14	1	0.14
1960	3.20	3	0.51	9	1.54
1962	3.75	5	1.06	25	5.31
1964	4.40	7	1.71	49	11.99

averages	1957	2,69
sums	168	36,5

Figure 10.49. *Details of calculations*

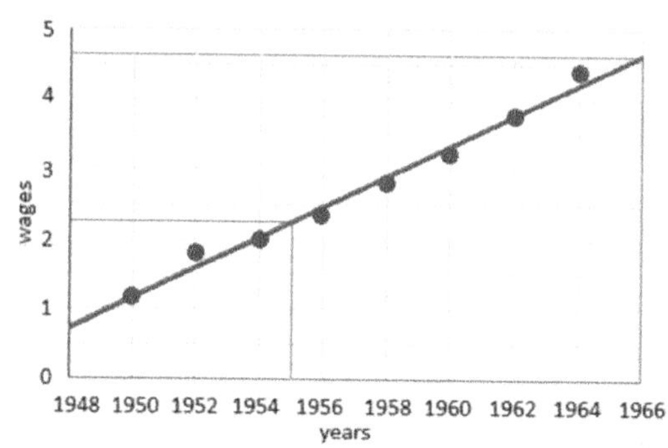

Figure 10.50. *Linear adjustment. For a color version of this figure, see www.iste.co.uk/cochard/mathematics1.zip*

10.6. Chronological series

10.6.1. *Introduction*

A *chronological series*, or *temporal series*, is a statistical series that represents changes in particular magnitude over time. Chronological series are commonly used to analyze economic, social or scientific phenomena.

In this section, a chronological series is used to illustrate variations in the number of job applications in France from 1973 to 1978, expressed in thousands. These data sourced from an article published in *Le Monde* on April 24, 1979 (see Figure 10.51).

years	January	February	March	April	May	June
1973	420	400	380	360	340	330
1974	460	455	440	420	390	375
1975	760	770	755	755	745	746
1976	1020	975	930	900	850	810
1977	1075	1050	1025	1000	975	970
1978	1130	1100	1075	1075	1060	1040

years	July	August	September	October	November	December
1973	350	360	410	450	460	465
1974	400	430	525	620	690	720
1975	775	800	950	1025	1030	1025
1976	805	850	950	1020	1040	1040
1977	1015	1040	1175	1210	1175	1150
1978	1100	1159	1280	1350	1325	1320

Figure 10.51. *Change in the number of job applications*

The variation of a socio-economic phenomenon over time generally results from four fundamental components:

– The long-term movement or general *trend*: this represents the overall direction and evolution of the phenomenon over an extended period. It can be ascending (upward), descending (downward) or stable.

– Seasonal movement: this includes regular and recurring variations observed at specific times of the year. These variations are typically linked to seasonal factors such as holidays, agricultural cycles and festivals.

– Accidental variations: these are due to unforeseen and non-periodic events such as strikes, droughts, floods and wars. These variations are often regarded as temporary disturbances in the time series.

– Cyclical movement: this results from the periodicity of economic phases such as and recession. The importance and impact of cyclical movements can be subject to debate among economists.

The study of a chronological series involves separating and identifying its different components. In this section, the focus will be on analyzing the trend and seasonal movement (at the monthly level in the chosen example).

Two models can be used to decompose a time series:

1) The additive model: in this model, the different components are treated as distinct elements that add together. The notations used are as follows: y_i represents the raw data, y_i' represents the data corresponding to the general trend, s_i denotes the seasonal component and e_i accounts for accidental variations and cyclical movement. According to this model, the relationship is expressed as:

$$y_i = y_i' + s_i + e_i$$

In other words, the observed value y_i is the sum of the general trend y_i', the seasonal component s_i and the accidental variations e_i.

2) The multiplicative model: in this model, the components are treated as multiplicative factors that interact with each other. In this case, it is assumed that the seasonal component is proportional to the general trend. The relationship is given by:

$$y_i = y_i' + \alpha_i y_i' + e_i = (1 + \alpha_i) y_i' + e_i = c_i y_i' + e_i$$

where $c_i = (1 + \alpha_i)$ is called the seasonal coefficient.

Depending on the nature of the data, we might choose between the additive and multiplicative models. However, the multiplicative model is generally preferred for its ability to account for proportional variations.

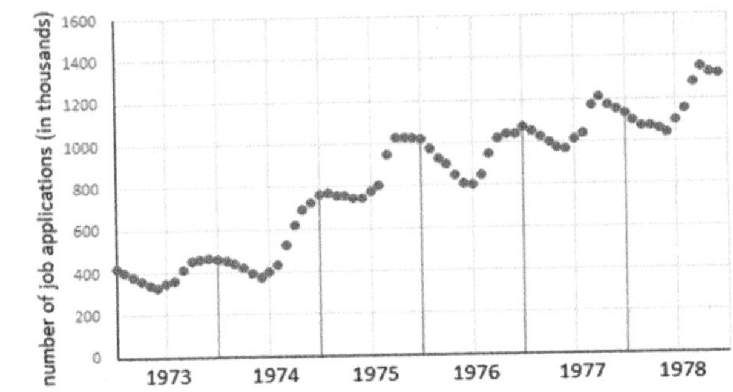

Figure 10.52. *Change year by year. For a color version of this figure, see www.iste.co.uk/cochard/mathematics1.zip*

In the study, it is possible to identify distinct *periods* within which *seasonal variations* can be observed. In this example, periods correspond to years and seasons to months. Figures 10.52 and 10.53 show the repeatability of seasons (months) within each period (year), helping to visualize seasonal variations and their regularity over time.

For the analysis, the term e_i, will be disregarded, simplifying the seasonal coefficient to:

$$c_i = \frac{y_i}{y'_i}$$

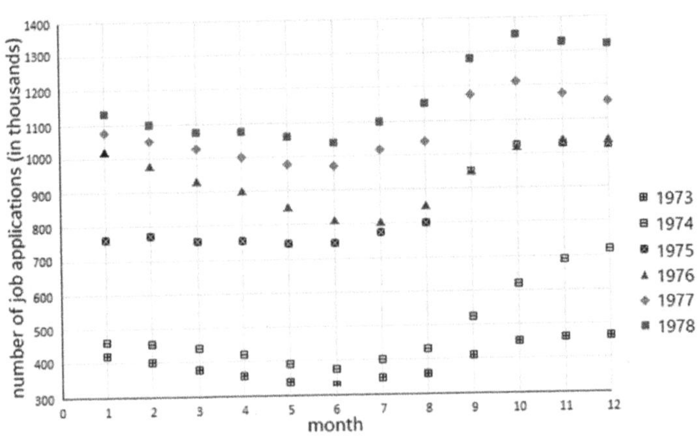

Figure 10.53. *Change month by month. For a color version of this figure, see www.iste.co.uk/cochard/mathematics1.zip*

Assuming that the seasonal component is independent of the period and only depends on the season, it is necessary to calculate 12 seasonal coefficients, each corresponding to a specific month.

Additionally, it is assumed that seasonal increases are offset by seasonal decreases, meaning that seasonal fluctuations balance out.

For an additive model, this results in the condition that the sum of the seasonal coefficients for all seasons of a given period is equal to zero: $\sum_j s_j = 0$, where the index "*j*" represents the seasons of a given period.

For a multiplicative model, the sum of the seasonal coefficients must equal the total number of seasons, i.e. $\sum_j c_j = (\text{number of seasons})$ (in this example, $\sum_j c_j = 12$). This hypothesis may not always be valid for every time series.

10.6.2. *Study of the general trend*

The general trend in a time can be obtained using one of the following methods:

- the scaled average method or the moving average method;
- analytical adjustment.

Each method has its advantages, and the choice depends on the nature of the data and the specific requirements of the analysis.

In the example, the scaled average method is first used to determine the general trend by calculating the averages for each year (see Figure 10.54). This method involves averaging over successive periods to smooth out seasonal fluctuations and highlight the general trend. The graphical representation in Figure 10.54 suggests a linear fit is appropriate for the overall trend, indicating an increasing trend that can be approximated by a straight line.

To simplify the data, month by month and year by year, values are assigned: -71 for January 1973, -69 for February 1973, -67 for March 1973, etc., up to +71 for December 1978. This adjustment results in an average m_x of 0, meaning that $X_i = x_i - m_x = x_i$.

years	averages
1973	393.75
1974	493.75
1975	844.67
1976	932.50
1977	1071.67
1978	1167.08

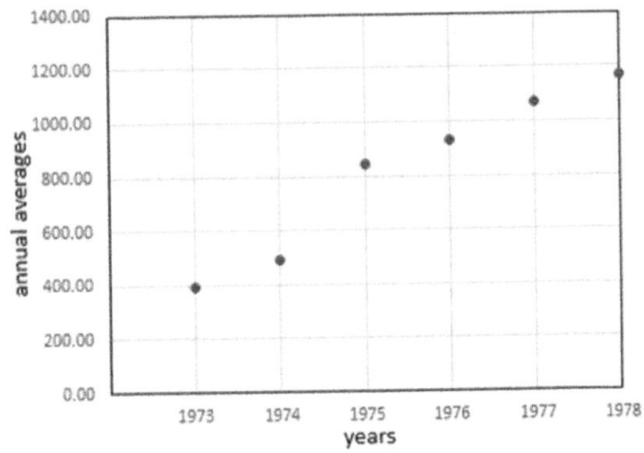

Figure 10.54. *Representation of change. For a color version of this figure, see www.iste.co.uk/cochard/mathematics1.zip*

$x_i = X_i$	y_i	Y_i	X_i^2	X_iY_i
-71	420	-397.36	5041.00	28212.64
-69	400	-417.36	4761.00	28797.92
-67	380	-437.36	4489.00	29303.19
-65	360	-457.36	4225.00	29728.47
-63	340	-477.36	3969.00	30073.75
-61	330	-487.36	3721.00	29729.03
-59	350	-467.36	3481.00	27574.31
-57	360	-457.36	3249.00	26069.58
-55	410	-407.36	3025.00	22404.86
-53	450	-367.36	2809.00	19470.14
-51	460	-357.36	2601.00	18225.42
-49	465	-352.36	2401.00	17265.69

$x_i = X_i$	y_i	Y_i	X_i^2	X_iY_i
-47	460	-357.36	2209.00	16795.97
-45	455	-362.36	2025.00	16306.25
-43	440	-377.36	1849.00	16226.53
-41	420	-397.36	1681.00	16291.81
-39	390	-427.36	1521.00	16667.08
-37	375	-442.36	1369.00	16367.36
-35	400	-417.36	1225.00	14607.64
-33	430	-387.36	1089.00	12782.92
-31	525	-292.36	961.00	9063.19
-29	620	-197.36	841.00	5723.47
-27	690	-127.36	729.00	3438.75
-25	720	-97.36	625.00	2434.03

a)

$x_i = X_i$	y_i	Y_i	X_i^2	X_iY_i
-23	760	-57.36	529.00	1319.31
-21	770	-47.36	441.00	994.58
-19	755	-62.36	361.00	1184.86
-17	755	-62.36	289.00	1060.14
-15	745	-72.36	225.00	1085.42
-13	745	-72.36	169.00	940.69
-11	775	-42.36	121.00	465.97
-9	800	-17.36	81.00	156.25
-7	950	132.64	49.00	-928.47
-5	1025	207.64	25.00	-1038.19
-3	1030	212.64	9.00	-637.92
-1	1025	207.64	1.00	-207.64

$x_i = X_i$	y_i	Y_i	X_i^2	X_iY_i
1	1020	202.64	1.00	202.64
3	975	157.64	9.00	472.92
5	930	112.64	25.00	563.19
7	900	82.64	49.00	578.47
9	850	32.64	81.00	293.75
11	810	-7.36	121.00	-80.97
13	805	-12.36	169.00	-160.69
15	850	32.64	225.00	489.58
17	950	132.64	289.00	2254.86
19	1020	202.64	361.00	3850.14
21	1040	222.64	441.00	4675.42
23	1040	222.64	529.00	5120.69

b)

$x_i = X_i$	y_i	Y_i	X_i^2	X_iY_i
25	1075	257.64	625.00	6440.97
27	1050	232.64	729.00	6281.25
29	1025	207.64	841.00	6021.53
31	1000	182.64	961.00	5661.81
33	975	157.64	1089.00	5202.08
35	970	152.64	1225.00	5342.36
37	1015	197.64	1369.00	7312.64
39	1040	222.64	1521.00	8682.92
41	1175	357.64	1681.00	14663.19
43	1210	392.64	1849.00	16883.47
45	1175	357.64	2025.00	16093.75
47	1150	332.64	2209.00	15634.03

$x_i = X_i$	y_i	Y_i	X_i^2	X_iY_i
49	1130	312.64	2401.00	15319.31
51	1100	282.64	2601.00	14414.58
53	1075	257.64	2809.00	13654.86
55	1075	257.64	3025.00	14170.14
57	1060	242.64	3249.00	13830.42
59	1040	222.64	3481.00	13135.69
61	1100	282.64	3721.00	17240.97
63	1150	332.64	3969.00	20956.25
65	1280	462.64	4225.00	30071.53
67	1350	532.64	4489.00	35686.81
69	1325	507.64	4761.00	35027.08
71	1320	502.64	5041.00	35687.36

c)

Figure 10.55. *Detailed calculations*

Detailed calculations of the linear fit are provided in Figure 10.55(a)–(c).

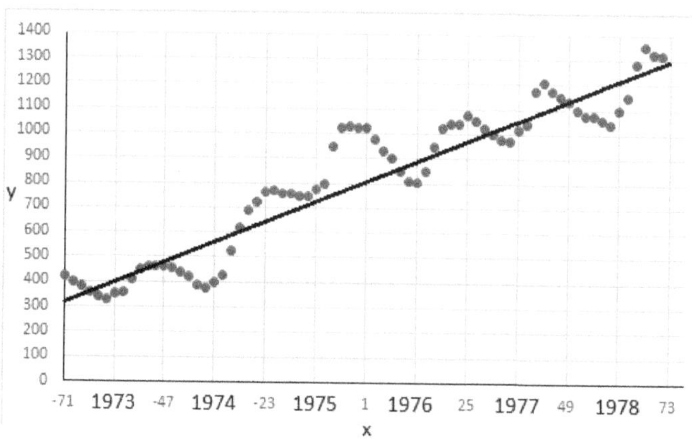

Figure 10.56. *Adjustment line for the general trend. For a color version of this figure, see www.iste.co.uk/cochard/mathematics1.zip*

From the tables in Figure 10.55, the following values are deduced: $m_x = 0$, $m_y = 817.22$, the next sum of squares $\sum_i X_i^2 = 124{,}392$ and the next sum of products $\sum_i X_i Y_i = 849{,}630$. Thus, the equation of the adjustment line, as shown in Figure 10.56, is: $y = 6.83x + 817.22$.

10.6.3. *Study of seasonal variations*

To analyze seasonal variations of a phenomenon, several methods can be employed. In this study, two methods will be highlighted.

a) Monthly average method

Calculate the average for each month across all years. This yields the monthly averages. Compute the annual average, which is the average of these monthly averages. The "seasonal coefficient" is then determined by the formula:

$$\text{Seasonal coefficient} = \frac{\text{Seasonal average}}{\text{Total average}}$$

In this approach, the overall period is defined as the year, and the season is represented by the month (see Figure 13.57).

By following these steps, the hypothesis that the sum of seasonal coefficients should equal 12 can be verified, i.e. $\sum_j c_j = 12$.

month	Jan	Feb	Mar	Apr	May	Jun	Jul	Aug	Sep	Oct	Nov	Dec
monthly average	810.83	791.67	767.50	751.67	726.67	711.67	740.83	771.67	881.67	945.83	953.33	953.33
seasonal coefficient	0.99	0.97	0.94	0.92	0.89	0.87	0.91	0.94	1.08	1.16	1.17	1.17

Figure 10.57. *Determination of seasonal coefficients*

The monthly seasonal coefficients obtained can be applied in two different ways:

1) Seasonal adjustment:

– Objective: to remove the seasonal component from the raw data.

– Method: divide the raw data by the corresponding monthly seasonal coefficients. This process yields seasonally adjusted data, which highlights the underlying trend and helps to analyze the general trend of the phenomenon without seasonal effects.

– Formula:

$$\text{Seasonal adjusted data} = \frac{\text{Raw data}}{\text{Monthly seasonal coeifficient}}$$

2) Seasonal reconstruction:

– Objective: to model or reconstruct the original data based on the identified general trend and seasonal effects.

– Method: multiply the trend values by the seasonal coefficient for each period. This method combines the general trend with the seasonal component to approximate the original data. Although simple, this method provides a basic representation of seasonal variations.

– Formula:

$$\text{Reconstructed data} = \text{Trend values} \times \text{Seasonal coeifficient}$$

b) Method of reporting to the general trend

In this method, the general trend is assumed to have been determined, for instance, through linear adjustment. The focus here is on understanding how the raw data deviates from this general trend over time. The steps involved are as follows:

1) Computing the ratio: for each month and year, compute the ratio between the raw data and the general trend data. This ratio reflects the seasonal component of the data relative to the trend. The formula used is as follows:

$$\text{Seasonal ratio}_i = \frac{\text{Raw data}_i}{\text{Trend data}_i}$$

2) Results: organize these ratios in a table or graph to visualize the seasonal deviations. This table or graph will provide insights into how much the raw data deviates from the general trend in each season.

For each month and year, calculate the ratio between the raw data and the general trend data, as shown in the table in Figure 10.58.

year		Jan	Feb	Mar	Apr	May	Jun	Jul	Aug	Sep	Oct	Nov	Dec
1973	raw data	420	400	380	360	340	330	350	360	410	450	460	465
	trend	332.29	345.95	359.61	373.27	386.93	400.59	414.25	427.91	441.57	455.23	468.89	482.55
	ratio	1.26	1.16	1.06	0.96	0.88	0.82	0.84	0.84	0.93	0.99	0.98	0.96
1974	raw data	460	455	440	420	390	375	400	430	525	620	690	720
	trend	496.21	509.87	523.53	537.19	550.85	564.51	578.17	591.83	605.49	619.15	632.81	646.47
	ratio	0.93	0.89	0.84	0.78	0.71	0.66	0.69	0.73	0.87	1.00	1.09	1.11
1975	raw data	760	770	755	755	745	746	775	800	950	1025	1030	1025
	trend	660.13	673.79	687.45	701.11	714.77	728.43	742.09	755.75	769.41	783.07	796.73	810.39
	ratio	1.15	1.14	1.10	1.08	1.04	1.02	1.04	1.06	1.23	1.31	1.29	1.26
1976	raw data	1020	975	930	900	850	810	805	850	950	1020	1040	1040
	trend	824.05	837.71	851.37	865.03	878.69	892.35	906.01	919.67	933.33	946.99	960.65	974.31
	ratio	1.24	1.16	1.09	1.04	0.97	0.91	0.89	0.92	1.02	1.08	1.08	1.07
1977	raw data	1075	1050	1025	1000	975	970	1015	1040	1175	1210	1175	1150
	trend	987.97	1001.63	1015.29	1028.95	1042.61	1056.27	1069.93	1083.59	1097.25	1110.91	1124.57	1138.23
	ratio	1.09	1.05	1.01	0.97	0.94	0.92	0.95	0.96	1.07	1.09	1.04	1.01
1978	raw data	1130	1100	1075	1075	1060	1040	1100	1150	1280	1350	1325	1320
	trend	1151.89	1165.55	1179.21	1192.87	1206.53	1220.19	1233.85	1247.51	1261.17	1274.83	1288.49	1302.15
	ratio	0.98	0.94	0.91	0.90	0.88	0.85	0.89	0.92	1.01	1.06	1.03	1.01
	averages	**1.11**	**1.06**	**1.00**	**0.96**	**0.90**	**0.87**	**0.88**	**0.91**	**1.02**	**1.09**	**1.09**	**1.07**

Figure 10.58. *Monthly seasonal coefficients*

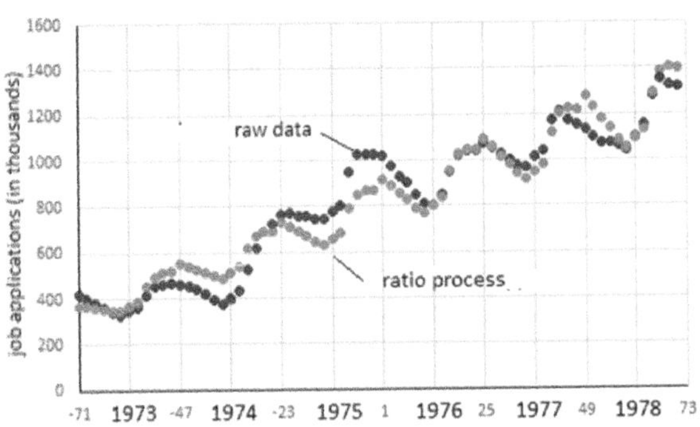

Figure 10.59. *Ratio process. For a color version of this figure, see www.iste.co.uk/cochard/mathematics1.zip*

The last row of the table shows the monthly seasonal coefficients, assumed to be independent of the years. By multiplying these seasonal coefficients with the general trend data, the model displayed in Figure 10.59 is produced and compared to the raw data.

Ratio chain process:

Calculate the ratios $\frac{y_t}{y'_{t-1}}$. For example:

$$\frac{\text{January 1973 applications}}{\text{1973 applications}} = \frac{400}{420} = 0.95,$$

$$\frac{\text{March 1973 applications}}{\text{1973 applications}} = \frac{380}{400} = 0.95,$$

etc.

	J/D-1	F/J	M/F	A/M	M'/A	J'/M'	J"/J'	A/J"	S/A	O /S	N/O	D/N
ratios greater than 1								1.08	1.22	1.18		
							1.07	1.06	1.19	1.10		
							1.06	1.05	1.14	1.08	1.11	
							1.06	1.03	1.13	1.07	1.02	1.04
	1.06			1.00			1.05	1.03	1.12	1.05	1.02	1.01
	1.03	1.01		1.00		1.00	1.04	1.02	1.11	1.03	1.00	1.00
	1.00	0.99	0.98	0.98	0.99	0.99	0.99				0.98	1.00
ratios less than 1	0.99	0.98	0.98	0.97	0.99	0.98					0.97	1.00
	0.98	0.97	0.98	0.95	0.98	0.97						0.98
		0.96	0.97	0.95	0.94	0.96						
		0.95	0.95		0.94	0.95						
			0.95		0.93							
averages	1.01	0.98	0.97	0.97	0.96	0.98	1.04	1.04	1.15	1.09	1.02	1.00

Figure 10.60. *Highlighting seasonal variations*

Record these ratios in a table, highlighting ratios greater than 1 and ratios less than 1, after sorting them column by column, as shown in Figure 10.60. This approach allows for clearer identification of seasonal phenomena.

Modify these results by relating them to January as follows:

$$\frac{F}{J} = 0.98 \qquad \frac{J"}{J} = \frac{J"}{J'} \times \frac{J'}{J} = 0.90$$

$$\frac{M}{J} = \frac{M}{F} \times \frac{F}{J} = 0.95 \qquad \frac{A'}{I} = \frac{A'}{J"} \times \frac{J"}{J} = 0.94$$

$$\frac{A}{J} = \frac{A}{M} \times \frac{M}{J} = 0.92 \qquad \frac{S}{J} = \frac{S}{A'} \times \frac{A'}{J} = 1.08$$

$$\frac{M'}{J} = \frac{M'}{A} \times \frac{A}{J} = 0.88 \qquad \frac{O}{J} = \frac{O}{S} \times \frac{S}{J} = 1.18$$

$$\frac{J'}{J} = \frac{J'}{M'} \times \frac{M'}{J} = 0.86 \qquad \frac{N}{J} = \frac{N}{O} \times \frac{O}{J} = 1.20$$

$$\frac{D}{J} = \frac{D}{N} \times \frac{N}{J} = 1.20$$

End the ratio chain by calculating $\frac{D}{J} \times \frac{J}{D_{-1}} = 1.22$. To eliminate the influence of the long-term trend, multiply each average by a factor k to achieve a final value of 1. This yields the adjusted series, as shown in Figure 10.61. Compute k^{12} as follows:

$$k^{12} = \frac{1}{1.22} = 0.82.$$

Thus:

$$k = e^{\frac{1}{12} \times Ln(0.82)} = 0.9837.$$

This adjustment results in the modified ratios presented in Figure 10.61.

	J/D-1	F/J	M/J	A/J	M'/J	J'/J	J"/J	A'/J	S/J	O/J	N/J	D/J	avg
ratios	1.22	0.98	0.95	0.92	0.88	0.86	0.90	0.94	1.08	1.18	1.20	1.20	
multiplicator	k^{12}	k	k^2	k^3	k^4	k^5	k^6	k^7	k^8	k^9	k^{10}	k^{11}	
modified ratios	1.00	0.96	0.91	0.88	0.83	0.80	0.82	0.84	0.95	1.02	1.02	1.01	0.92
seasonal coefficients	Jan	Feb	Mar	Apr	May	Jun	Jul	Aug	Sep	Oct	Nov	Dec	
	1.00	1.05	1.00	0.95	0.90	0.87	0.89	0.91	1.04	1.11	1.11	1.09	0.99

Figure 10.61. *Modified ratios*

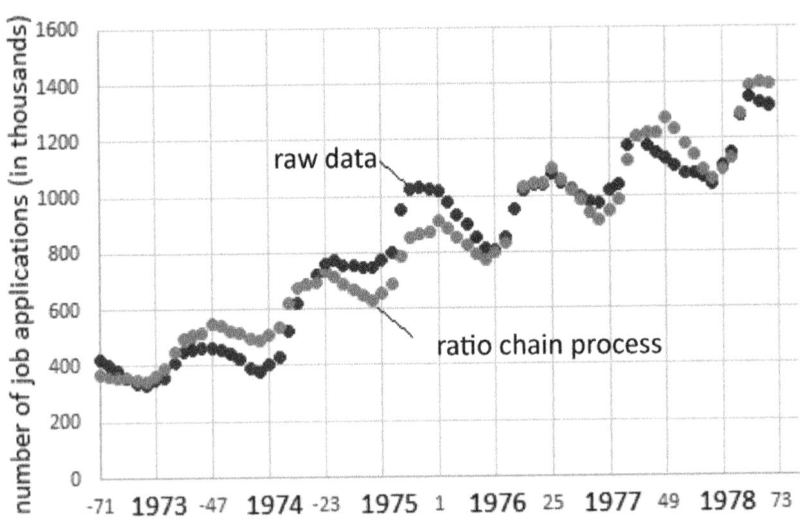

Figure 10.62. *Ratio chain process. For a color version of this figure, see www.iste.co.uk/cochard/mathematics1.zip*

The average of the modified ratios is 0.92. Normalizing this average to 1 yields the seasonal coefficients listed in the final row of the table, as shown in Figure 10.60. Figure 10.62 presents a comparison between the models obtained with the raw data, showing that the result is closest to that derived from the ratios to the general trend.

10.7. Covariance and correlation

10.7.1. *Regression lines*

To perform a linear fit on two-dimensional statistical series, consider the set of triples (x_i, y_j, n_{ij}), where x_i is a value of the first character, y_j is a value of the second character and n_{ij} represents the frequency or count associated with the pair (x_i, y_j). Such a series is typically displayed in a crosstab format (see example 10.9) and can be visualized using a scatter plot, as shown in Figure 10.24.

To fit a line to this scatter plot, we can seek a linear relationship of the form $y = f(x)$, or $x = g(y)$.

In this section, the focus is on linear adjustments using straight lines. For a linear fit, the goal is to find a line of the form $y = ax + b$, where a and b are the parameters to be determined.

– For the curve $y = f(x) = ax + b$, which represents the straight line $D_{x/y}$:

Parameters in original coordinates:

$$a = \frac{\sum_i x_i y_i - n m_x m_y}{\sum_i x_i^2 - n m_x^2} \quad \text{and} \quad b = \frac{m_y \sum_i x_i^2 - m_x \sum_i x_i y_i}{\sum_i x_i^2 - n m_x^2}$$

Parameters in centered coordinates:

$$(X_i = x_i - m_x \quad \text{and} \quad Y_i = y_i - m_y),$$

Then:

$$a = \frac{\sum_i X_i Y_i}{\sum_i X_i^2} \quad \text{and} \quad b = m_y - m_x \frac{\sum_i X_i Y_i}{\sum_i X_i^2}$$

– For the curve $x = g(y) = a'y + b'$, which represents the straight line $D_{x/y}$, a simple exchange between x and y in the previous relationships provides the following.

Parameters in original coordinates:

$$a' = \frac{\sum_i x_i y_i - n m_x m_y}{\sum_i y_i^2 - n m_y^2} \quad \text{and} \quad b' = \frac{m_x \sum_i y_i^2 - m_y \sum_i x_i y_i}{\sum_i y_i^2 - n m_y^2}$$

Parameters in centered coordinates:

$$a' = \frac{\sum_i X_i Y_i}{\sum_i Y_i^2} \quad \text{and} \quad b' = m_x - m_y \frac{\sum_i X_i Y_i}{\sum_i Y_i^2}$$

The equation of this line $x = g(y) = a'y + b'$ can be rewritten as:

$$y = \frac{x}{a'} - \frac{b'}{a'}$$

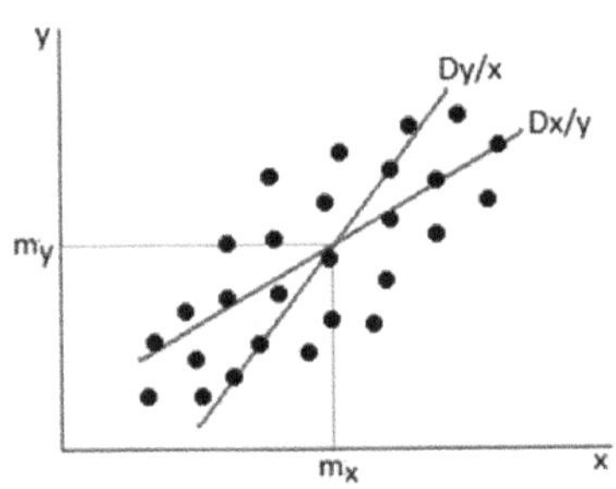

Figure 10.63. *Linear regression lines. For a color version of this figure, see www.iste.co.uk/cochard/mathematics1.zip*

These two lines $D_{y/x}$ and $D_{x/y}$ are called (linear) *regression lines.* They are, in general, distinct, but both pass through the "middle point" (m_x, m_y) (see Figure 10.63).

EXAMPLE 10.31.– To determine the regression lines for the statistical series from example 10.9, detailed calculations are provided in Figure 10.64. For each class, represented by its center, the frequency c of each pair (x_i, y_i) must be considered. For instance, the pair (3.5, 7.5) appears 19 times, so $c = 19$. This requires adjusting the average and sums to ensure accurate results.

From Figure 10.63, the calculated values for a, b, a' and $b'$are:

$$a = \frac{\sum_i X_i Y_i}{\sum_i X_i^2} = \frac{7589.49}{2267.99} = 0.17 \qquad \text{and} \qquad b = m_y - m_x a = 16.64$$

$$a' = \frac{\sum_i X_i Y_i}{\sum_i Y_i^2} = \frac{7589.49}{69473.75} = 0.002 \qquad \text{and} \qquad b' = m_x - m_y a' = 4.38$$

Thus, the equations of the regression lines are:

$$\mathrm{D}_{y/x}: \ y = 0.17x + 16.64 \qquad \textit{and} \qquad D_{x/y}: y = 500x - 2{,}190$$

These regression lines are shown in Figure 10.65.

	x	[3, 4[	[4, 5[	[5, 6]	
y	centers	3.5	4.5	5.5	marg.dist.
[0, 15[	7.5	19	7	1	27
[15, 20[	17.5	32	21	12	65
[20, 25[	22.5	3	18	28	49
[25, 30]	27.5	0	0	1	1
	marg. dist.	54	46	42	142

x	y	c	X	Y	X²	Y²	XY
3.5	7.5	19	-0.92	-9.89	0.84	97.90	9.06
3.5	17.5	32	-0.92	0.11	0.84	0.01	-0.10
3.5	22.5	3	-0.92	5.11	0.84	26.07	-4.67
3.5	27.5	0	-0.92	10.11	0.84	102.12	-9.25
4.5	7.5	7	0.08	-9.89	0.01	97.90	-0.84
4.5	17.5	21	0.08	0.11	0.01	0.01	0.01
4.5	22.5	18	0.08	5.11	0.01	26.07	0.43
4.5	27.5	0	0.08	10.11	0.01	102.12	0.85
5.5	7.5	1	1.08	-9.89	1.18	97.90	-10.73
5.5	17.5	12	1.08	0.11	1.18	0.01	0.11
5.5	22.5	28	1.08	5.11	1.18	26.07	5.54
5.5	27.5	1	1.08	10.11	1.18	102.12	10.96
averages: 4.42	17.39			sums	8.09	678.30	1.37

Figure 10.64. *Details of calculations*

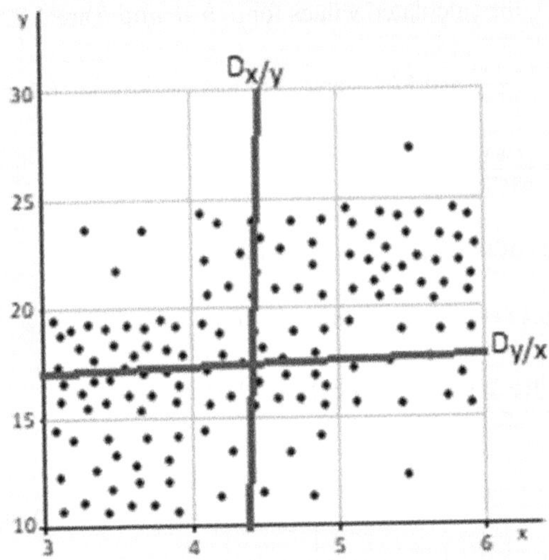

Figure 10.65. *Regression lines. For a color version of this figure, see www.iste.co.uk/cochard/mathematics1.zip*

10.7.2. *Linear correlation*

The difference between the regression lines indicates the strength of the correlation, in this case, linear. If the lines nearly overlap, it suggests a high correlation. Conversely, if the lines are nearly perpendicular, it implies a weak or even zero correlation. To obtain a more precise measure of correlation, calculate the product $a \times a'$:

$$a \times a' = \frac{(\sum_i X_i Y_i)^2}{(\sum_i X_i^2)(\sum_i Y_i^2)}$$

This quantity is always non-negative, leading to Schwarz's inequality:

$$\left(\sum_i X_i Y_i\right)^2 \geq \left(\sum_i X_i^2\right)\left(\sum_i Y_i^2\right)$$

The correlation is maximum when the regression lines are perfectly aligned, which implies:

$$a = \left(\frac{1}{a'}\right) \quad \text{or} \quad a \times a' = 1.$$

In this case, Schwarz's equality holds:

$$\left(\sum_i X_i Y_i\right)^2 = \left(\sum_i X_i^2\right)\left(\sum_i Y_i^2\right)$$

Correlation is zero when the lines are perpendicular, which implies

$$a \times \left(\frac{1}{a'}\right) = -1 \quad \text{or} \quad (a \times a') = -a^2.$$

Since $a \times a' \geq 0$, the only possible value for $a \times a'$ in this case is zero.

However, instead of using $a \times a'$ directly, the square root of this product is used as the measure of correlation, known as *the linear correlation coefficient*, denoted by *r*:

$$r = \frac{\sum_i X_i Y_i}{\sqrt{\sum_i X_i^2}\sqrt{\sum_i Y_i^2}} \quad \text{with} \quad -1 \leq r \leq +1$$

The range of variation of the correlation coefficient *r* is between -1 and +1. Extreme values of *r* correspond to a strong correlation, while a value of 0 indicates no correlation (see Figure 10.66).

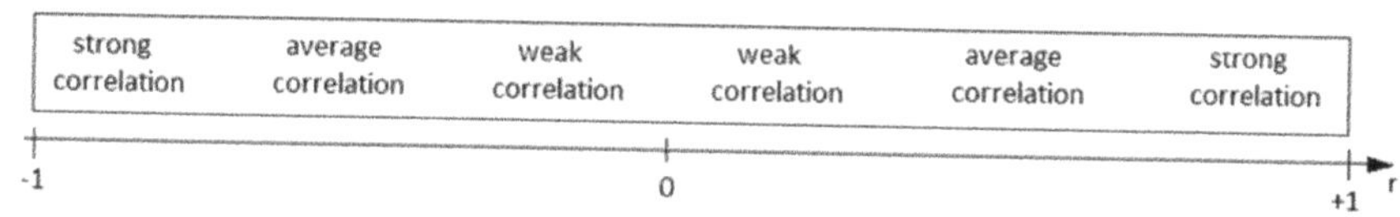

Figure 10.66. *Overview of r values*

The *linear correlation coefficient r* can also be expressed using the *covariance* of *x* and *y* as well as their standard deviations $\sigma(x)$ and $\sigma(y)$.

Recall that the variances $v(x)$ and $v(y)$ are calculated as follows:

$$v(x) = \sigma^2(x) = \frac{1}{n}\sum_i (x_i - m_x)^2 = \frac{1}{n}\sum_i X_i^2$$

$$v(y) = \sigma^2(y) = \frac{1}{n}\sum_i (y_i - m_y)^2 = \frac{1}{n}\sum_i Y_i^2$$

The *covariance* of x and y is defined as:

$$cov(x,y) = \frac{1}{n}\sum_i (x_i - m_x)(y_i - m_y) = \frac{1}{n}\sum_i X_i Y_i$$

The linear correlation coefficient r can then be written in the following form:

$$r = \frac{cov(x,y)}{\sigma(x)\sigma(y)}$$

EXAMPLE 10.32.– Returning to the calculations from example 10.31, the linear correlation coefficient can be derived using the formula:

$$r = \sqrt{aa'}$$

where $a = 0.17$ and $a' = 0.002$. Substituting these values yields:

$$r = \sqrt{0.17 \times 0.002} = 0.018$$

This result indicates a correlation that is practically zero. Figure 10.64 shows that the regression lines are nearly perpendicular, reinforcing the observation of a very weak correlation.

For further context, the covariance between x and y, and standard deviations $\sigma(x)$ and $\sigma(y)$ are calculated as follows:

$$cov(x,y) = \frac{1.37}{142} = 0.0096$$

$$\sigma(x) = \sqrt{\frac{8.09}{142}} = 0.2387$$

$$\sigma(y) = \sqrt{\frac{678.3}{142}} = 2.1856$$

These calculations confirm that the linear correlation coefficient $r = \frac{cov(x,y)}{\sigma(x)\sigma(y)} \approx 0.018$ reflects a negligible correlation between x and y.

The linear correlation coefficient has two key properties: invariance by translation and invariance by change of scale.

– Translation invariance: when x_i is transformed into $x_i' = x_i + \alpha$ and y_i into $y_i' = y_i + \beta$, the means m_x and m_y are shifted to $m_x' = m_x + \alpha$ and $m_y' = m_y +$

β, respectively. Consequently, X_i transforms into $X_i' = x_i' - m_x' = x_i - m_x = X_i$ and $Y_i' = Y_i$. This results in no change to the expression of r.

– Invariance by change of scale: when x_i is transformed into $x_i' = \alpha x_i$ and y_i into $y_i' = \beta y_i, (\alpha,\ \beta > 0)$, the means transforms to $m_x' = \alpha m_x$ and $m_y' = \beta m_y$. Thus, X_i becomes $X_i' = x_i' - m_x' = \alpha x_i - \alpha m_x = \alpha X_i$ and $Y_i' = \beta Y_i$.

Consequently:

$$r' = \frac{\sum_i X'_i Y'_i}{\sqrt{\sum_i X'^2_i}\sqrt{\sum_i Y'^2_i}} = \frac{\alpha\beta \sum_i X_i Y_i}{\alpha\sqrt{\sum_i X_i^2}\,\beta\sqrt{\sum_i Y_i^2}} = r$$

Important warning: although correlation can suggest a mutual influence between two traits or even imply a cause-and-effect relationship, it is essential to consider the limitations of correlation analysis:

a) Determining causality: it is generally challenging to ascertain with certainty which of the two variables, x or y, is the cause and which is the effect, unless it is clearly evident from the context or experimental design.

b) Spurious correlations: it is crucial to be aware of "accidental" or even far-fetched correlations. A strong correlation does not necessarily imply a real causal relationship. For an entertaining look at some surprising correlations, the reader may visit the website: *https://tylervigen.com/spurious-correlations*.

c) Third-variable influence: it is also important to recognize that x and y might be related to a third variable. For instance, while a strong correlation might exist between sunglasses sales and ice cream sales, the underlying factor could be temperature. Higher temperatures may lead to increased sales of both sunglasses and ice cream, without a direct causal link between the two variables.

11

Probability Laws and Simulation

CONCEPTS COVERED IN THIS CHAPTER.–

This chapter serves as a bridge between the previous chapters on probability (Chapter 9) and statistics (Chapter 10).

After covering the fundamental concepts of random variables – both discrete and continuous – various types of event occurrence modeling, known as probability distributions (laws), are explored.

These descriptions are described in terms of probability expressions, mathematical expectation and standard deviation, with a focus on common distributions such as the uniform, binomial, Poisson, exponential and normal distributions.

The field of computer science has facilitated the development of simulation techniques that recreate real phenomena on a computer. These simulations rely on random sampling according to specific probability distributions, which are modeled using generating programs.

This chapter concludes with the application of simulation to a classic problem in queuing theory, providing a detailed examination of the M/M/1 queue model.

References: [PHE 77, COC 19].

11.1. Random variables

Suppose that an experiment can produce several results under identical conditions, assuming for the moment that their number is finite and denoted by $A, B, \ldots, K$. The appearance of these results is random, and it is convenient to associate the value of a *random variable X* with each result.

Mathematics for Digital Science 1,
by Gérard-Michel COCHARD and Mhand HIFI.

For example:

$X = 1$ for event A,

$X = 2$ for event B,

$X = 3$ for event C,

etc.

The probabilities of occurrence of events are then denoted as follows:

$p(\text{event A}) = p(X = 1),$

$p(\text{event B}) = p(X = 2),$

etc.

The probability $p(X)$ can be considered a function of the variable X. This function, $p(X)$, is known as a *probability law* and characterizes the distribution of possible results of the random experiment.

EXAMPLE 11.1.– A jar contains 2 white balls and 4 black balls. Apart from their color, the balls are identical. Two balls are drawn at random. The possible events are as follows:

A: 2 white balls.

B: 1 white ball and 1 black ball.

C: 2 black balls.

The random variable X is defined such that:

X: 1 for event A;

X: 2 for event B;

X: 3 for event C.

To determine the probabilities of these events in terms of X, note that the jar contains a total of 6 balls.

For event A (2 white balls), there is only one way to draw 2 white balls, which is by drawing the two white balls available in the jar.

To find the probability of event A (i.e. $X = 1$):

$$p(X = 1) = \left(\frac{\text{number of draws favorable to A}}{\text{number of possible draws}}\right)$$

$$p(X = 1) = \left(\frac{\text{number of ways to draw 2 white balls}}{\text{number of ways to draw 2 balls from 6 balls}}\right)$$

$$p(X = 1) = \left(\frac{C_2^2}{C_6^2}\right)$$

Like $C_6^2 = \frac{6\,!}{(6-2)!\,2\,!} = \frac{6\times5}{2} = 15$ and $C_2^2 = 1$, then $p(X = 1) = \frac{1}{15}$.

For event B (i.e. $X = 2$: 1 white ball and 1 black ball), the number of ways to choose a white ball and a black ball from the available balls needs to be determined.

There are 2 white balls and 4 black balls. The number of ways to choose 1 white ball from 2 and 1 black ball from 4 is calculated as follows:

Number of ways to choose a white ball and a black ball = (number of ways to choose 1 white ball from 2) ×(number of ways to choose 1 black ball from 4)

Number of ways to choose 1 white ball from 2 and 1 black ball from 4 =$C_1^2 \times C_1^4 = 2 \times 4 = 8$

Therefore, the probability of event B (i.e. $X = 2$) is:

$$p(X = 2) = \left(\frac{\text{nunber of draws favorable to B}}{\text{number of possible draws}}\right)$$

$$p(X = 2) = \left(\frac{\text{nunber of ways to draw 1 white ball and 1 black ball}}{\text{number of ways to draw 2 balls from 6 balls}}\right)$$

$$p(X = 2) = \left(\frac{C_1^2 \times C_1^4}{C_6^2}\right)$$

As $C_6^2 = 15$ and $C_1^2 \times C_1^4 = 8$, then $p(X = 2) = \frac{8}{15}$.

For event C (i.e. $X = 3$: 2 black balls), there is only one way to draw 2 black balls, which involves drawing both black balls from the jar.

To find the probability of event C:

$$p(X=3)=\left(\frac{\text{number of draws favorable to C}}{\text{total number of possible draws}}\right)$$

$$p(X=3)=\left(\frac{\text{number of ways to draw 2 black balls from 4 black balls}}{\text{number of ways to draw 2 balls from 6 balls}}\right)$$

$$p(X=3)=\left(\frac{C_4^2}{C_6^2}\right)$$

As $C_6^2 = 15$ and $C_4^2 = \frac{4\,!}{(4-2)!\,2\,!} = \frac{4\times 3}{2} = 6$, then $p(X=3) = \frac{6}{15}$.

Confirming that the sum of probabilities for all events equals 1: $p(X=1) + p(X=2) + p(X=3) = 1$.

The probability law $p(X)$ is shown in Figure 11.1. This is a discrete probability law because the random variable X takes on discrete values.

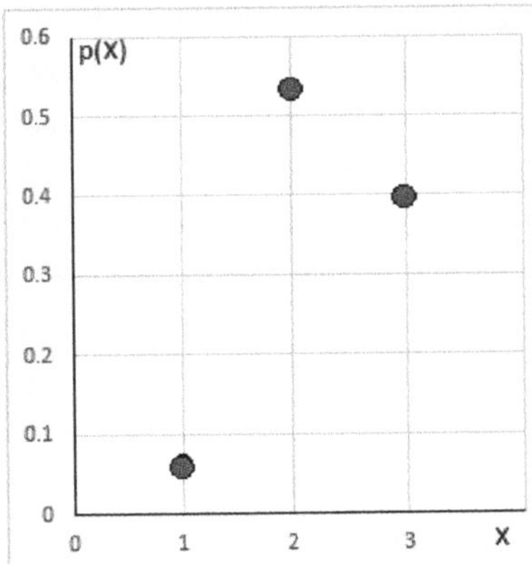

Figure 11.1. *Graphical representation of the probability law. For a color version of this figure, see www.iste.co.uk/cochard/mathematics1.zip*

In the previous example, the values taken by the random variable X serve primarily to "count" the events, such as indicating whether 2 white balls, 1 white ball and 1 black ball or 2 black balls are drawn. In this case, X simply labels the outcomes of the random experiment. However, a random variable can also take on values with specific, meaningful interpretations. For instance, consider example 11.2, where the values of the random variable are not just labels but represent quantities with inherent significance.

EXAMPLE 11.2.– A dice game works as follows: the die is thrown four times. The player's gain or loss is determined by the number of times a "1" appears.

The player wins 2 euros each time a "1" appears and loses 2 euros otherwise. For 4 rolls of the die, there are 5 possible situations, each representing the player's gain or loss. These outcomes are used to define the value of a random variable X, which reflects the player's net gain or loss.

Case 1: "1" never comes out. The player loses 2 euros four times, leading a total loss of 8 euros. Thus, the value of the random variable is therefore $X = -8$.

Case 2: "1" appears exactly once. The player wins 2 euros once but loses 6 euros, resulting in a net loss of 4 euros. Therefore, $X = -4$.

Case 3: "1" appears exactly twice. The player wins 2 euros and loses 2 euros, resulting in no net gain or loss. Thus, $X = 0$.

Case 4: "1" appears exactly three times. The player wins 6 euros and loses 2 euros, resulting in a net gain of 4 euros. Hence, $X = 4$.

Case 5: "1" appears four times. The player wins 8 euros, resulting in a net gain of 8 euros. Thus, $X = 8$.

The random variable X thus takes 5 discrete values: $-8; -4, 0, +4$ *and* $+8$. When rolling a six-sided die, the probability of rolling a "1" is effectively $\frac{1}{6}$, while the probability of rolling a number other than "1" is $\frac{5}{6}$. To simplify the notations, let p represent the probability of rolling a "*1*" (i.e. $p = \frac{1}{6}$) and q represent the probability of rolling any other number (i.e. $q = (1-p) = \frac{5}{6}$).

Calculate the probabilities corresponding to the different situations as follows:

Probability that no "1" appears (therefore multiplying the probability q four times by itself) comes out during the four throws of the die:

$$p(X = -8) =$$
$$(\text{Number of ways no "1" appears in throws}) \times$$
$$(\text{Probability of not rolling a "1" in each throw})^4$$

$$p(X = -8) = C_4^0 \times (q.q.q.q)$$

$$p(X=-8)=q^4=\left(\frac{5}{6}\right)^4=\frac{625}{1{,}296}$$

Probability of rolling *a "1" exactly once* during the four rolls of the die:

$p(X=-2)$ = (Number of ways to roll a "1" exactly once in 4 throws) × (Product of the probabilities that "1" comes out 1 time and that it does not come out three times).

$$p(X=-2)=p.C_4^1\times(q.q.q)$$

$$p(X=-2)=4\cdot p\cdot q^3=4\cdot\frac{1}{6}\cdot\left(\frac{5}{6}\right)^3=\frac{500}{1{,}296}$$

Probability of rolling *a "1" exactly twice* during the four rolls of the die:

$p(X=0)$ = (Number of ways to roll a "1" exactly two times out of 4 throws) × (Probability of rolling a "1" twice) × (Probability of not rolling a "1" twice).

$$p(X=0)=C_4^2\times(p.p.)\times(q.q)$$

$$p(X=0)=6\cdot p^2\cdot q^2=6\cdot\left(\frac{1}{6}\right)^2\cdot\left(\frac{5}{6}\right)^2=\frac{150}{1{,}296}$$

Probability for *a 1* to roll *exactly three times* during the four rolls of the die:

$p(X=4)$ = (number of times to see 1 come out three times out of 4 throws) × (probability that 1 comes out three times) × (probability that 1 does not come out).

$$p(X=4)=C_4^3(p.p.p)\times(q)$$

$$p(X=4)=4\ p^3\ q=4\ \left(\frac{1}{6}\right)^3\ \frac{5}{6}=\frac{20}{1{,}296}$$

Probability of rolling a "1" *exactly four times* during the four rolls of the die:

$p(X=8)$ = (Number of ways to roll a "1" exactly four times out of 4 throws) × (Probability of rolling a "1" in each throw).

$$p(X=8)=C_4^4\times(p.p.p.p)$$

$$p(X=8)=p^4=\left(\frac{1}{6}\right)^4=\frac{1}{1{,}296}$$

The corresponding probability law is graphically represented in Figure 14.2.

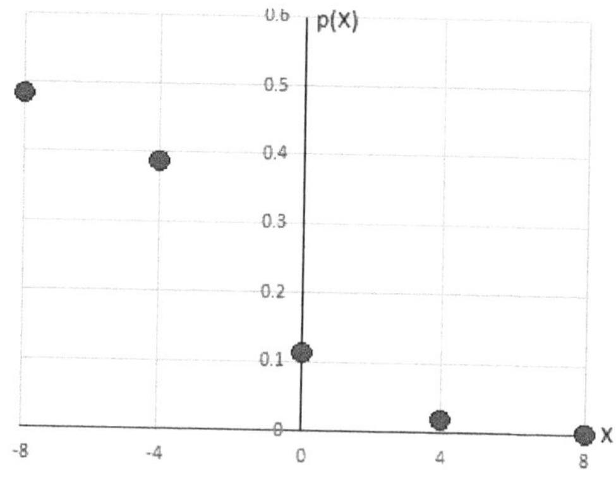

Figure 11.2. *Graphical representation of the probability law. For a color version of this figure, see www.iste.co.uk/cochard/mathematics1.zip*

Note that, in general, the probability is given by the following expression:

$$P(X = k) = C_n^k p^k q^{n-k},$$

where C_n^k is the binomial coefficient, representing the number of ways to choose k objects from n total objects. This formula is applicable within the framework of the binomial probability distribution, also known as Bernoulli's law. This formula will be used for calculations later on.

A random variable can take real values, which can be either discrete or continuous. In the previous examples, the values were discrete, indicating that the probability law is discrete.

For continuous random variables, the probability is determined using the concept of probability density. Consider the example of selecting a random point on a line segment AB. To locate the points of the segment, the line containing the segment is oriented as an axis, with an origin defining the abscissa x. The position of any point on the segment is given by its abscissa x.

– The abscissa of the endpoints A and B of the segment are a and b (where $a < b$).

– A point M of the segment has an abscissa x such that $a \leq x \leq b$.

– Choosing a point at random is the equivalent to selecting x randomly between a and b.

Consider X the random variable with abscissa x as its values (see Figure 11.3). To characterize the distribution of possible values of X, the concept of probability density is introduced.

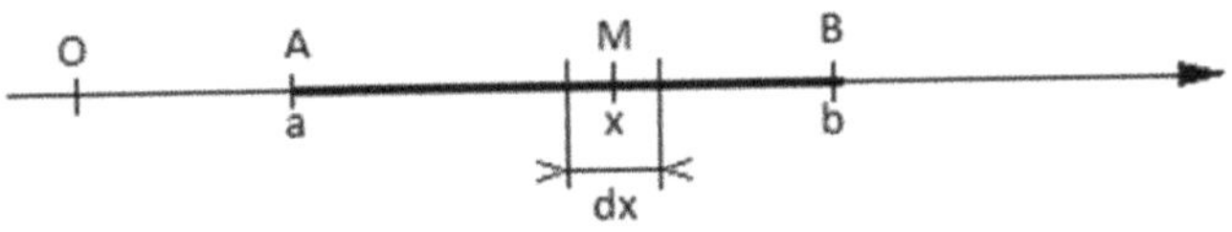

Figure 11.3. *Problem situation*

A segment contains an infinite number of points, so the variable X has an infinite range of values. Choosing an abscissa point x represents a choice among an infinite set, which results in a probability of zero. Therefore, the approach must be modified to not consider the probability of finding a point "*exactly at* x", but rather the probability of finding a point within the interval dx as it will be greater for larger dx and smaller for smaller dx. Hence, there should be a proportional relationship between dx and the probability, denoted as dp (by analogy with dx), i.e.

$$dp(X = x) = f(x)\,dx \quad \text{or} \quad \frac{dp}{dx}(X = x) = f(x)$$

This notation expresses the relationship between the interval dx and the probability dp. The "proportionality coefficient" $f(x)$ can generally depend on the value x. The term $f(x)$ represents the *probability density*

In the case of a continuous random variable where each point is equally likely, $f(x)$ does not depend on x and is therefore constant.

The probability density function must satisfy the following conditions:

– It must be non-negative: $f(x) \geq 0$.

– It must be continuous over the interval $[a, b]$.

– The integral of the probability density function over the interval $[a, b]$ must be equal to:

$$\int_a^b f(x)dx = 1$$

This condition has a graphical interpretation: the area under the curve $y = f(x)$, between the x-axis and the vertical lines $x = a$ and $x = b$, must be equal to 1 (see Figure 11.4).

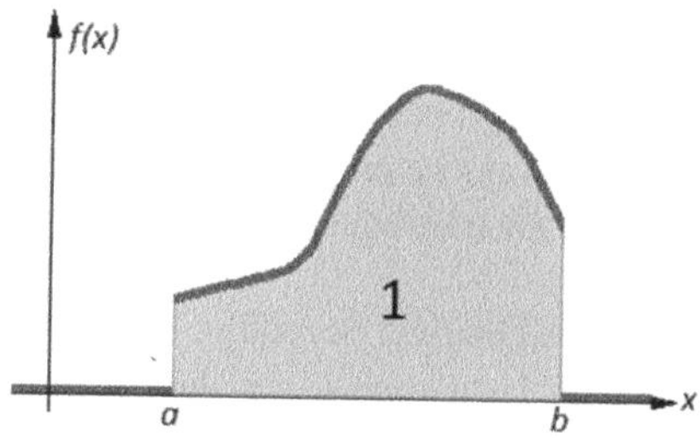

Figure 11.4. *Probability density. For a color version of this figure, see www.iste.co.uk/cochard/mathematics1.zip*

If a and b are finite, then $f(x) = 0$ for $x < a$ and $x > b$. In the example where $f(x) = k$ is a constant, the integral $\int_a^b f(x)dx$ simplifies to $k(b-a)$. This leads to:

$$f(x) = k = \frac{1}{b-a}$$

This result is shown in Figure 11.5.

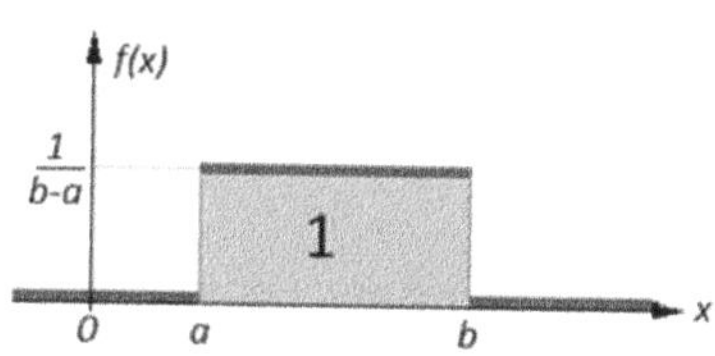

Figure 11.5. *Probability density. For a color version of this figure, see www.iste.co.uk/cochard/mathematics1.zip*

Returning to the general case where the probability density is a function of x. In the continuous case, the expression $p(X = x)$ is replaced by $p(\alpha \leq X \leq \beta)$, which is represented as follows:

$$p(\alpha \leq X \leq \beta) = \int_{\alpha}^{\beta} f(x)dx$$

EXAMPLE 11.3.– Let the function be $f(x) = \alpha e^{(-|x|)}$.

Can this function be a probability density? If so, what should the value of the coefficient α be? What is the probability that X is between $\frac{1}{2}$ and 1?

Recall that for a function to be a probability density, it must satisfy the following conditions: the function must always be positive, $f(x) \geq 0$ for all x, and the integral of the function over its domain must be equal to 1.

Firstly, $f(x) \geq 0$ for all x if α is positive. Secondly, $f(x)$ is continuous on $\mathbb{R}$ and is therefore a good candidate to be a probability density function. To satisfy the second condition, determine α by ensuring that the integral of all the function over its entire domain is equal to 1:

$$1 = \int_{-\infty}^{+\infty} f(x)dx = \alpha[\int_{-\infty}^{0} e^{x}dx + \int_{0}^{+\infty} e^{-x}dx] = 2\alpha$$

Hence, $\alpha = \frac{1}{2}$.

The graphical representation of $f(x)$ is given in Figure 11.6.

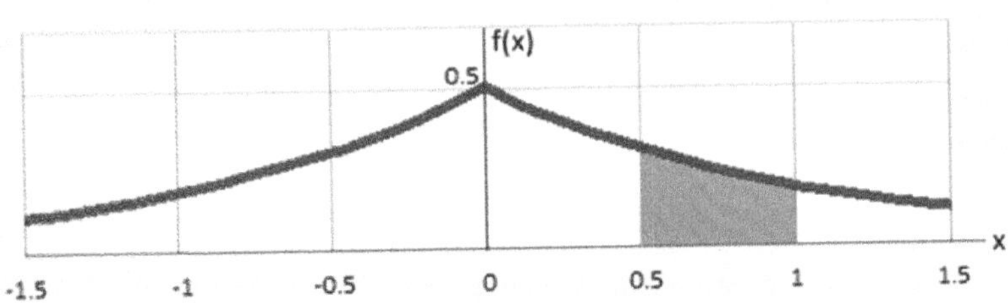

Figure 11.6. *Probability density. For a color version of this figure, see www.iste.co.uk/cochard/mathematics1.zip*

What is the probability that $\frac{1}{2} \leq X \leq 1$?

$$p\left(\frac{1}{2} \leq X \leq 1\right) = \int_{0.5}^{1} f(x)dx = \frac{1}{2}\int_{0.5}^{1} e^{-x}dx \approx 0.12$$

11.2. Mathematical expectation, variance and standard deviation

Definitions similar to those seen in statistics are used, but in probability theory, the term "*mathematical expectation*" is employed instead of "*average*". The mathematical expectation is an average weighted by the probability values.

Given a random variable X associated with a probability law $p(X)$, the *average* or *mathematical expectation* $E(X)$ is defined differently depending on whether the variable is discrete or continuous:

If X is a discrete variable taking n values x_i with probability $p(x_i)$

$$E(X) = \sum_{i=1}^{n} p(x_i)x_i$$

If X is a continuous variable with probability density $f(x) \geq 0$ over the interval (a, b), then

$$E(X) = \int_a^b f(x)xdx$$

Mathematical expectation has linearity properties: for two random variables X and Y, and real numbers α and β,

$$E(\alpha X + \beta Y) = \alpha E(X) + \beta E(Y)$$

Moreover, if X and Y are independent variables, the joint probability $p(X = x_i \text{ and } Y = y_j)$ can be written as $p(X = x_i)p(Y = y_j)$. Consequently,

$$E(XY) = \sum_{i,j} p(X = x_i \; and \; Y = y_j)x_i y_j$$

$$= \sum_i p(X = x_i)x_i \sum_j p(Y = y_j)y_j = E(X)E(Y)$$

Consider X as a random variable with mathematical expectation $E(X)$. The *variance* $v(X)$ is defined as the weighted average of the squares of the deviations from the mathematical expectation:

If the variable is discrete and takes n values x_i with probability $p(x_i)$

$$v(X) = \sum_{i=1}^{n} p(x_i)[x_i - E(X)]^2$$

If the variable is continuous with probability density $f(x) \geq 0$ on [a, b]

$$v(X) = \int_a^b f(x)[x - E(X)]^2 dx$$

The standard deviation $\sigma(X)$ is the square root of the variance:

$$\sigma(X) = \sqrt{v(X)}$$

These definitions align with those in statistics. A convenient expression for calculating the variance is provided by the Huygens-Koenig theorem:

$$v(X) = E(X^2) - [E(X)]^2$$

This theorem is straightforward to prove. For a discrete variable, the proof is as follows. By definition,

$$v(X) = \sum_{i=1}^{n} p(x_i)x_i - E(X)]^2$$
$$= \sum_{i=1}^{n} p(x_i)x_i^2 - 2E(X)\sum_{i=1}^{n} p(x_i)x_i + [E(X)]^2 \sum_{i=1}^{n} p(x_i)$$

Expanding this,

$$\sum_{i=1}^{n} p(x_i)x_i^2 = E(X^2) \quad \sum_{i=1}^{n} p(x_i)x_i = E(X) \quad \sum_{i=1}^{n} p(x_i) = 1$$

it follows that:

$$v(X) = E(X^2) - 2E(X)E(X) + [E(X)]^2 = E(X^2) - [E(X)]^2$$

This result is often remembered as "mean of the square minus square of the mean", ensuring that the square of the mean is correctly subtracted from the mean of the squares. If unsure, verify that the variance should not be negative.

When X and Y are independent variables, the variance of their sum is simply the sum of their variances. This can be shown as follows:

$$v(X+Y) = E((X+Y)^2) - [E(X+Y)]^2$$
$$= E(X^2 + 2XY + Y^2) - [E(X) + E(Y)]^2$$

$$= E(X^2) + 2E(XY) + E(Y^2) - -2E(X)E(Y) - [E(Y)]^2$$

$$= E(X^2) - [E(X)]^2 + E(Y^2) - [E(Y)]^2 + 2E(X)E(Y) - 2E(X)E(Y)$$

Therefore,

$$v(X+Y) = v(X) + v(Y)$$

EXAMPLE 11.4.– For the example of balls in a jar (example 11.1):

Given:

$$E(X) = 1 \times \frac{1}{15} + 2 \times \frac{8}{15} + 3 \times \frac{6}{15} = \frac{5}{3} \approx 1.67$$

$$E(X^2) = 1 \times \frac{1}{15} + 4 \times \frac{8}{15} + 9 \times \frac{6}{15} = \frac{29}{5} \approx 5.8$$

The variance $v(X)$ is calculated as:

$$v(X) = E(X^2) - [E(X)]^2 = \frac{136}{45} \approx 3.02$$

The standard deviation $\sigma(X)$ is the square root of the variance:

$$\sigma(X) = \sqrt{v(X)} \approx 1.74$$

EXAMPLE 11.5.– Returning to the example of choosing a point on a segment defined by the interval $[a, b]$.

Recall that the probability density function is: $f(x) = \frac{1}{b-a}$.

Expected value $E(X)$:

$$E(X) = \int_a^b \frac{x dx}{b-a} = \frac{1}{2(b-a)}(b^2 - a^2) = \frac{b+a}{2}$$

which is a predictable result.

Expected value of X^2, $E(X^2)$:

Calculating the standard deviation requires calculating $E(X^2)$ and variance.

$$E(X^2) = \int_a^b \frac{x^2 dx}{b-a} = \frac{(b^3-a^3)}{3(b-a)} = \frac{b^2+ab+a^2}{3}$$

Variance $v(X)$:

The variance is calculated as follows:

$$v(X) = E(X^2) - [E(X)]^2 = \frac{(b-a)^2}{12}$$

Standard deviation $\sigma(X)$:

Finally, the standard deviation is computed as follows:

$$\sigma(X) = \frac{b-a}{2\sqrt{3}}$$

EXAMPLE 11.6.– Given the probability distribution from example 11.2:

$$p(X=-8) = \frac{625}{1{,}296},\ p(X=-4) = \frac{500}{1{,}296},\ p(X=0) = \frac{150}{1{,}296},\ p(X=4) = \frac{20}{1{,}296},$$
$$p(X=8) = \frac{1}{1{,}296}.$$

Expected value $E(X)$:

$$E(X) = \frac{-8\times 625 - 4\times 500 + 0\times 150 + 4\times 20 + 8\times 1}{1{,}296} = -\frac{6{,}912}{1{,}296} \approx -5.33$$

This is the amount in euros that the player loses on average.

Expected value of X^2, $E(X^2)$:

$$E(X^2) = \frac{64\times 625 + 16*500 + 0*150 + 16*20 + 64*1}{1{,}296} = \frac{48{,}384}{1{,}296} \approx 37.33$$

Variance $v(X)$:

The variance can therefore be calculated as follows:

$$v(X) = E(X^2) - [E(X)]^2 \approx 8.92$$

Standard deviation $\sigma(X)$:

Finally, the standard deviation is given by:

$$\sigma(X) \approx 2.99$$

EXAMPLE 11.7.– Given the continuous variable X with the probability density function $f(x) = \frac{1}{2}e^{(-|x|)}$, the calculations for the expected value and variance are as follows (example 11.3):

Expected value $E(X)$:

$$E(X) = \int_{-\infty}^{+\infty} f(x)x dx = \frac{1}{2}\int_{-\infty}^{0} xe^{x} dx + \frac{1}{2}\int_{0}^{+\infty} xe^{-x} dx$$

For the first integral, change x to $-x$:

$$E(X) = -\frac{1}{2}\int_{0}^{+\infty} xe^{-x} dx + \frac{1}{2}\int_{0}^{+\infty} xe^{-x} dx = 0,.$$

The expected value $E(X)$ is 0, which aligns with the symmetry of $f(x)$.

Expected value of X^2, $E(X^2)$:

In the same way, $E(X^2) = \int_{-\infty}^{+\infty} f(x)x^2 dx = \frac{1}{2}\int_{-\infty}^{0} x^2 e^{x} dx + \frac{1}{2}\int_{0}^{+\infty} x^2 e^{-x} dx$

By changing x to $-x$ in the first integral, $E(X^2) = \int_{0}^{+\infty} x^2 e^{-x} dx$ and integrating twice by parts, the following result is obtained:

$$E(X^2) = 2$$

Variance $v(X)$ and standard deviation $\sigma(X)$:

The variance:

$$v(X) = 2$$

and the standard deviation $\sigma(X) = \sqrt{2}$.

11.3. Distribution function

The *distribution function* of a probability law of a random variable X is defined as:

$$F(x) = p(X \le x)$$

The cumulative distribution function provides the probabilities that the random variable X takes on a value less than or equal to x.

– $F(x)$ ranges between 0 and 1, as it represents the cumulative probability from the smallest value of X (which could be $-\infty$) up to x.

– Since the total probability is 1 and all probabilities are non-negative, $F(x)$ is always an increasing function.

– For a discrete random variable, $F(x)$ is a "step" function, meaning it has jumps at the points where the random variable X takes specific values.

– For a continuous random variable, $F(x)$ is a continuous function, reflecting the continuous nature of the underlying probability density.

EXAMPLE 11.8.– For the discrete variable X in example 11.1, which takes the values 1, 2 and 3 with probabilities $\frac{1}{15}$, $\frac{8}{15}$ and $\frac{6}{15}$, respectively.

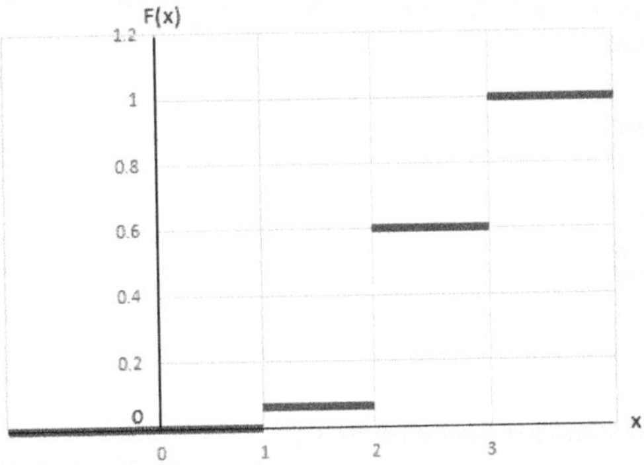

Figure 11.7. *Distribution function. For a color version of this figure, see www.iste.co.uk/cochard/mathematics1.zip*

The cumulative distribution function $F(x)$ can be determined as follows:

If $x < 1$, then $F(x) = 0$.

If $1 \leq x < 2$, then $F(x) = \frac{1}{15}$.

If $2 \leq x < 3$, then $F(x) = \frac{1}{15} + \frac{8}{15} = \frac{9}{15} = \frac{3}{5}$.

If $x \geq 3$, then $F(x) = \frac{1}{15} + \frac{8}{15} + \frac{6}{15} = 1$.

This representation is shown in Figure 11.7.

EXAMPLE 11.9.– For the continuous random variable X with the probability density function $f(x) = \frac{1}{2}e^{-|x|}$, the cumulative distribution function $F(x)$ is computed as follows:

$$F(x) = p(X \leq x) = \int_{-\infty}^{x} f(t)dt$$

The following two cases can be distinguished:

If $x < 0$, then $F(x) = \frac{1}{2}\int_{-\infty}^{x} e^t dt = \frac{e^x}{2}$.

If > 0, then $F(x) = \frac{1}{2}\int_{-\infty}^{0} e^t\, dt + \frac{1}{2}\int_{0}^{x} e^{-t}\, dt = \frac{1}{2} + \frac{1-e^{-x}}{2} = 1 - \frac{e^{-x}}{2}$.

This function shows the cumulative probability up to a point x and is shown graphically in Figure 11.8.

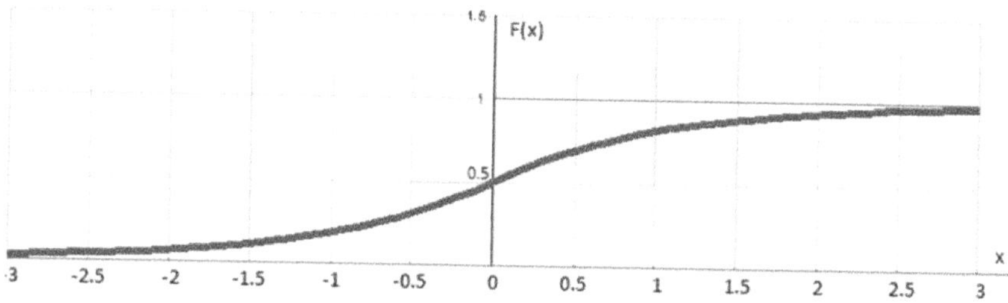

Figure 11.8. *Distribution function. For a color version of this figure, see www.iste.co.uk/cochard/mathematics1.zip*

11.4. Usual probability laws

Some probability laws frequently encountered in experiments include the uniform probability law, the binomial (or Bernoulli) law, Poisson's law, the exponential law and the normal (or Gaussian) law.

11.4.1. *Uniform law*

The uniform law is one of the simplest probability laws and is used to model continuous random variables with probability density $f(x)$ that is *constant* over a given interval (a, b) and zero outside this interval. This law ensures that each value within the interval has an equal probability of being observed (see Figure 11.5). The density function is denoted as follows: $f(x) = C$ (where C is a constant) with the condition on the given interval (a, b):

$$\int_{a}^{b} f(x)dx = 1$$

This implies that the constant C can be calculated as follows: $C = \frac{1}{b-a}$.

It is then straightforward to deduce (as calculated in example 11.5):

– mathematical expectation: $E(X) = \frac{a+b}{2}$;

– variance: $v(X) = \frac{(a-b)^2}{12}$;

– standard deviation: $\sigma(X) = \frac{b-a}{2\sqrt{3}}$.

If the interval (a, b) is (0, 1), then

$$E(X) = \frac{1}{2} \qquad v(X) = \frac{1}{12} \qquad \sigma(X) = \frac{1}{2\sqrt{3}}$$

11.4.2. Binomial or Bernoulli's law

The binomial law is frequently used to model repeated experiments (trials) of the same type with binary results. It is based on a *Bernoulli scheme* and is associated with discrete random variables. In this law, it is assumed that the trials are identical and independent of each other. This law was illustrated in example 11.2., where a specific case of the binomial law was studied, and probabilities of success in a certain number of repeated trials were calculated.

In each trial of a binomial experiment, there are two possible results: the event of interest, with an elementary probability p, and its complement, with a probability $q = 1 - p$. When n trials are conducted, the number of times the event of interest occurs is denoted by k. This value k is assigned to a random variable X, which can take discrete values from 0 to n. Thus, in the context of the binomial law, X represents the number of successes (occurrences of the event of interest) in the n trials. It is a discrete random variable that can take values between 0 and n, inclusive.

The probability law associated with the Bernoulli diagram is expressed by:

$$p(X = k) = C_n^k p^k q^{n-k}$$

where C_n^k is the binomial coefficient defined by:

$$C_n^k = \frac{n!}{k!\,(n-k)!}$$

Recall Newton's binomial formula on this subject.

$$(p+q)^n = \sum_{k=0}^{n} C_n^k p^k q^{n-k}$$

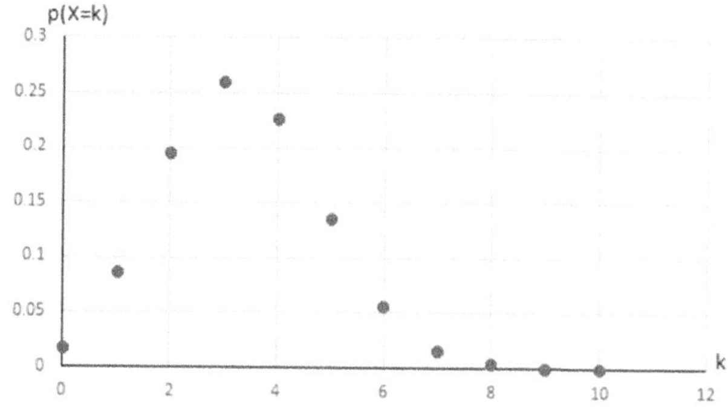

Figure 11.9. *Binomial law. For a color version of this figure, see www.iste.co.uk/cochard/mathematics1.zip*

This formula guarantees us that the sum of all the probabilities of the binomial law is equal to 1. Since $q = 1 - p$, it follows that $(p+q)^n = 1$, and therefore,

$$\sum_{k=0}^{n} C_n^k p^k q^{n-k} = \sum_{k=0}^{n} p(X=k) = 1$$

Figure 11.9 provides the graphical representation of this law for $p = \frac{1}{3}$ and $n = 10$.

Note that this law depends only on two parameters: n and p. Therefore, it is sometimes denoted as $B(n, p)$.

To calculate the mathematical expectation $E(X)$:

$$E(X) = \sum_{k=0}^{n} p(X=k)k$$

The first term of the sum, for $k = 0$, is zero, so:

$$E(X) = \sum_{k=1}^{n} p(X = k)k = \sum_{k=1}^{n} \frac{n!\,k}{k!\,(n-k)!} p^k q^{n-k}$$
$$= \sum_{k=1}^{n} \frac{n!}{(k-1)!\,(n-k)!} p^k q^{n-k}$$

$$= n \sum_{k=1}^{n} \frac{(n-1)!}{(k-1)!\,(n-k)!} p^k q^{n-k}$$
$$= n \sum_{k=1}^{n} \frac{(n-1)!}{(k-1)!\,[(n-1)-(k-1)]!} p^k q^{n-k}$$

$$= n \sum_{k=1}^{n} C_{n-1}^{k-1} p^k q^{n-k}$$

Change the index by letting $h = (k - 1)$ and use Newton's binomial formula. This leads to the following:

$$E(X) = n \sum_{h=0}^{n-1} C_{n-1}^{h} p^{k+1} q^{n-1-h} = np \sum_{h=0}^{n-1} C_{n-1}^{h}\, p^h q^{n-1-h} = np(p+q)^{n-1}$$
$$= np(1)^{n-1}$$

It follows that the expectation *E(x)* is simply: $E(X) = np$.

To calculate the variance $v(X)$, an alternative approach can be used. Consider X_i as the random variable for an individual test. Here, $X_i = 1$ if the expected event occurs and $X_i = 0$ if the opposite event occurs. Thus, X_i follows Bernoulli's law with a success probability p. The expected value is $E(X_i) = 1 \times p + 0 \times q = p$, and the variance is $v(X_i) = p \times (1-p)^2 + q \times (0-p)^2 = pq(p+q) = pq$.

The variable X of the Bernouilli scheme (n identical and independent trials) can be considered as the sum of the variables $X_1, X_2, \ldots, X_n$, each corresponding to a single trial. The variables X_i are independent. Therefore:

$$E(X) = E(X_1 + X_2 + \cdots + X_n) = E(X_1) + E(X_2) + \cdots + E(X_n)$$
$$= p + p + \cdots + p = np$$

This confirms the previous result. For the variance, since the variables X_i are independent, the variance of X is:

$$v(X) = v(X_1 + X_2 + \cdots + X_n) = v(X_1) + v(X_2) + \cdots + v(X_n)$$

Each X_i has the same variance pq, so:

$$v(X) = pq + pq + \cdots + pq = npq$$

11.4.3. *Poisson's law*

Poisson's law approximates the binomial distribution when the probability p is small (e.g. $p < 0.1$) and the number of trials n is large. The probability mass function of Poisson's distribution is given by:

$$p(X = k) = e^{-m}\frac{m^k}{k!}$$

where m is a positive real number. To verify that the sum of all the probabilities equals 1:

$$\sum_{k=0}^{\infty} p(X = k) = e^{-m}\left\{1 + m + \frac{m^2}{2!} + \frac{m^3}{3!} + \cdots.\right\} = e^{-m}e^{m} = 1$$

This confirms that the sum of all probabilities is indeed 1. Poisson's distribution depends on a single parameter, m. To find the mathematical expectation:

$$E(X) = \sum_{k=0}^{\infty} e^{-m}\frac{m^k}{k!}k = me^{-m}\sum_{k=1}^{\infty}\frac{m^{k-1}}{(k-1)!}$$

rewriting k as $k = h + 1$, where $h = k - 1$:

$$E(X) = me^{-m}\sum_{h=0}^{\infty}\frac{m^h}{h!} = me^{-m}e^{m} = m$$

The parameter m is therefore the mathematical expectation of the variable X. To calculate the variance, first compute $E(X^2)$:

$$E(X^2) = e^{-m}\sum_{k=1}^{\infty}\frac{m^k}{k!}k^2 = me^{-m}\sum_{k=1}^{\infty}\frac{m^{k-1}}{(k-1)!}k$$

Let $h = k - 1$. Then,

$$E(X^2) = me^{-m}\sum_{h=0}^{\infty}\frac{m^h}{h!}(h+1)$$
$$= me^{-m}\{\sum_{h=1}^{\infty}\frac{m^h}{(h-1)!}+\sum_{h=0}^{\infty}\frac{m^h}{h!}\} = me^{-m}\{\sum_{h=1}^{\infty}\frac{m^h}{(h-1)!}$$
$$+ e^m\}$$

In the remaining sum, let $l = h - 1$:

$$E(X^2) = me^{-m}\left\{m\sum_{l=0}^{\infty}\frac{m^l}{l!} + e^m\right\} = me^{-m}[me^m + e^m] = m(m+1).$$

Therefore, the variance is:

$$v(X) = E(X^2) - [E(X)]^2 = m(m+1) - m^2 = m.$$

The variance, like the mathematical expectation, is equal to m. The standard deviation is:

$$\sigma(X) = \sqrt{m}$$

Figure 11.10 graphically shows Poisson's law for $m = 1, m = 4, m = 10$. Tables provide the value of $p(X = k)$ (see section 11.8.1. Appendix 1 for details).

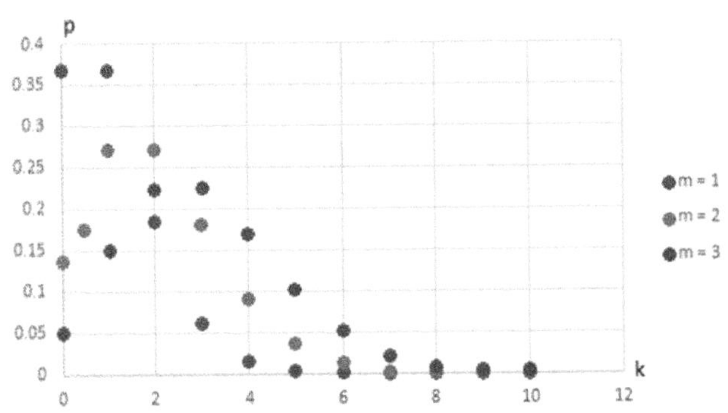

Figure 11.10. *Poisson's law. For a color version of this figure, see www.iste.co.uk/cochard/mathematics1.zip*

11.4.4. *Exponential law*

The exponential law is a law for a continuous random variable with a probability density function given by:

$$f(x) = \lambda e^{-\lambda x} \quad \text{for } x \geq 0 \quad \text{and} \quad f(x) = 0 \text{ for } x < 0$$

where λ is a positive real number.

To verify that the total probability is equal to 1, calculate the integral of the probability density function over its entire range:

$$\int_0^{+\infty} \lambda e^{-\lambda x}\, dx = -\left[e^{-\lambda x}\right]_0^{+\infty} = 1$$

Thus, the total probability is indeed equal to 1.

Analogous to Poisson's law, the exponential law depends on only one parameter, λ, whose meaning becomes clear when calculating the following mathematical expectation $E(X)$:

$$E(X) = \int_0^{+\infty} \lambda e^{-\lambda x}\, x dx = \lambda \int_0^{+\infty} x e^{-\lambda x}\, dx$$

Integrating by parts, with $u = x$ and $= \lambda e^{-\lambda x}$, results in:

$$E(X) = \int_0^{+\infty} e^{-\lambda x} dx = \frac{1}{\lambda}.$$

Thus, λ is the inverse of the mathematical expectation.

To calculate the variance $v(X)$:

$$E(X^2) = \lambda \int_0^{+\infty} x^2\, e^{-\lambda x} dx$$

Integrating twice by parts yields:

$$E(X^2) = \frac{2}{\lambda^2}$$

From this, the variance is:

$$v(X) = E(X^2) - [E(X)]^2 = \frac{1}{\lambda^2}.$$

The standard deviation is:

$$\sigma(X) = \frac{1}{\lambda}.$$

Figure 11.11 shows the probability density function for $\lambda = 0.5$.

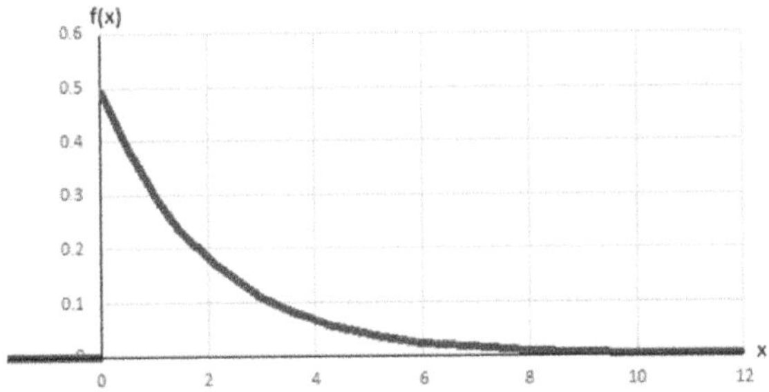

Figure 11.11. *Exponential law. For a color version of this figure, see www.iste.co.uk/cochard/mathematics1.zip*

The exponential law is characteristic of *memoryless* systems, sometimes referred to as the *law of amnesia*. Consider two instants on the time axis: t and $t + h$ $(h \geq 0)$. A random variable X is memoryless if it satisfies the Markov property:

$$p(X > t + h \mid X > t) = p(X > h),$$

This means the probability for the time being greater than $t + h$ does not depend on t (no memory of the past state). Using the conditional probability, this simplifies to:

$$p(X > t + h) = p(X > t).p(X > t + h \mid X > t)$$

Given the memoryless property:

$$p(X > t + h) = p(X > t).p(X > h) \qquad [P]$$

With the properties of the exponential function:

$$p(X > t) = \int_t^{+\infty} \lambda e^{-\lambda x} dx = e^{-\lambda t} \quad \Leftrightarrow \quad p(X > h)$$

$$= \int_h^{+\infty} \lambda e^{-\lambda x} dx = e^{-\lambda h}$$

Also:

$$p(X > t + h) = \int_{t+h}^{+\infty} \lambda e^{-\lambda x} dx = \lambda e^{-\lambda(t+h)} = e^{-\lambda t} e^{-\lambda h}$$

This confirms that the exponential law satisfies the memoryless property [P]. Furthermore, it can be demonstrated (accepted as true) that a memoryless random variable follows the exponential law.

To illustrate this concretely, consider a scenario where individuals arrive in a queue at an average rate of $\lambda = 4$ individuals per minute, with arrivals following Poisson's law. In this case, the time θ between two consecutive arrivals has an average of $\frac{1}{\lambda} = \frac{1}{4}$ minute. This time θ follows an exponential law with parameter $\lambda = 4$, corresponding to an average of $\frac{1}{4}$ minute.

Poisson's law and the exponential law are complementary and play significant roles in queue simulation. Poisson's law models the number of arrivals within a given time frame, while the exponential law describes the time between consecutive arrivals. Together, these laws provide a comprehensive framework for analyzing and simulating queue behavior.

Furthermore, a relationship exists between the exponential law and Poisson's law. The following table illustrates this relationship:

Occurrence of n events in the interval $[0,+\infty[$	Interval θ of time between two consecutive events
Poisson's law: $p_n(t) = e^{-\lambda t}\frac{(\lambda t)^n}{n!}$	Exponential law: $p(\theta) = \lambda e^{-\lambda\theta}$
Law for a discrete random variable	Law for a continuous random variable
λ is the average occurrence rate of events	*$1/\lambda$* is the average of the time intervals separating two consecutive events

Table 11.1. *Comparison of Poisson's law and exponential law*

11.4.5. *Normal law*

The normal law, also known as Gauss' law or Laplace–Gauss law, is one of the best-known distributions in statistics and applies to continuous random variables. Its probability density function is given by:

$$f(x) = \frac{1}{\sigma\sqrt{2\pi}} e^{-\frac{(x-m)^2}{2\sigma^2}}$$

where m represents the mean (mathematical expectation) and σ represents the standard deviation. The normal distribution is commonly denoted as $N(m, \sigma^2)$.

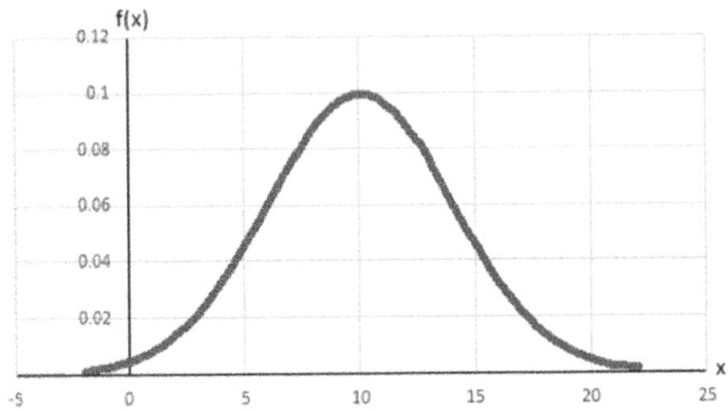

Figure 11.12. *Normal law. For a color version of this figure, see www.iste.co.uk/cochard/mathematics1.zip*

The graphical representation of this probability density features a distinctive "bell" curve shape. Figure 11.12 shows this for $m = 10$ and $\sigma = 4$.

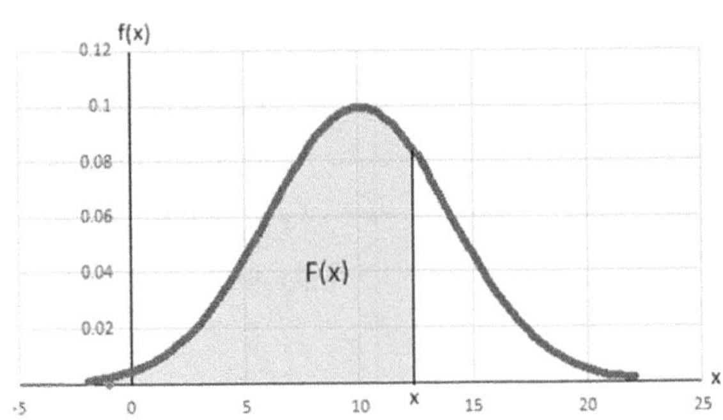

Figure 11.13. *p (X < x). For a color version of this figure, see www.iste.co.uk/cochard/mathematics1.zip*

The probability that $X < x$ is given by the following cumulative distribution function of the normal distribution:

$$p(X \leq x) = \int_{-\infty}^{x} f(x)dx$$

This probability is shown in Figure 11.13. In this graphical representation:

– on the one hand, $p(X \leq x)$ is represented by the area under the curve $y = f(x)$. This area is bounded by the curve, the x-axis and the vertical line at x (i.e. $y = x$);

– on the other hand, $p(X \leq x)$ corresponds to the value of the cumulative distribution function $F(X)$, which gives the probability that the random variable X takes on a value less than or equal to x.

In particular, the following properties can be verified:

a) Sum of the probabilities: the total area under the probability density function is 1. This can be shown using the mathematical result:

$$\int_{-\infty}^{+\infty} e^{-u^2} du = \sqrt{\pi}$$

By substituting

$$u = \frac{(x - m)}{\sigma\sqrt{2}} \quad \text{and} \quad du = \frac{dx}{\sigma\sqrt{2}}$$

The integral becomes:

$$\int_{-\infty}^{+\infty} f(x)dx = \frac{1}{\sigma\sqrt{2\pi}} \int_{-\infty}^{+\infty} e^{-u^2} \sigma\sqrt{2}\, du = \frac{1}{\sqrt{\pi}} \int_{-\infty}^{+\infty} e^{-u^2} du$$

Simplifying gives:

$$\int_{-\infty}^{+\infty} f(x)dx = 1$$

b) Mathematical expectation: the mathematical expectation M is indeed m. To calculate it, use the following formula:

$$M = \int_{-\infty}^{+\infty} xf(x)dx = \frac{1}{\sigma\sqrt{2\pi}} \int_{-\infty}^{+\infty} x\,e^{-\frac{(x-m)^2}{2\sigma^2}} dx$$

Changing variables as before:

$$M = \frac{1}{\sigma\sqrt{2\pi}} \int_{-\infty}^{+\infty} (m + u\sigma\sqrt{2})e^{-u^2}\,\sigma\sqrt{2}du$$

$$= \frac{m}{\sqrt{\pi}}[\int_{-\infty}^{+\infty} e^{-u^2}\,du + \sigma\sqrt{2}\int_{-\infty}^{+\infty} u\,e^{-u^2}du]$$

The second term in the above expression involves integrating an odd function over $]-\infty;\,+\infty[$, which results in zero. Therefore, $M = m$.

c) The standard deviation is indeed σ. To calculate the variance $v(X)$:

$$v(X) = \int_{-\infty}^{+\infty} f(x)(x-m)^2dx = \frac{1}{\sigma\sqrt{2\pi}} \int_{-\infty}^{+\infty} e^{-\frac{(x-m)^2}{2\sigma^2}} (x-m)^2dx$$

Let $u = \frac{x-m}{\sigma\sqrt{2}}$, so $du = \frac{dx}{\sigma\sqrt{2}}$. The integral becomes:

$$v(X) = \frac{1}{\sigma\sqrt{2\pi}} \int_{-\infty}^{\circ} e^{-u^2} 2\sigma^2\,u^2\sigma\sqrt{2}du = \frac{2\sigma^2}{\sqrt{\pi}} \int_{-\infty}^{+\infty} u^2\,e^{-u^2}du$$

Integrate by parts by setting $U = u, dV = ue^{-u^2}du$.

Thus, $dU = du$ and $V =-\frac{1}{2}e^{-u^2}$ which leads to:

$$v(X) = \frac{2\sigma^2}{\sqrt{\pi}}\left[-\frac{u^2}{2}e^{-u^2}\right]_{-\infty}^{+\infty} + \frac{1}{2}\int_{-\infty}^{+\infty} e^{-u^2}du\} = \frac{\sigma^2}{\sqrt{\pi}}\sqrt{\pi} = \sigma^2$$

Thus, σ is indeed the standard deviation.

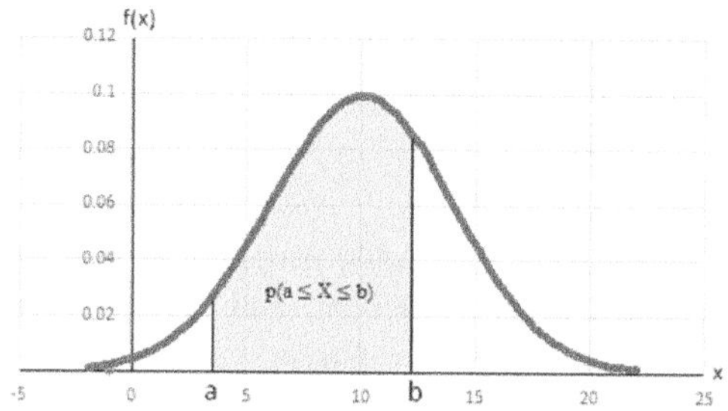

Figure 11.14. $p(a \leq X \leq b)$. *For a color version of this figure, see www.iste.co.uk/cochard/mathematics1.zip*

The probability that X is between two values is obtained as follows:

$$p(a \leq X \leq b) = \int_a^b f(x)dx = 1 - \int_{-\infty}^{a} f(x)dx - \int_{b}^{+\infty} f(x)dx = 1 - p(X \leq a) - p(X \geq b) = p(X \leq b) - p(X \leq a)$$

Thus:

$$p(a \leq X \leq b) = p(X \leq b) - p(X \leq a)$$

This approach can be applied to expressions of the type $p(X \leq x)$ (see Figure 11.14). In particular, the following values are interesting:

$$p(m - \sigma \leq X \leq m + \sigma) = 0.68$$

$$p(m - 2\sigma \leq X \leq m + 2\sigma) = 0.95$$

$$p(m - 3\sigma \leq X \leq m + 3\sigma) = 0.997$$

Tables of the normal distribution (see Appendix 2) provide these quantities for the standard normal distribution, which is introduced below.

The standard normal distribution corresponds to $m = 0$ and $\sigma = 1$:

$$\varphi(t) = \frac{1}{\sqrt{2\pi}} e^{-\frac{t^2}{2}}$$

It is obtained by changing variables, transitioning to standardized coordinates:

$$t = \frac{x - m}{\sigma}$$

EXAMPLE 11.10.– For a population of 2,500 people, where the height is measured in cm, the average height is $m = 169$ cm, and the height follows a normal distribution with a standard deviation $\sigma = 5.6$ cm.

To find the probability that an individual's height is less than 155 cm:

Calculate the z-score: $t = \frac{(155-169)}{5.6}$

Using the standard normal distribution tables, we find:

$$p(T < -2.5) = 1 - p(T < 2.5) = 1 - 0.9938 = 0.0062.$$

Thus, the probability that an individual's height is less than 155 cm is 0.0062.

To find the probability that an individual's height is greater than 172 cm:

Similarly, calculate the z-score:

$$t = \frac{(172-169)}{5.6} = 0.536$$

Using the standard normal distribution tables, find:

$$p(X > 172) = p(T > t) = 1 - p(T < t)$$

$$p(X > 172) = p(T > 0.536) = 1 - p(T < 0.536)$$

From the tables: $p(T < t) \approx 0.7019$, so

$$p(X > 172) = p(T > 0.536) = 1 - p(T < 0.536) = 1 - 0.7019 = 0.2981$$

Therefore, the probability that an individual's height is greater than 172 cm is approximately 0.2981.

To find the probability that an individual's height is between 160 and 178 cm:

Calculate the z-scores for the endpoints, according to the interval $160 \leq X \leq 178$:

$$p(160 \leq X \leq 178) = p(X \leq 178) - p(X \leq 160).$$

Thus:

$$t_1 = \frac{(178-169)}{5.6} = 1.607 \quad \text{and} \quad t_2 = \frac{(160-169)}{5.6} = -1.607.$$

Using the standard normal distribution tables, we have:

$$p(T \leq t_1) \approx 0.9525$$

$$p(T \leq t_2) = p(T > t_1) = 1 - p(T \leq t_1) \approx 0.0475.$$

Therefore, the probability that the height is between 160 and 178 cm is:

$$p(160 \leq X \leq 178) = 0.9525 - 0.0475 = 0.905$$

Thus, the probability that an individual's height is between 160 and 178 cm is approximately 0.905.

To conclude on normal distribution, it is crucial to highlight its *universal character*, as demonstrated by the *central limit theorem*. This theorem asserts that, regardless of the original distribution of a population, the distribution of the sample mean approaches a normal distribution as the sample size becomes large enough. This property reveals the widespread nature of the normal distribution across various fields and scenarios, reinforcing its significance in statistical analysis and real-world applications.

CENTRAL LIMIT THEOREM.– Consider X_i to be a series of independent and identically distributed random variables, each following the same probability law with mean μ and standard deviation σ. Let the random variable X be defined by:

$$X = \frac{1}{n}\sum_{i=1}^{n} X_i$$

Then, when n is sufficiently large, the variable X approximately follows a normal distribution with mean μ and standard deviation $\sigma/\sqrt{n}$.

EXAMPLE 11.11.– Let us provide a concrete justification for the central limit theorem.

Consider rolling a fair die and focus on the result "1". The probability of rolling a "1" (success) is $p = \frac{1}{6}$, and the probability of not rolling a "1" (failure) is $q = \frac{5}{6}$. For this Bernoulli trial:

– the mean (μ) is $p = \frac{1}{6}$;

– the standard deviation (σ) is given by $\sqrt{pq}$. Thus:

$$\sigma = \sqrt{\frac{1}{6} \times \frac{5}{6}} = \sqrt{\frac{5}{36}} = \frac{\sqrt{5}}{6}$$

A single throw of a die corresponds to a random variable X_i. For n throws, each random variable X_i follows the same probability law defined previously. For a sufficiently large n, such as $n = 100$, the central limit theorem predicts that the variable $X = \frac{1}{n}\sum_{i=1}^{n} X_i$ follows a normal law with mean $\mu = \frac{1}{6} \approx 0.1667$ and standard deviation $\sigma = \frac{\sqrt{5}}{6\sqrt{n}}$. For $n = 100$, this standard deviation is approximately 0.03727.

In practice, simulating this with a computer involves performing several series of 100 dice throws and computing the average for each series. Repeating this series of 100 throws 100 times yields an empirical mean of approximately 0.165 and a standard deviation of about 0.039, which are very close to the values predicted by the central limit theorem.

The distribution of these averages closely resembles the Gaussian (normal) curve, confirming that even with 100 series of 100 throws each, the results align closely with the prediction of the central limit theorem.

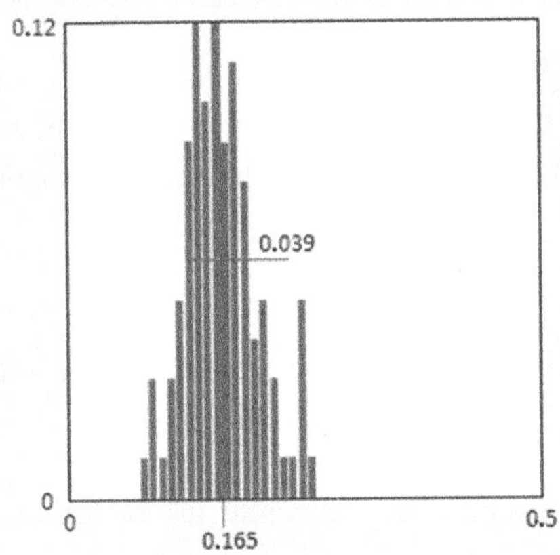

Figure 11.15. *Simulation result. For a color version of this figure, see www.iste.co.uk/cochard/mathematics1.zip*

It is important to note that the central limit theorem makes no assumptions about the specific probability law followed by each random variable X_i. The theorem applies regardless of the underlying distribution, provided that the variables are independent and identically distributed with a finite mean and variance.

11.5. General information on simulation

Simulation is a popular technique for reproducing real-world events using a computer (a modern approach, as earlier simulations were performed manually). This method uses randomness, which might seem paradoxical at first because randomness is typically viewed as inherently unpredictable. The idea of incorporating chance into a simulation might suggest a mastery over randomness, which is often thought to be impossible. And yet…

11.5.1. *Weak law of large numbers*

A fundamental mathematical result underlying simulation techniques is the Law of Large Numbers. Intuitively, this law suggests that a statistical result should approximate the notion of probability. It forms the basics of statistics, which, by analyzing large samples (or, in mathematical terms, large *numbers*), aims to estimate the probability of events. But what exactly constitutes "*large*" in this context?

Two fundamental theorems justify this intuition: the *Weak Law of Large Numbers* (originally developed by Jacques Bernoulli) and the *Strong Law of Large Numbers*. To understand these laws, it is necessary to define what is meant by the *convergence* of statistical frequency towards probability.

The Weak Law of Large Numbers, on the contrary, asserts that the probability of the empirical average deviating from the theoretical average becomes increasingly smaller as the sample size grows. In other words, with a sufficiently larger sample, the empirical average will "almost surely" approach the theoretical average.

The Strong Law of Large Numbers, on the contrary, asserts that the probability of the empirical average deviates from the theoretical average becomes increasingly smaller as the sample size grows. In other words, with a sufficiently sample, the empirical average "almost surely" approaches the theoretical average.

In the case of rolling a die, it is expected that if the number of times "1" appears is averaged over a large number of rolls, the result should approximate $\left(\frac{1}{6}\right)$, which is the probability p of rolling a "1". To formalize this, introduce the random variable X_i, which takes the value "1" if "1" appears on the i-th throw, and 0 otherwise.

Define the statistical frequency f_n as the average of the results obtained from n throws:

$$f_n = \frac{1}{n}\sum_{i=1}^{n} x_i$$

where x_i is the result of the i-th throw.

The statistical frequency f_n represents the proportion of times of "1" appears among the n throws. According to the Law of Large Numbers, as the number of throws n becomes large, this statistical frequency should approach the theoretical probability $p = \frac{1}{6}$. Thus, with an increasing number of throws, f_n should get closer and closer to the theoretical probability of the rolling a "1". This convergence is a key result of probability theory and justifies the expectation that the frequency of occurrence of "1" converges to $\frac{1}{6}$ when the number of throws is sufficiently large.

To provide a practical mathematical result related to this convergence, we consider the Bienaymé–Chebyshev inequality. Consider a random variable X with mean μ and standard deviation σ, taking discrete values $x_1., x_2, \ldots., x_n$ with corresponding probabilities $p_1, p_2, \ldots., p_n$. Its variance is given by:

$$v(X) = \sigma^2 = \sum_i p_i (x_i - \mu)^2$$

Now, focus on the event A defined by $|X - \mu| \geq \varepsilon$, where "$\varepsilon$" is a positive real number. Let $i(A)$ denote the indices i that correspond to this event. It follows that:

$$v(X) = \sigma^2 = \sum_i p_i (x_i - \mu)^2 \geq \sum_{i(A)} p_{i(A)} (x_{i(A)} - \mu)^2$$
$$\geq \sum_{i(A)} p_{i(A)} \varepsilon^2 = \varepsilon^2 \sum_{i(A)} p_{i(A)}$$

As a result

$$p(|X - \mu| \geq \varepsilon) = \sum_{i(A)} p_{i(A)} \leq \frac{\sigma^2}{\varepsilon^2}$$

Hence, the Bienaymé–Chebyshev inequality is:

$$p(|X - \mu| \geq \varepsilon) \leq \frac{\sigma^2}{\varepsilon^2}$$

Consider the following random variable

$$\bar{X}_n = \frac{1}{n}\sum_{i=1}^{n} X_i,$$

defined as the average of a series of independent random variables X_i with the same probability law (mean μ and standard deviation σ). For this variable, the mathematical expectation and variance are given by:

$$E(\bar{X}_n) = \frac{1}{n}\sum_{i=1}^{n} E(X_i) = \frac{n\mu}{n} = \mu \qquad v(\bar{X}_n) = \frac{1}{n^2}\sum_{i=1}^{n} v(X_i) = \frac{n\sigma^2}{n^2} = \frac{\sigma^2}{n}$$

Applying the Bienaymé–Chebyshev inequality to $\underline{X}_n$ yields

$$p(|\underline{X}_n - \mu| \geq \varepsilon) \leq \frac{\sigma^2}{n\varepsilon^2}$$

The complement of the event $|\underline{X}_n - \mu| \geq \varepsilon$ is $|\underline{X}_n - \mu| < \varepsilon$. Therefore,

$$1 - p(|\underline{X}_n - \mu| < \varepsilon) \leq \frac{\sigma^2}{n\varepsilon^2} \quad \text{or} \quad p(|\underline{X}_n - \mu| < \varepsilon) \geq 1 - \frac{\sigma^2}{n\varepsilon^2}$$

As $\varepsilon > 0$ and n tends to infinity (∞), $\frac{\sigma^2}{n\varepsilon^2}$ approaches 0. Thus, the probability $p(|\underline{X}_n - \mu| < \varepsilon)$ approaches 1, indicating that the probability of $\underline{X}_n$ being within ϵ of μ approaches certainty. This result demonstrates the weak law of large numbers.

> **WEAK LAW OF LARGE NUMBERS.–** Consider a sequence of independent and identically distributed random variables, each with mean μ and standard deviation σ. Let $\bar{X}_n$ be defined as the average of thee first n random variables in the sequence:
>
> $$\bar{X}_n = \frac{1}{n}\sum_{i=1}^{n} X_i$$
>
> Then, for every $\varepsilon > 0$, as n tends to infinity (∞), the probability $p(|\bar{X}_n - \mu| < \varepsilon)$ tends to 1.

This theorem implies that it is highly improbable for the difference $|\bar{X}_n - \mu|$ to be greater than ε when n is large. In simpler terms, $\bar{X}_n$ converges to μ in probability as n approaches infinity. It is important to note that this result, initially established by Bernoulli, concerns the limit of probability.

For practical applications, consider an experiment in which a particular event A is expected. For this experiment, associate a random variable X_i (where the index i accounts for different trials of the experiment) that takes the value 1 if event A occurs and the value 0 if event A does not occur. The mathematical expectation of the variable X_i is defined as follows:

$$E(X_i) = 1 \times p(A) + 0 \times p(\bar{A}) = p(A)$$

If the experiment is repeated n times and the statistical frequency f_n, which corresponds to the empirical mean $\bar{X}_n$, is calculated, the weak law of large numbers indicates that as n becomes sufficiently large, it is very likely that f_n will be close to $E(X_i)$, which is the theoretical probability $p(A)$. In other words, the statistical frequency and the theoretical probability should be similar for very large n, reflecting the practical principle of the weak law of large numbers.

However, it is important to note that the weak law of large numbers does not claim that the statistical frequency f_n converges to the probability $p(A)$ as n becomes very large. Instead, it states that as n becomes very large, the probability that the absolute difference between the statistical frequency f_n and the theoretical probability $p(A)$ is less than a given ϵ tends to 1. This means that f_n and $p(A)$ will be close for a sufficiently large n.

11.5.2. *Some simulations*

In this section, several examples will be presented to illustrate how a simulation operates in practice.

EXAMPLE 11.12.– Take the famous example of estimating the value of π devised by Buffon (1707–1788). This naturalist and mathematician designed an experiment to estimate π as follows:

The experiment involves throwing a needle of length L onto a floor composed of parallel strips with a uniform width d. The number of needle throws is denoted by N, and the number of times the needle crosses the gaps between the strips is denoted by n. Buffon demonstrated (see Appendix 3) that the probability of the needle crossing a gap between the strips is given by $\frac{2L}{\pi d}$ when $d > L$. Therefore, for a large number of throws N, the ratio $\frac{n}{N}$ should approach. This allows for the approximation value of π using the formula: $\pi \approx \frac{2LN}{nd}$.

Buffon conducted this experiment manually, as computers did not exist in his time. Today, computers can be used to perform this simulation. For instance, with the parameter $L = 6, d = 3$, and $N = 1{,}000$, a single simulation might yield an approximation of $\pi = 3.02543$. To improve accuracy, it is necessary to repeat the experiment many times and average the results obtained.

EXAMPLE 11.13.– Here is another example illustrating the use of chance. Suppose the goal is to estimate the area of a surface bounded by a closed curve, such as a disk centered at the origin with radius R. Theoretically, the area of the disk is given by $A = \pi R^2$. Now, consider that this disk is inscribed in a square with side length $2R$, as shown in Figure 11.16.

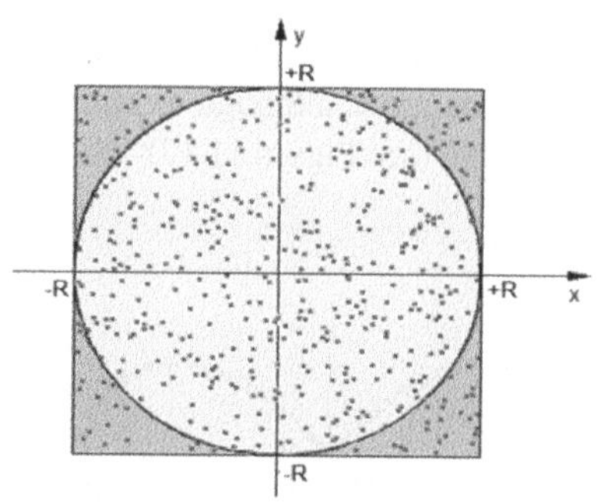

Figure 11.16. *Areas for throwing points. For a color version of this figure, see www.iste.co.uk/cochard/mathematics1.zip*

The area of the square is also known and is given by: $= (2R)^2 = 4R^2$.

To estimate the area of the disk, a computer can be used to randomly and uniformly choose points (x, y) within the square. Specifically, values for x and y will be generated such that $0 \leq x \leq 2$ and $0 \leq y \leq 2$.

Suppose N points are randomly thrown, and n of those points are identified as being inside the disk. The points that lie inside the disk must satisfy the condition $x^2 + y^2 < R^2$. Given that the distribution of points within both the square and the disk is uniform, the number of points inside the disk is proportional to the area of the disk. Thus, $\frac{n}{N} = \frac{A}{B}$, where A is the area of the disk and B is the area of the square.

Since $B = 4R^2$, the area of the disk A can be calculated as:

$$A = \frac{nB}{N} = \frac{nR^2}{N}$$

This provides an "experimental" value of the area of a disk. Given that the theoretical area of the disk is $4R^2$, the "experimental" value of π can be approximated by:

$$\pi \approx \frac{n}{N}$$

For example, with $N = 500$ points, an experiment yields an approximate value of π = 3.152. Repeating the experiment, a second time results in π = 3.04, and a third time yields π = 3.216. This process is continued up to 10 times. Taking the average of the results gives π ≈ 3.114, which is a reasonably close approximation to the "theoretical" value of π.

EXAMPLE 11.14.– Calculate the area of the surface limited by a Bernoulli lemniscate. This is a well-known curve, although its name might not be familiar: it is the shape of the infinity symbol (see Figure 11.17). The equation of this curve is:

$$(x^2 + y^2)^2 = a^2(x^2 - y^2)$$

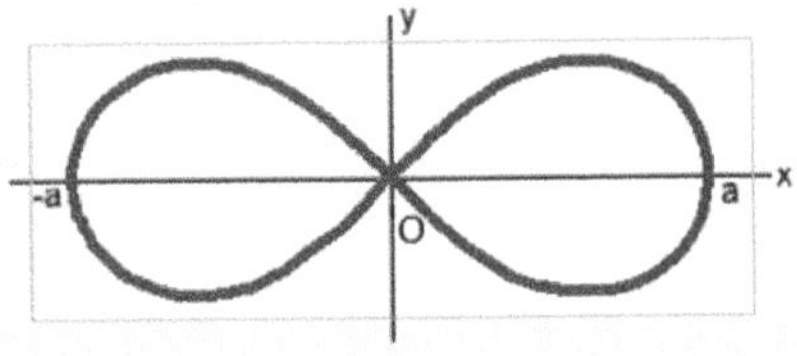

Figure 11.17. *Bernoulli's lemniscate. For a color version of this figure, see www.iste.co.uk/cochard/mathematics1.zip*

This figure is symmetrical with respect to the origin. To find the total area, first calculate the area in the first quadrant (see Figure 11.18) and then multiply by 4.

To generate points, use a uniform distribution to obtain two random numbers, x and y, such that $x \in [0, a]$, and $y \in [0, a/2]$. This ensures that the points lie within a rectangle R with sides a and $a/2$. Repeat this process N times and count the number n of points that fall within the area of the surface, which is shown in dark gray. The points within this region must satisfy the following conditions:

$$(x^2 + y^2)^2 < a^2(x^2 - y^2)$$

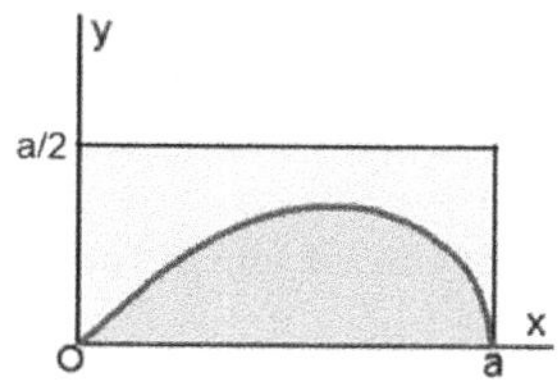

Figure 11.18. *Definition of zones for simulation. For a color version of this figure, see www.iste.co.uk/cochard/mathematics1.zip*

Since the distribution of points is uniform, let A be the area of interest. The proportion of points inside the region to the total number of points is given by:

$$\frac{(A/4)}{\frac{a^2}{2}} = \frac{n}{N}$$

Solving for A:

$$A = 2a^2\frac{n}{N}$$

This formula enables the calculation of the area A from n, N, and a.

By choosing a sufficiently large N, the result should approximate $A \approx a^2$, which corresponds to the expected theoretical result. This also implies that $A/4$ is half the area of the rectangle R.

EXAMPLE 11.15.– Consider the following game. A player starts with 1 euro. They roll a die, and if the result is even, they win 1 euro; if the result is odd, they lose 1 euro. The game continues until the player has either 0 euros or 5 euros. The goal is to determine the average duration of the game, expressed in the number of dice throws. To estimate this, a simulation program can be used, which can be represented by the following algorithm:

```
Input N // number of parts
S = 0 // accumulation of durations j obtained in the games
For i = 1 to N
      j = 0
      M = 1 // initial amount
      While (M > 0 and M < 5) // while M  0 and M  5
            j = j + 1
            X = unif(1, 6) // uniform law
            If (X even)
                  M = M + 1 // increase in amount
```

```
            Otherwise
                M = M - 1 // decrease in amount
            End if
        EndWhile
        S = S + j // accumulation of durations
EndFor
R = S/N // average of game durations
```

The function $unif(1,6)$ generates a random integer between 1 and 6, according to a uniform distribution. To implement the simulation, the algorithm must be translated into a programming (see below). It is advisable to use a sufficiently large number *N*, for example, at least 10,000, to obtain reliable results. Below is an example of Python code to implement this simulation, with different executions varying the value of N.

```
import random
# Entering the number of games to play / launch
N = int(input("Enter the value of N = ")) # We use the variable S
to calculate the accumulation of durations j obtained in the games
S = 0
for i in range(1, N + 1):
    j = 0
    M = 1 # Initial availability, here 1 euro
    while M > 0 and M < 5:
        j += 1
        X  =  random.randint(1,  6)  #  Generating  a  random  number
between
        1 and 6
          if X % 2 == 0: M += 1 # Increase in amount
          else: M -= 1 # Decrease in amount
          S += j # We calculate the accumulation of durations
          R = S / N # We calculate the average of the durations
# Displaying the game average result
print("The average duration of the game (in number of dice rolls) =
", R)
```

With $N = 10{,}000$, the following results were obtained from three executions:

For execution 1:

Entering $N = 10{,}000$

Average game duration (in number of dice rolls) = 4.0174

For execution 2:

Entering $N = 10{,}000$

Average game duration (in number of dice rolls) = 4.0677

For execution 3:

Entering $N = 10{,}000$

Average game duration (in number of dice rolls) = 4.0131

In the same way, if we wish to increase the number N, by setting $N = 10^6$, we obtain the following results:
For execution 1:
Entering $N = 1{,}000{,}000$
Average game duration (in number of dice rolls) = 4.004777
For execution 2:
Entering $N = 1{,}000{,}000$
Average game duration (in number of dice rolls) = 4.005863
For execution 2:
Entering $N = 1{,}000{,}000$
Average game duration (in number of dice rolls) = 4.005863
From the two values associated with N ($N = 10^4$and $N = 10^6$), we can note a better approximation for the largest value of N.

The examples provided above illustrate several characteristics of numerical simulation:

a) A computer is used for its practicality in performing calculations and, more importantly, for avoiding real experiments. In all examples, random numbers are generated, and favorable and possible cases are counted.

b) Statistical frequency is approximated to probability when the conditions of the law of large numbers are met.

c) It becomes evident that obtaining significant results requires a large number of experiments (and averaging the results). This approach aligns with the concept of statistical frequency approaching probability (as supported by the law of large numbers) and justifies the use of a computer for handling repetitive tasks.

d) Generating random numbers with a computer, which is inherently deterministic, presents a challenge. This case will be examined in the next section.

The process of generating random numbers to simulate events is commonly referred to as the *Monte Carlo method*, named after the casino in Monte Carlo city.

11.5.3. *Generation of random numbers*

As showed in the previous examples, simulation relies on the use of random numbers to generate artificial samples.

Before the advent of computers, and even today, tables of random numbers derived from physical phenomena governed by chance, such as radioactive decay, were used. A well-known example from this era is a table produced by the Rand Corporation in 1955. This table was generated by observing the results of an electronic device functioning similarly a casino roulette. These tables were used to extract sequences of random numbers. For instance, consider the Rand Corporation table containing 100,000 digits:

58360937587312385305968851265431298745609854325789043268721095010285730265742892

These numbers can then be grouped into pairs to obtain random numbers ranging from 00 to 99:

58 36 09 37 58 73 12 38 53 05 96 88 51 26 54 31 29 87 45 60 98 54 32 57 89 04 32 68 72 10 95 01 02 85 73 02 65 74 28 92

Alternatively, to obtain random numbers ranging from 000 to 999, the numbers can be grouped into triples:

583 609 375 873 123 853 059 688 512 654 312 987 456 098 543 257 890 432 687 210 950 102 857 302 657 428

In 1939, Kendall and Babington-Smith created a random number table by photographing the numbers on a roulette wheel.

Today, computers are primarily used to generate random numbers, although these are actually pseudo-random, as they are computed rather than truly random. Programming languages provide functions to generate pseudo-random numbers. For example, in JavaScript, the *Math.random*() function returns a number between 0 (inclusive) and 1 (exclusive). To obtain a number between a and b ($b > a$), the following expression can be used: $(b - a) * Math.random() + a$. In PHP, the rand() function generates an integer between 0 and getrandmax() by default. By providing arguments, such as $rand(a, b)$, it returns an integer between a and b (inclusive).

There are several methods for generating these pseudo-random numbers. All of these methods produce numbers following a uniform distribution. Let us explore two of these methods, among many others, and also a third, less conventional one.

Mid-square process

This process involves choosing an initial number u_0 composed of $2m$ digits. The number is squared, resulting in a number with $4m$ digits. From this result, the

$2m$ central digits are extracted to obtain the number u_1. The process is then repeated with u_1 to get u_2, etc. Here is an example illustrating this process:

$u_0 = 7{,}637$ $u_0^2 = 58{,}323{,}769$
$u_1 = 3{,}237$ $u_1^2 = 10{,}478{,}169$
$u_2 = 4{,}781$ $u_2^2 = 22{,}857{,}961$
$u_3 = 8{,}579$ $u_3^2 = 73{,}599{,}241$
$u_4 = 5{,}992$ $u_4^2 = 35{,}904{,}064$
$u_5 = 9{,}040$ $u_5^2 = 81{,}721{,}600$
etc.

This generates the following sequence: 763732374781857959929040 ….

Fibonacci process

Start with three numbers u_0, u_1 and k, where k is larger than the first two numbers. Generate a sequence of random numbers using the following procedure:

$$u_{n+2} = u_{n+1} + u_n + ak \quad \text{where} \quad a = \begin{cases} 0 \; if \; u_{n+1} + u_n \leq k \\ -1 \; if \; u_{n+1} + u_n > k \end{cases}$$

For example, let $u_0 = 3{,}765$, $u_1 = 1{,}529$, and $k = 1{,}6245$. The calculations proceed as follows:

$u_2 = 1{,}529 + 3{,}765 = 05{,}294$
$u_3 = 5{,}294 + 1{,}529 = 06{,}823$
$u_4 = 6{,}823 + 5{,}294 = 12{,}117$
$u_5 = 12{,}117 + 6{,}823 - 16{,}245 = 02{,}695$
$u_6 = 2{,}695 + 12{,}117 = 14{,}812$
$u_7 = 14{,}812 + 2{,}695 - 16{,}245 = 01{,}262$
etc.

Next, eliminate the first and last digit of each result, yielding the following sequence:

376152529682211269481126….

Decimal places of π

The irrationality of π implies that the sequence of its decimal places shows no periodicity and appears completely disordered. While this suggests that this sequence of decimal places could be considered a sequence of random numbers, proving this rigorously remains a challenge. Nonetheless, this sequence is often used in simulations and computations However, in practice, computers typically do not use this sequence directly.

As of now, 62,800 billion decimal places of π are known. Here are the first 100 of these known decimal places.

π = 3.1415926535897932384626433832795028841971693993751058209749445923078164062862089986280348253421170679

11.6. Generating programs

Generating random numbers that follow a specific probability distribution is essential for conducting simulations. These random numbers will form samples for the simulation. Initially, consider common probability distributions, where the cumulative distribution function is particularly useful. Next, address the general case for any probability distribution. The importance of the distribution function can be illustrated for both discrete and continuous variables (see Figure 11.19).

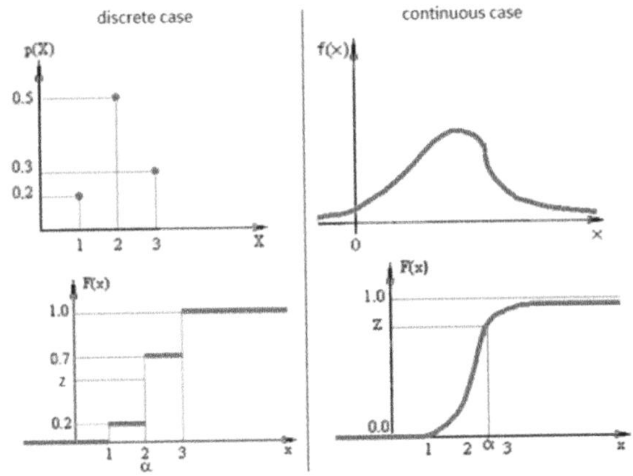

Figure 11.19. *Using the distribution function. For a color version of this figure, see www.iste.co.uk/cochard/mathematics1.zip*

The method is as follows:

1) Uniformly choose a number z between 0 and 1.

2) Use the distribution function $F(x)$ to determine the number α that follows the corresponding probability law (i.e. such that $F(\alpha) = z$).

This approach allows for the generation of random samples that conform to the desired probability distribution. The key problem is in inverting the function $F(x)$, since α is obtained from $\alpha = F^{-1}(z)$.

11.6.1. *Usual laws*

Previously, it was demonstrated how computers can generate random numbers (in fact, pseudo-random numbers) following a uniform distribution. This will serve as the basis for generating random numbers according to other probability laws.

Binomial law

For the binomial distribution, given a value of z (from the uniform distribution), it corresponds to the number α (following the binomial distribution) such that

$$\sum_{k=0}^{\alpha} C_n^k p^k q^{n-k} \leq z \leq \sum_{k=0}^{\alpha+1} C_n^k p^k q^{n-k}$$

Poisson's laws

Poisson's distribution describes a discrete random variable. Given a random number z (from the uniform distribution), the number α (following Poisson's distribution) is determined by the condition

$$\sum_{k=0}^{\alpha} e^{-m} \frac{m^k}{k!} \leq z \leq \sum_{k=0}^{\alpha+1} e^{-m} \frac{m^k}{k!}$$

Exponential law

The exponential distribution is for a continuous variable with a probability density function given by

$$f(x) = \lambda e^{-\lambda x}$$

The cumulative distribution function is

$$F(x) = \int_{-\infty}^{x} \lambda e^{-\lambda t} dt = 1 - e^{-\lambda x} \quad \text{if } x \geq 0 \quad \text{and } F(x) = 0 \text{ otherwise}$$

The inversion is straightforward. For a number z (uniformly distributed between 0 and 1), the value α (following the exponential distribution) is given by:

$$\alpha = F^{-1}(z) = -\frac{1}{\lambda} Ln(1 - z)$$

Normal law

There are several methods for simulating a normal distribution. One common approach is described here. Directly inverting the distribution function is difficult. Therefore, alternative algorithms are used. One simple method is:

$$\alpha = \sigma \left[\sum_{i=1}^{12} z_i - 6 \right] + m$$

where z_i are numbers generated from a uniform distribution between 0 and 1, and m and σ are the mean and standard deviation of the normal distribution to be simulated.

11.6.2. *Any probability law*

Consider a practical situation where the probability distribution is unknown, but statistical data is available. Imagine a counter where customers arrive, and measurements of the number of arriving customers are taken every 15 minutes. After conducting 1,000 measurements, the resulting frequencies of customer arrivals are displayed in Figure 11.20. These frequencies are provided to two decimal places.

Number of arrivals	0	1	2	3	4	5	6	7	8	9	10	11	12	13
Frequency	0	0.01	0.12	0.23	0.22	0.11	0.08	0.08	0.07	0.03	0.03	0.01	0.01	0

Figure 11.20. *Any probability law*

Assume that this statistic approximates the probability distribution of customer arrivals, which is unknown. The average number of arrivals is $\lambda = 4.77$ arrivals per quarter hour. Suppose that measurements were also taken of the number of customers served per quarter hour. A total of 100 measurements provide the statistical results (table) shown in Figure 11.21.

Number of services	0	1	2	3	4	5	6	7	8	9
Frequency	0	0.04	0.08	0.1	0.15	0.2	0.24	0.15	0.04	0

Figure 11.21. *Any probability law*

Assume that this statistic represents the probability distribution of services (or departures). The average number of services is $m = 4.91$ per quarter hour. To study this waiting system and estimate, through simulation, the average queue length of customers waiting to be served, complete the previous two tables by associating a range of two-digit numbers with each frequency, as shown in Figure 11.22. The size of each range should be proportional to the frequency. For instance, for an arrival count of 3 with a frequency of 0.23, multiplying by 100 gives an integer of 23. Given that 12 was the last value for the previous arrival count, create a range of 13 numbers from 13 to 35.

Numbers of arrivals	0	1	2	3	4	5	6	7	8	9	10	11	12	13
Frequency	0	0.01	0.12	0.23	0.22	0.11	0.08	0.08	0.07	0.03	0.03	0.01	0.01	0
Numbers	-	0	01-12	13-35	36-57	58-68	69-76	77-84	85-91	92-94	95-97	98	99	-

Numbers of services	0	1	2	3	4	5	6	7	8	9
Frequency	0	0.04	0.08	0.1	0.15	0.2	0.24	0.15	0.04	0
Numbers	-	00-03	04-11	12-21	22-36	37-56	57-80	81-95	96-99	-

Figure 11.22. *Using number ranges*

The meaning of these numbers is as follows. Suppose two random sequences are generated using a computer:

Sequence S1: 90 27 80 39 52 98 24 79 …
Sequence S2: 72 71 67 53 43 97 30 98 …

Period	S1	S2	Number of arrivals	Number of services	File length
1	90	72	8	6	2
2	27	71	3	6	0
3	80	67	7	6	1
4	39	53	4	5	0
5	52	43	4	5	0
6	98	97	11	8	3
7	24	30	3	4	0
8	79	98	7	8	0

Figure 11.23. *Arrivals and departures table*

– For the first quarter-hour, draw 90 for arrivals and 72 for departures. According to the arrivals table, the number 90 corresponds to eight arrivals table, while 72 corresponds six departures.

– For the second quarter-hour, draw 27 for arrivals and 71 for departures. In the arrivals table, 27 corresponds to three arrivals, and 71 corresponds to six departures.

– Continuing in this manner, create a table showing, arrivals and departures for each quarter-hour, as shown in Figure 11.23.

In this table, by calculating the difference between the number of arrivals and the number of services (i.e. departures), the length of the queue is obtained. If the difference is positive, it represents the length of the queue; if negative, the queue length is zero.

With a sufficiently large number of values for S1 and S2 (i.e. a large sample size), the average queue lengths can be determined, providing the desired result. To ensure a reliable result, consider the following precautions:

1) Number of periods: ensure that there are a sufficiently large number of periods, at least 100. It is also recommended to repeat the simulation multiple times and average the results to obtain a more accurate estimate.

2) Edge effects: in the given example, the waiting system is not "open-ended". For instance, the service only during specific hours, such as an 8-hour workday. It is therefore important to account for "edge" effects. To address these, discard results from the beginning and end of the simulation period retaining only those that reflect the system's stable, or "steady-state", regime.

11.7. M/M/1 waiting system

Waiting phenomena are prevalent in daily life (such as waiting at counters or experiencing traffic jams), and they significantly impact economic, social and financial aspects. Waiting systems are studied through various models that share certain common characteristics.

A waiting system typically consists of a queue where customers wait and one or more stations that perform the required services (see Figure 11.24).

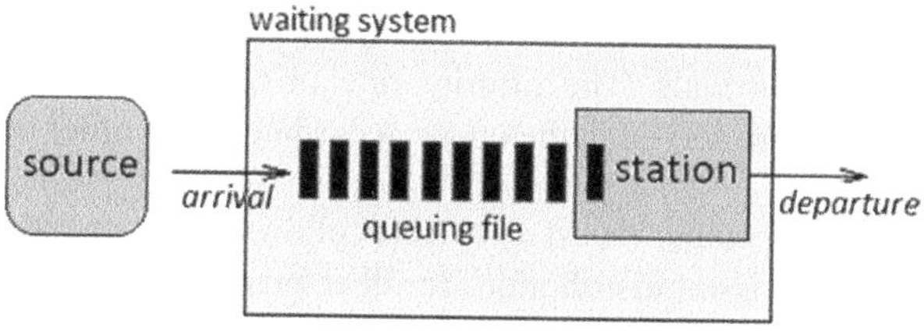

Figure 11.24. *M/M/1 waiting system*

In a queuing system, individuals enter the queue from a source, which can be either infinite (open systems) or finite (closed systems). They advance in line according to a specific order of priority and are directed to a station where they receive individual service. Once the service is complete, they leave the station. The arrival and departure processes are governed by probabilistic laws, which may differ from one another.

The interesting case is that of equilibrium, where the queue neither grows indefinitely nor remains perpetually empty. The waiting system is subject to a process of arrivals (births) and departures (deaths), and understanding this balance is crucial for comprehending the overall functioning of the system.

Two key parameters, which are often random, influence the behavior of the system over time:

– Arrival rate: the frequency (or rate) of arrival of individuals.

– Service duration: the duration of service.

Arrival frequency represents the rate at which individuals enter the waiting system. It may vary depending on various factors such as time of day, day of the week or other relevant variables. Duration of service refers to the time required to serve each individual before they leave the station. This duration may also be subject to random variations.

One of the simplest waiting systems is the M/M/1 system. The first "*M*" indicates the exponential distribution of arrival times, meaning the time intervals between consecutive arrivals (which therefore follow an exponential distribution). The second "*M*" signifies that the duration of services also follows an exponential distribution. The number "1" indicates that there is a single service station, which implies the existence of a single queue.

Additionally, it is assumed that the waiting system is open, meaning that there is a continuous flow of arrivals. The priority rule in the queue is FIFO (First-In, First-Out): individuals are served in the queue according to the order of their arrival.

It is important to remember that if a variable follows an exponential distribution, its inverse follows a Poisson distribution. In other words, if the time between two consecutive arrivals follows an exponential distribution with a mean of $\frac{1}{\lambda}$, then the frequency of arrivals follows a Poisson distribution with a mean λ. Similarly, if the duration of a service follows an exponential distribution with a mean of $\frac{1}{\mu}$, then the frequency of departures or services follows a Poisson distribution with a mean of μ.

Thus, the M/M/1 system can be analyzed from two equivalent perspectives, as shown in Figure 11.25.

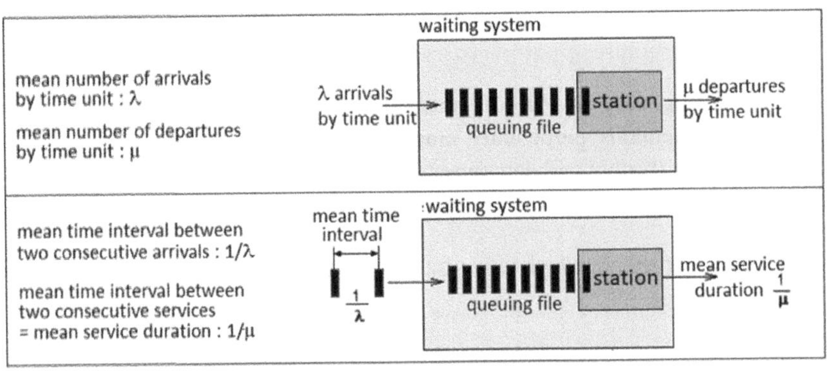

Figure 11.25. *Two equivalent ways of seeing a waiting system*

11.7.1. *Theoretical study*

Consider the time interval from t to $t + h$, where h is a very small positive duration (time). It can be hypothesized that during this interval h, at most one arrival or departure occurs (the "rare events" hypothesis). Therefore, if there are n individuals in the system at $t + h$, there are only four possible scenarios for the system at time t:

– n individuals at time t: no arrivals and no departures or one arrival and one departure during the duration (interval) h.

– $n - 1$ individuals at time t: one arrival during the interval h.

– $n + 1$ individuals at time t: one departure during the interval h.

Let $p_n(t)$ denote the probability of having n individuals in the system at time t. To describe the evolution of the system at time $t + h$, the compound probabilities can be used

$$p_n(t+h) = p_n(t)\rho_{n,n}(h) + p_{n-1}(t)\rho_{n-1,n}(h) + p_{n+1}(t)\rho_{n+1,n}(h)$$

where $\rho_{n,m}(h)$ represents the probability of transitioning from a state with n individuals to a state with m individuals during the interval h. This probability expression accounts for all three possible cases. The process is represented as a Markov chain where the states are the number of individuals in the waiting system (see Figure 11.26).

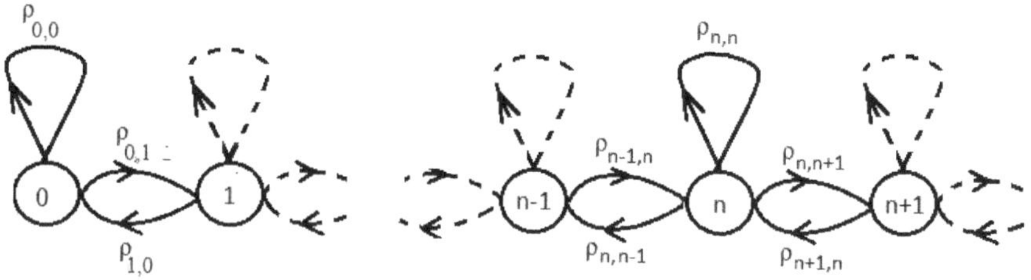

Figure 11.26. *Markov process*

The probability $\rho_{n,n}(h)$ is approximately greater as h becomes larger. This leads to the following approximations:

$$\rho_{n,n+1}(h) = \lambda_n h + O(h) \quad \text{and} \quad \rho_{n,n-1}(h) = \mu_n h + O(h)$$

Since the sum of the transition probabilities must be equal to 1, it follows that:

$$\rho_{n,n}(h) = 1 - \rho_{n,n+1}(h) - \rho_{n,n-1}(h) = 1 - \lambda_n h - \mu_n h + O(h)$$

where $O(h)$ represents terms that become negligible (tend to 0) as h approaches 0. The parameter λ_n is the "birth rate" and generally depends on the current state n. For simplicity, λ_n is assumed to be constant (so $\lambda_n = \lambda$). Similarly, μ_n is the death rate also assumed to be constant (so $\mu_n = \mu$).

With these notations, the probability $p_n(t+h)$ for $n > 0$ is expressed as:

$$p_n(t+h) = p_n(t)(1 - \lambda h - \mu h) + p_{n-1}(t)\lambda h + p_{n+1}(t)\mu h + O(h)$$

Rearranging and dividing by h gives:

$$\frac{p_n(t+h)-p_n(t)}{h}=-p_n(t)(\lambda+\mu)+p_{n-1}(t)\lambda+p_{n+1}(t)\mu+O(h)$$

Taking the limit as h approaches 0, the term $O(h)$ becomes negligible, and the equation simplifies to:

$$\frac{dp_n(t)}{dt}=-p_n(t)(\lambda+\mu)+p_{n-1}(t)\lambda+p_{n+1}(t)\mu$$

For $n=0$, $p_0(t+h)=p_0(t)(1-\lambda h)+p_1(t)\mu h+O(h)$ and

$$\frac{dp_0(t)}{dt}=-p_0(t)\lambda+p_1(t)\mu$$

The primary focus is on the equilibrium state, where the system reaches a steady-state. In this case, the probabilities associated with the different states of the system no longer depend on time:

$$0=-p_n(\lambda+\mu)+p_{n-1}\lambda+p_{n+1}\mu$$

$$0=-p_0\lambda+p_1\mu$$

The second equation allows p_1 to be expressed in terms of p_0:

$$p_1=\frac{\lambda}{\mu}p_0=\Psi p_0 \quad \text{with} \quad \Psi=\frac{\lambda}{\mu}$$

Using the first equation and applying induction:

$$p_2=\frac{1}{\mu}[p_1(\lambda+\mu)-p_0\lambda]=p_1(\Psi+1)-p_0\Psi=p_0\Psi^2$$

And so forth:

$$p_n=p_0\Psi^n.$$

To determine p_0, use the normalization condition:

$$\sum_i p_i=1=p_0[1+\Psi+\Psi^2+\cdots]$$

Assuming $\Psi < 1$ (this will be justified later):

$$1 = p_0 \frac{1}{1 - \Psi}$$

Thus:

$$p_0 = 1 - \Psi \quad \text{and} \quad p_n = (1 - \Psi)\Psi^n \quad \text{for } n > 0$$

At this stage, the relationship is between the parameters λ and μ from the previous calculations, and the average rates of arrivals and departures become clear. These parameters represent the average rates themselves. For instance (see example 11.15.), if $p_k(t)$ represents the probability of having k arrivals from time 0 to time t, this probability follows Poisson's law in the M/M/1 system:

$$p_k(t) = e^{-\lambda t} \frac{(\lambda t)^k}{k!}$$

The expected number of arrivals $E(k)$ is given by:

$$E(k) = \sum_{k=0}^{\infty} p_k(t)t = e^{-\lambda t} \sum_{k=1}^{\infty} \frac{(\lambda t)^k}{(k-1)!} = (\lambda t)\, e^{-\lambda t} \sum_{k=1}^{\infty} \frac{(\lambda t)^{k-1}}{(k-1)!}$$
$$= (\lambda t) e^{-\lambda t} e^{\lambda t} = \lambda t$$

By differentiating this with respect to λt and multiplying by λt, the result is:

$$\lambda = \frac{E(k)}{t}$$

The parameter λ therefore represents the mean arrival rate, i.e. the mean number of arrivals per unit of time. Similarly, μ represents the mean departure rate.

From the relationship obtained for the probability p_n, the following results can be derived (see section 11.8.4, Appendix 4):

- The average number of individuals in the waiting system is $n_S = \frac{\Psi}{1-\Psi}$.
- The average number of people in the queue is $n_F = \frac{\Psi^2}{1-\Psi}$.
- The average time spent by an individual in the system is $t_S = \frac{1}{\mu-\lambda}$.
- The average time spent by an individual in the queue is $t_F = \frac{\Psi}{\mu-\lambda}$.

It is observed that $t_S - t_F = \frac{1}{\mu}$, which corresponds to the average duration of a service.

The assumption that $\lambda < \mu$ is justified because, for the queue to remain stable and not grow indefinitely, the arrival rate must be less than the departure rate. Thus, on average, $\lambda < \mu$.

EXAMPLE 11.16.– A bank office in a small town has only one teller. The arrival rate is three customers per hour. The average service duration is 15 minutes (or ¼ hour). The time between two consecutive arrivals follows an exponential distribution with parameter λ (hence, the average time between arrivals is $\left(\frac{1}{\lambda}\right)$), and the service duration follows an exponential distribution with parameter μ (hence, the average service time is $\left(\frac{1}{\mu}\right)$). This represents an M/M/1 waiting system with $\lambda = 3$ and $\mu = 4$. The waiting system is "viable" because $\lambda < \mu$ or $\Psi < 1$.

Theoretical results indicate the following:

– The average number of customers in the bank is $n_S = \frac{\Psi}{1-\Psi} = 3$ customers.

– The average number of customers in the queue is $n_F = \frac{\Psi^2}{1-\Psi} = 2.25$ customers.

– The average time spent by a customer in the bank is $t_S = \frac{1}{\mu-\lambda} = 1$ hour.

– The average time spent by a customer in the queue is $t_F = \frac{\Psi}{\mu-\lambda} = \frac{3}{4}$ hour = 45 minutes.

It is verified that $t_S - t_F = 15$ minutes.

11.7.2. *Simulation*

The basic data are the average arrival rate λ and the average departure (or service) rate μ, assuming an M/M/1 system. Figure 11.27 describes a typical situation with nine elements. N individuals, numbered 1, 2, 3, ... N, are successively entered into the waiting system. For each individual $k > 1$, the following are defined or calculated:

– The time difference between two successive arrivals, $I[k]$ is generated randomly according to an exponential law with parameter λ.

– The arrival date, $da[k]$, is calculated as the sum of the previous individual's arrival date, $da[k-1]$, and $I[k]$. For the first individual, $da[1]$ is set to 0. Thus, $da[k] = da[k-1] + I[k]$, and for $k = 1, da[1] = 0$.

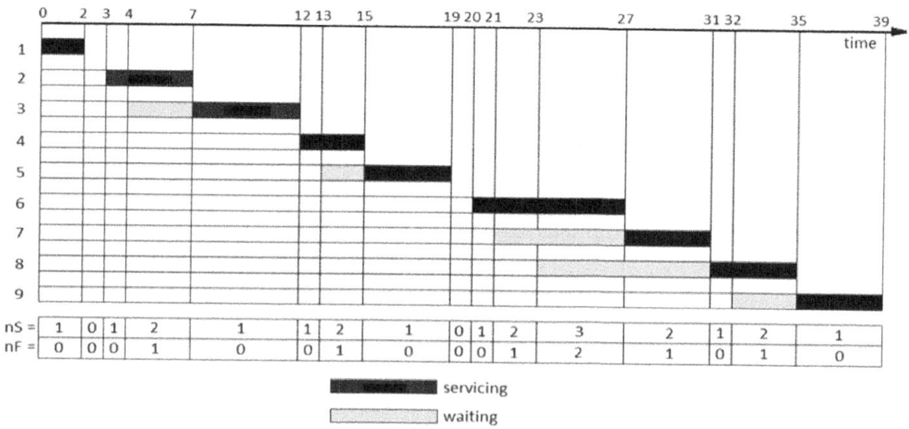

Figure 11.27. n_S, n_F: *number of individuals in the system and in the queue*

– The duration of the service, $D[k]$, is randomly generated according to an exponential law with parameter μ.

– The start date of service, $ds[k]$, determined by taking the greater value between the departure date of the previous individual, $dd[k-1]$, and the individual's arrival date k, $da[k]$. If $da[k] > dd[k-1]$, the station is free, and the service can start immediately. If $da[k] < dd[k-1]$, service will begin only after the departure of the $(k-1)$th individual. In other words, $ds[k] = \max\{dd[k-1], da[k]\}$, where $dd[k-1]$ is the departure date of the $(k-1)$th individual.

– The departure date, $dd[k]$, is calculated by adding the duration of service, $D[k]$, to the service start date, $ds[k]$. In other words, $dd[k] = ds[k] + D[k]$.

– The time spent by the individual k in the queue, $tF[k]$, is calculated as the difference between the service start date, $ds[k]$, and the arrival date, $da[k]$. In other words, $tF[k] = ds[k] - da[k]$.

– The time spent by the individual k in the waiting system, $tS[k]$, is calculated as the difference between the departure date, $dd[k]$, and the arrival date, $da[k]$. In other words, $tS[k] = dd[k] - da[k]$.

For the first individual, who is first to enter the waiting system, the initial values are $da[1] = 0$, $ds[1] = 0$, $dd[1] = ds[1] + D[1]$, $tF[1] = 0$, and $tS[1] = dd[1] - da[1]$.

From the tables $tF[k]$ and $tS[k]$, the average values t_F and t_S can be calculated as follows:

$$t_F = \frac{1}{N}\sum_{k=1}^{N} tF[k] \quad \text{and} \quad t_S = \frac{1}{N}\sum_{k=1}^{N} tS[k]$$

Summarize the algorithm that handles the timing of events and the calculation of t_F and t_S:

```
ttf=0 // total queue time
tts=0 // total system waiting time
For k = 1 to N
     If k = 1
             I[1]=0
             da[1]=0
             ds[1]=0
             D[1]=genexp(m)
             dd[1]=D[1]
             tf[1]=0 // wait time
             tS[1]=dd[1]=da[1] // presence time in the system
             ttf=ttf+tf[1]
             tts=tts+ts[1]
     Otherwise
             I[k]=genexp(l)
             da[k]=da[k-1]+I[k]
             ds[k]= max{dd[k-1], da[k]}
             D[k]=genexp(m)
             dd[k]=ds[k]+D[k]
             tf[k]=ds[k]-da[k] // wait time
             tS[k]=dd[k]-da[k] // presence time in the system
             ttf=ttf+tf[k]
             tts=tts+ts[k]
     End if
EndFor
```

In this algorithm, t_F and t_S represent the average waiting time in the queue and the average presence time spent in the system, respectively.

The algorithm uses the function $genexp(x)$, which generates a random number following the exponential distribution:

$$y = \text{genexp}(x) = -\frac{1}{\lambda} Ln(1-z) \quad \text{where} \quad z = \text{unif}(0,1)$$

where $z = \text{unif}(0,1)$. The function unif(0,1). returns a random number uniformly distributed between 0 and 1.

Calculating the average number n_F of individuals in the queue and the average number n_S of individuals in the waiting system is less straightforward. This is because the durations during which there are values for the number of individuals in the queue and in the system need to be considered such as durations 2, 1, 1 and 3, as shown in Figure 11.27. The key dates are provided in the tables $da[k], dd[k]$ and $ds[k]$.

These three lists can be merged while maintaining chronological order and noting whether each date corresponds to an arrival, a, a start of service, s, or a departure, d. Multiple events can occur at the same date (e.g. date 12 in Figure 11.27, which represents an arrival, a start of service and a departure). In such cases, events are processed in the order: arrival (da), departure (dd) and start of service (ds). Identical dates should be included and considered as separate events.

For example, from Figure 11.27, the tables in Figure 11.28 show sorted dates from da, ds and dd, with the corresponding events listed in line E.

To determine the average number of intervals, represent the consecutive time intervals with their length, starting and ending events, as shown in Figure 11.29. In these tables, a, s and d denote an arrival, a start of service and a departure, respectively. Note that there are intervals of zero length corresponding to identical dates of different events.

da	0	3	4	12	13	20	21	23	32
ds	0	3	7	12	15	20	27	31	35
dd	2	7	12	15	19	27	31	35	39

L	0	0	2	3	3	4	7	7	12	12	12	13	15	15	19	20	20	21	23	23	27	27	31	31	32	35	35	39
E	a	s	d	a	s	a	d	s	d	a	s	a	d	s	d	a	s	a	a	s	d	s	d	s	a	d	s	d

Figure 11.28. *Timelines*

To determine the number nS of individuals present in the system during an interval, add 1 each time an interval begins with an arrival and subtract 1 each time an interval begins with a departure.

interval	0-0	0-2	2-3	3-3	3-4	4-7	7-7	7-12	12-12	12-12	12-13	13-15	15-15
duration d	0	2	1	0	1	3	0	5	0	0	1	2	0
events	as	sd	da	as	sa	ad	ds	sd	da	as	sa	ad	ds

interval	15-19	19-20	20-20	20-21	21-23	23-27	27-27	27-31	31-31	31-32	32-35	35-35	35-39
duration d	4	1	0	1	2	4	0	4	0	1	3	0	4
events	sd	da	as	sa	aa	ad	ds	sd	ds	sa	ad	ds	sd

Figure 11.29. *Consecutive time intervals*

interval	0-0	0-2	2-3	3-3	3-4	4-7	7-7	7-12	12-12	12-12	12-13	13-15	15-15
duration d	0	2	1	0	1	3	0	5	0	0	1	2	0
events	as	sd	da	as	sa	ad	ds	sd	da	as	sa	ad	ds
nS	1	1	0	1	1	2	1	1	0	1	1	2	1
nF	1	0	0	1	0	1	1	0	0	1	0	1	1

interval	15-19	19-20	20-20	20-21	21-23	23-27	27-27	27-31	31-31	31-32	32-35	35-35	35-39
duration d	4	1	0	1	2	4	0	4	0	1	3	0	4
events	sd	da	as	sa	aa	ad	ds	sd	ds	sa	ad	ds	sd
nS	1	0	1	1	2	3	2	2	1	1	2	1	1
nF	0	0	1	0	1	2	2	1	1	0	1	1	0

Figure 11.30. *Completed tables*

Likewise, to determine the number of individuals nF in the queue during an interval, add 1 each time an interval begins with an arrival and subtract 1 each time an interval begins with a start of service. Figure 11.30 shows the completed tables.

The weighted averages are then calculated using the following formulas:

$$n_S = \frac{\sum_k nS[i] \times d[i]}{\sum_i d[i]} \qquad n_F = \frac{\sum_k nF[i] \times d[i]}{\sum_i d[i]}$$

where i denotes an interval, $d[i]$ is its duration, $nS[i]$ is the number of individuals in the system, and $nF[i]$ is the number of individuals in the queue.

The algorithm below follows the one given above, with the tables $da, dd,$ and ds now being known.

```
L = concatenation of lists da, dd, ds
For k=1 to N
      Le[k]="a"
      Le[k+N]="d"
      Le[k+2*N]="s"
```

```
    EndFor
    //LL= sorted list of L according to the values of L[h] for h=1 to
    3*N
    //LLe = sorted list of LLe in the same order as LL
    // list of time intervals
    i = 0 // number of intervals between two consecutive events
    nS=0 // number of individuals present in interval i
    nF=0 // number of individuals waiting in interval i
    T[i][1]=0 // low value of the 1st interval
    T[i][2]=0 // high value of the 1st interval
    T[i][3]=0 // duration of the 1st interval
    T[i][4]="a" // event of the start of the 1st interval
    T[i][5]="s" // end of 1st interval event
    T[i][6]=0//number of individuals in system during the i interval
    T[i][7]=0 //number of individuals waiting during
    the interval i
    For h = 1 to 3*N-1
        i=i+1 // interval number
        T[i][1]=LL[h] // low value of the interval
        T[i][2]=LL[h+1] // high value of the interval
        T[i][3]=T[i][h+1]-T[i][h] //interval duration
        T[i][4]=LLe[h] // event for low value
        T[i][5]=LLe[h+1] // event for the high value
        If T[i][4]="a" // an individual enters the system
               nS=nS+1
               nF=nF+1
        End if
        T[i][6]=nS // number of presents during the interval
        T[i][7]=nF // number of patients during the interval
        If T[i][4]="d" // an individual leaves the system
               nS=nS-1
        End if
        T[i][6]=nS
        If T[i][4]="s" // an individual leaves the queue
               nF=nF-1
        End if
        T[i][7]=nF
    EndFor
    M=i
    //calculation of averages
    TnS=0
    TnF=0
    For i=1 to M
      TnS=TnS+T[i][6]*T[i][3]//cumulative products duration×individuals
presents
```

```
    TnF=TnF+T[i][7]*T[i][3]//cumulative products duration×individuals
waiting
  EndFor
  mnS=TnS/LL[3*N] // average number of individuals present in the
  system
  mnF=TnF/LL[3*N] // average number of individuals waiting
```

EXAMPLE 11.17.– Concretely, by using a program based on the previously described algorithms and repeating the experiment 20 times with 110 individuals in an M/M/1 type waiting system, with an arrival rate of $\lambda = 3$ arrivals per unit time and a service rate of $\mu = 4$ services per unit time, the results are similar to those presented in the table in Figure 11.31.

To ensure a steady state, the first five and last five individuals are eliminated, resulting in a sample of 100 individuals.

The benefit of conducting a large number of experiments (potentially more than 20) is evident, as results can vary significantly between experiments. The primary interest is in the average values obtained. Comparing these values to the theoretical values for an M/M/1 queue, as given in example 11.16 ($t_S = 1$, $t_F = 0.75$, $n_S = 3$ and $n_F = 2.25$), shows that the obtained values are relatively close, with for t_S and t_F aligning more closely than n_S and n_F.

Calculating the standard deviations σ_{t_S}, σ_{t_F}, σ_{n_S}, and σ_{n_F}, from the table in Figure 11.31, reveals that the theoretical values t_S, t_F, n_S and n_F fall within the following ranges:

$$t'_S - \sigma_{t_S} \leq t_S \leq t'_S + \sigma_{t_S}$$

$$t'_F - \sigma_{t_F} \leq t_F \leq t'_F + \sigma_{t_F}$$

$$n'_S - \sigma_{n_S} \leq n_S \leq n'_S + \sigma_{n_S}$$

$$n'_F - \sigma_{n_F} \leq n_F \leq n'_F + \sigma_{n_F}$$

where t'_S, t'_F, n'_S and n'_F are the values obtained through simulation.

experiments	t_S	t_F	n_S	n_F
1	1.14	0.90	3.70	2.91
2	0.60	0.16	0.96	0.38
3	0.74	0.49	2.34	1.56
4	1.08	0.82	2.75	2.10
5	0.81	0.54	2.55	1.70
6	0.67	0.44	2.17	1.45
7	0.72	0.47	2.38	1.56
8	0.63	0.39	1.95	1.24
9	0.55	0.29	1.45	0.78
10	1.28	1.00	3.74	2.93
11	2.17	1.87	6.58	5.67
12	0.60	0.35	1.81	1.08
13	2.61	2.34	9.25	8.28
14	1.09	0.83	3.25	2.50
15	0.88	0.64	2.57	1.88
16	1.33	1.10	4.43	3.66
17	1.25	0.98	3.98	3.13
18	1.83	1.54	6.16	5.21
19	0.50	0.27	1.44	0.80
20	0.42	0.20	1.21	0.59
averages	**1.05**	**0.78**	**3.23**	**2.47**

Figure 11.31. *Results*

11.8. Appendices

11.8.1. *Appendix 1: Table for Poisson's law*

Poisson's law table

$$p(X = k) = e^{-m}\frac{m^k}{k!}$$

k \ m	0.5	1	1.5	2	2.5	3	3.5	4	4.5	5	5.5
0	0.6065	0.3679	0.2231	0.1353	0.0821	0.0498	0.0302	0.0183	0.0111	0,0067	0.0041
1	0.3033	0.3679	0.3347	0.2707	0.2052	0.1494	0.1057	0.0733	0.0500	0.0337	0.0225
2	0.0758	0.1839	0.2510	0.2707	0.2565	0.2240	0.1850	0.1465	0.1125	0.0842	0.0618
3	0.0126	0.0613	0.1255	0.1804	0.2138	0.2240	0.2158	0.1954	0.1687	0.1404	0.1133
4	0.0016	0.0153	0.0471	0.0902	0.1336	0.1680	0.1888	0.1954	0.1898	0.1755	0.1558
5	0.0002	0.0031	0.0141	0.0361	0.0668	0.1008	0.1322	0.1563	0.1708	0.1755	0.1714
6	0.0000	0.0005	0.0035	0.0120	0.0278	0.0504	0.0771	0.1042	0.1281	0.1462	0.1571
7	0.0000	0.0001	0.0008	0.0034	0.0099	0.0216	0.0385	0.0595	0.0824	0.1044	0.1234
8	0.0000	0.0000	0.0001	0.0009	0.0031	0.0081	0.0169	0.0298	0.0463	0.0653	0.0849
9	0.0000	0.0000	0.0000	0.0002	0.0009	0.0027	0.0066	0.0132	0.0232	0.0363	0.0519
10	0.0000	0.0000	0.0000	0.0000	0.0002	0.0008	0.0023	0.0053	0.0104	0.0181	0.0285
11	0.0000	0.0000	0.0000	0.0000	0.0000	0.0002	0.0007	0.0019	0.0043	0.0082	0.0143
12	0.0000	0.0000	0.0000	0.0000	0.0000	0.0001	0.0002	0.0006	0.0016	0.0034	0.0065
13	0.0000	0.0000	0.0000	0.0000	0.0000	0.0000	0.0001	0.0002	0.0006	0.0013	0.0028
14	0.0000	0.0000	0.0000	0.0000	0.0000	0.0000	0.0000	0.0001	0.0002	0.0005	0.0011
15	0.0000	0.0000	0.0000	0.0000	0.0000	0.0000	0.0000	0.0000	0.0001	0.0002	0.0004
16	0.0000	0.0000	0.0000	0.0000	0.0000	0.0000	0.0000	0.0000	0.0000	0.0000	0.0001

11.8.2. *Appendix 2: Table for normal law*

Table for the reduced normal law

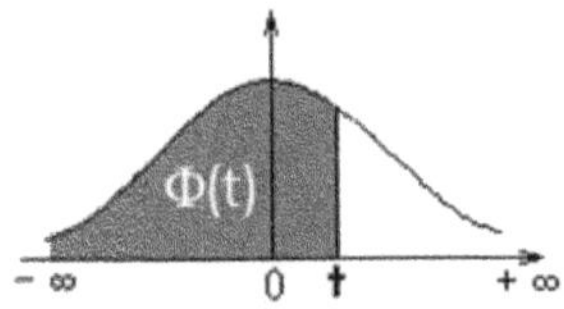

$$\Phi(t) = \frac{1}{\sqrt{\pi}} \int_{-\infty}^{t} e^{-u^2/2}\, du$$

t	0.00	0.01	0.02	0.03	0.04	0.05	0.06	0.07	0.08	0.09
0.0	0.5000	0.5040	0.0800	0.5120	0.5160	0.5199	0.5239	0.5279	0.5319	0.5359
0.1	0.5398	0.5438	0.5478	0.5517	0.5557	0.5596	0.5636	0.5675	0.5714	0,5753
0.2	0.5793	0.5832	0.5871	0.5910	0.5948	0.5987	0.6026	0.6064	0.6103	0.6141
0.3	0.6179	0.6217	0.6255	0.6293	0.6331	0.6368	0.6406	0.6443	0.6480	0.6517
0.4	0.6554	0.6591	0.6628	0.6664	0.6700	0.6736	0.6772	0.6808	0.6844	0.6879
0.5	0.6915	0.6950	0.6985	0.7019	0.7054	0.7088	0.7123	0.7157	0.7190	0.7224
0.6	0.7257	0.7290	0.7324	0.7357	0.7389	0.7422	0.7454	0.7486	0.7517	0.7549
0.7	0.7580	0.7611	0.7642	0.7673	0.7704	0.7734	0.7764	0.7794	0.7823	0.7852
0.8	0.7881	0.7910	0.7939	0.7967	0.7995	0.8023	0.8051	0.8078	0.8106	0.8133
0.9	0.8159	0.8186	0.8212	0.8238	0.8264	0.8289	0.8315	0.8340	0.8365	0.8389
1.0	0.8413	0.8438	0.8461	0.8485	0.8508	0.8531	0.8554	0.8577	0.8599	0.8621
1.1	0.8643	0.8665	0.8686	0.8708	0.8729	0.8749	0.8770	0.8790	0.8810	0.8830
1.2	0.8849	0.8869	0.8888	0.8907	0.8925	0.8944	0.8962	0.8980	0.8997	0.9015
1.3	0.9032	0.9049	0.9066	0.9082	0.9099	0.9115	0.9131	0.9147	0.9162	0.9177
1.4	0.9192	0.9207	0.9222	0.9236	0.9251	0.9265	0.9279	0.9292	0.9306	0.9319
1.5	0.9332	0.9345	0.9357	0.9370	0.9382	0.9394	0.9406	0.9418	0.9429	0.9441
1.6	0.9452	0.9463	0.9474	0.9484	0.9495	0.9505	0.9515	0.9525	0.9535	0.9545
1.7	0.9554	0.9564	0.9573	0.9582	0.9591	0.9599	0.9608	0.9616	0.9625	0.9633
1.8	0.9641	0.9649	0.9656	0.9664	0.9671	0.9678	0.9686	0.9693	0.9699	0.9706
1.9	0.9713	0.9719	0.9726	0.9732	0.9738	0.9744	0.9750	0.9756	0.9761	0.9767
2.0	0.9772	0.9779	0.9783	0.9788	0.9793	0.9798	0.9803	0.9808	0.9812	0.9817
2.1	0.9821	0.9826	0.9830	0.9834	0.9838	0.9842	0.9846	0.9850	0.9854	0.9857
2.2	0.9861	0.9864	0.9868	0.9871	0.9875	0.9878	0.9881	0.9884	0.9887	0.9890
2.3	0.9893	0.9896	0.9898	0.9901	0.9904	0.9906	0.9909	0.9911	0.9913	0.9916
2.4	0.9918	0.9920	0.9922	0.9925	0.9927	0.9929	0.9931	0.9932	0.9934	0.9936
2.5	0.9938	0.9940	0.9941	0.9943	0.9945	0.9946	0.9948	0.9949	0.9951	0.9952
2.6	0.9953	0.9955	0.9956	0.9957	0.9959	0.9960	0.9961	0.9962	0.9963	0.9964
2.7	0.9965	0.9966	0.9967	0.9968	0.9969	0.9970	0.9971	0.9972	0.9973	0.9974
2.8	0.9974	0.9975	0.9976	0.9977	0.9977	0.9978	0.9979	0.9979	0.9980	0.9981
2.9	0.9981	0.9982	0.9982	0.9983	0.9984	0.9984	0.9985	0.9985	0.9986	0.9986

Table for large values of t

t	3.0	3.1	3.2	3.3	3.4	3.5	3.6	3.8	4.0	4.5
Φ(t)	0.99865	0.99904	0.99931	0.99952	0.99966	0.99976	0.99984	0.99993	0.99997	0.99999

NB : The table gives the values of Φ(t) for positive t. When t is negative we proceed as follows:

for t = 2.72 Φ(t) = 0.9967

for t = -2.72 Φ(t) = 1 - 0.9967 = 0.0033

11.8.3. *Appendix 3: Buffon's needle*

The probability that the needle will pass through a gap between the slats needs to be calculated. Assuming the parquet floor is perfect; the impact location is irrelevant since all the parquet slats have equivalent properties. Thus, the problem can be approached using two random variables. Considering horizontal slats and focusing on a specific slat (see Figure 11.32), the calculation proceeds as follows:

– The distance y of the needle's lowest point from the upper dividing line is the first random variable, taking values between 0 and d.

– The angle θ of the needle relative to the horizontal is the second random variable, taking values between 0 and π.

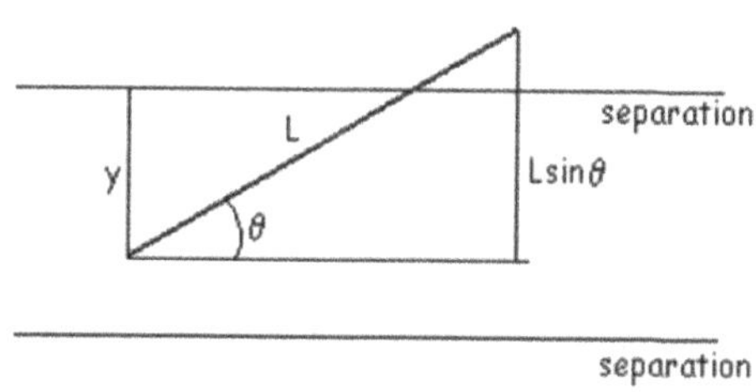

Figure 11.32. *Geometry of the problem*

It is straightforward to determine that the needle crosses the upper separation if $y < L\sin(\theta)$. This relationship can be shown graphically: Figure 11.33(a) shows the various possible points (light gray rectangle) when y and θ are chosen randomly, assuming the selection follows a uniform probability distribution.

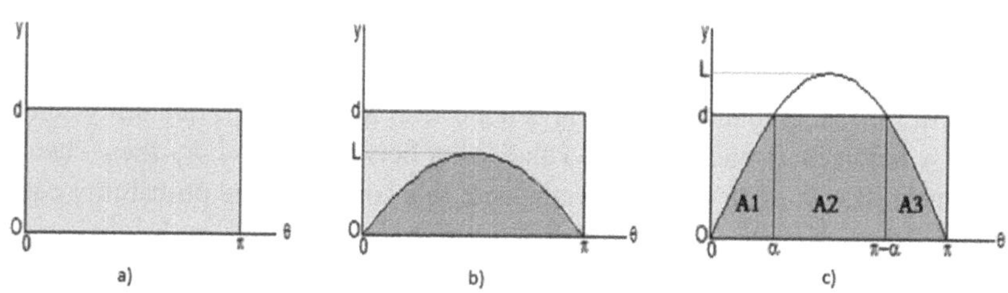

Figure 11.33. *Different scenarios*

Case ($d > L$) (see Figure 11.33(b)):

The curve $y < L\sin(\theta)$ forms a sinusoid arc, and the region below this curve (shown in dark gray) represents the situations where the needle intersects the separation. The probability of this occurrence is the ratio of the area of the dark gray region to the area of the rectangle (with part of the rectangle covered by the dark gray region). The area of the rectangle is ($d\pi$); the area of the dark gray portion is given by:

$$\int_0^{\pi} L\, sin\theta\, d\theta = 2\,L$$

Thus, the probability is:

$$p = \frac{2L}{\pi d} \qquad [11.1]$$

Case ($d < L$) (see Figure 11.33(c)):

The calculation is more involved. The dark gray area must be divided into three parts with areas A1, A2 and A3:

$$\mathrm{A1} = \int_0^{\alpha} Lsin(\theta)d\theta = L\big(1 - cos(\alpha)\big),\ \ \mathrm{A2} = d(\pi - 2\alpha)\ \ \text{and}\ \ \mathrm{A3} = \mathrm{A1}$$

where $\alpha = \arcsin\left(\frac{d}{L}\right)$. The resulting probability is then:

$$p = \frac{2L(1-\sqrt{1-\frac{d^2}{L^2}})+d[\pi-2arcssin\left(\frac{d}{L}\right)]}{\pi d} \qquad [11.2]$$

To perform the simulation (nowadays done with a computer), randomly select a value of y between 0 and d and a value of θ between 0 and π, then check if $y < L\sin(\theta)$. If, out of N trials, n trials meet this condition, the probability can be estimated using the law of large numbers approximating the probability to the statistical frequency ($f = n/N \approx p$):

– If $d > L$: $\pi = \frac{2LN}{dn}$

– If $d < L$: $\pi = \frac{2\,arcsin\left(\frac{d}{L}\right)-2\frac{L}{d}(1-\sqrt{1-\frac{d^2}{L^2}})}{1-\frac{n}{N}}$

11.8.4. *Appendix 4: Results for M/M/1*

Calculation of n_S

$$n_S = \sum_{n=0}^{\infty} n p_n = (1-\Psi)\sum_{n=0}^{\infty} n\Psi^n = (1-\Psi)\Psi(1+2\Psi+3\Psi^2+\cdots)$$
$$= (1-\Psi)\Psi\frac{d}{d\Psi}(\Psi+\Psi^2+\Psi^3+\cdots)$$

But since $\Psi < 1$,

$$1+\Psi+\Psi^2+\cdots = \frac{1}{1-\Psi} \quad \text{therefore} \quad \Psi+\Psi^2+\Psi^3+\cdots = \frac{\Psi}{1-\Psi}$$

Thus:

$$n_S = (1-\Psi)\Psi\frac{1}{(1-\Psi)^2} = \frac{\Psi}{1-\Psi}$$

Calculation of t_F

If an individual arrives in the system, waiting is necessary until all individuals who arrived before have been served. Given that there are on average n_S individuals in the system, the total time required to serve these individuals is equal to the product of n_S the average length of service time, which is $\left(\frac{1}{\mu}\right)$. Thus:

$$t_F = \frac{n_S}{\mu} = \frac{\Psi}{(1-\Psi)\mu} = \frac{\Psi}{\mu-\lambda}$$

Calculation of t_S

An individual spends time t_S in the system, which is the sum of the waiting time in the queue t_F and the average duration of a service $\left(\frac{1}{\mu}\right)$. Therefore,

$$t_S = t_F + \frac{1}{\mu} = \frac{\Psi}{\mu-\lambda} + \frac{1}{\mu} = \frac{\mu\Psi+\mu-\lambda}{\mu(\mu-\lambda)} = \frac{1}{\mu-\lambda}$$

Calculation of n_F

If there are n people in the system, then there are $(n - 1)$ people in the queue. To express the average number of people in the queue, n_F, use the following approach:

$$n_F = \sum_{n=2}^{\infty} (n-1)\, p_n$$

$$= (1-\Psi) \sum_{n=2}^{\infty} (n-1)\Psi^n = (1-\Psi)(\Psi^2 + 2\Psi^3 + 3\Psi^4$$
$$+ \cdots) = (1-\Psi)\Psi^2(1 + 2\Psi + 3\Psi^2 + \cdots)$$
$$= (1-\Psi)\Psi^2 \frac{d}{d\Psi}(\Psi + \Psi^2 + \Psi^3 + \cdots)$$

However:

$$\frac{d}{d\Psi}(\Psi + \Psi^2 + \Psi^3 + \cdots) = \frac{d}{d\Psi}\left(\frac{1}{1-\Psi}\right) = \frac{1}{(1-\Psi)^2}$$

Simplify to:

$$n_F = \frac{\Psi^2}{1-\Psi}$$

References

[BAD 02] BADRIKIAN J., *Codes correcteurs*, Ellipses, Paris, 2002.

[BEL 94] BELLOWIN S.M., CHESWICK W.R., *Firewalls et sécurité Internet*, Addison-Wesley, Paris, 1994.

[BOI 68] BOISSONNADE M., HALBIQUE, A.M., *Statistique et probabilités*, Ligel, Paris, 1968.

[BUC 98] BUCHMANN J., "La factorisation des grands nombres", *Pour la science*, n°251, Paris, available at: https://www.apprendre-en-ligne.net/crypto/rsa/251_088_096.pdf, 1998.

[COC 19] COCHARD G.-M., *Introduction to Stochastic Processes and Simulation*, ISTE Ltd, London and John Wiley & Sons, New York, 2019.

[CON 68] CONDAMINE M., VISSIO P., *Mathématique 1ère C et E*, Delagrave, Paris, 1968.

[ENG 76] ENGEL A., *Processus aléatoires pour les débutants*, Cassini, Paris, 1976.

[EXE 59] EXENER R., *Logic in Elementary Mathematics*, McGraw Hill, New York, 1959.

[FAU 73] FAURE R., LEMAIRE B., *Mathématiques pour l'informatique*, Gauthier Villars, Paris, 1973.

[FAU 79] FAURE R., *Précis de recherche opérationnelle*, Dunod/Bordas, Paris, 1979.

[GHE 08] GHERNAOUTI-HELIE S., *Sécurité informatique*, Dunod, Paris, 2008.

[GOM 13] GOMEZ J., *Codage et cryptographie*, Le monde est mathématique, Barcelona, 2013.

[GRA 78] GRANIER C., GUILBAUD B., *Statistique dans l'entreprise*, Foucher, Paris 1978.

[HUR 79] HURE F., *Traité expérimental de logique digitale*, Editions Techniques et Scientifiques Françaises, Paris, 1979.

[INA 99] INA, *Réseaux et protection numérique des documents*, Etudes INA, Paris, 1999.

[LAH 95] LAHECHE K., *Les codes en informatique : codes détecteurs et correcteurs d'erreurs*, Hermès, Paris, 1995.

[LEH 19] LEHNING H., *La bible des codes secrets*, Flammarion, Paris, 2019.

[LIP 83] LIPSCHUTZ S., *Mathématiques pour informaticiens*, Schaum – McGraw Hill, Paris, 1983.

[LOV 91] LOVINFOSSE J.-P., *Le piratage informatique*, Marabout, Paris, 1991.

[MAC 87] MACCHI C., GUIBERT J.-F., *Téléinformatique*, Dunod, Paris, 1987.

[MAS 76] MASIERI W., *Notions essentielles de statistique et calcul des probabilités*, Sirey, Paris, 1976.

[MEN 64] MENDELSON E., *Introduction to Mathematical Logic*, International Student Editions, New York, 1964.

[MUL 18] MULLER D., Les courbes elliptiques, available at: https://www.apprendre-en-ligne.net/crypto/moderne/elliptique.html, 2018.

[PHE 77] PHELIZON J.-F., *Informatique opérationnelle, tome II, modèles conduisant à la simulation*, Economica, Paris, 1977.

[POC 71] POCHARD H., *Nombres réels, nombres complexes, probabilités, calcul différentiel, calcul intégral, terminales C, D, E*, Gauthiers-Villars, Paris, 1971.

[PRA 23] PRADEAU Y., *La guerre des nombres premiers*, Flammarion, Paris, 2023.

[QUI 84] QUINE W.V.O., *Méthodes de logique*, Armand Colin, Paris, 1984.

[SIN 99] SINGH S., *Histoire des codes secrets*, Le livre de poche, Paris, 1999.

[VEL 19] VELU J., *Méthodes mathématiques pour l'informatique*, Dunod, Paris, 2019.

[WOO 80] WOOLARD B., *Conquérir la logique*, Dunod, Paris, 1980.

List of Authors

Gérard-Michel COCHARD
Eco-PRocédés Optimisation et Aide à la Décision (EPROAD, EA 4669)
Université de Picardie Jules Verne
Amiens
France

Mhand HIFI
Eco-PRocédés Optimisation et Aide à la Décision (EPROAD, EA 4669)
Université de Picardie Jules Verne
Amiens
France

Index

G, H

I, K

L, M

N, O

P, Q

R, S

T, U, V, W, X

Printed and bound by CPI Group (UK) Ltd, Croydon, CR0 4YY

13/09/2026

14961071-0007